Rummage tells the long story of British recycling: from dog hair spun into yarn to Second World War stretchers reused as railings, from Victorian gentlemen selling papier-mâché pianofortes to Puritan soldiers making kites from books.

In this fascinating and original new book, historian Emily Cockayne illuminates our relationship to rubbish and reveals how these attitudes have defined private and public life, built and destroyed businesses, shaped wars and created history. Starting with the hyper-consumerism of the 1990s, working backwards through world wars, Napoleonic wars and civil wars all the way to the Reformation, Cockayne exposes the hidden work that has gone into shaping the material world for successive generations.

Richly detailed, full of surprising stories and ingenious characters, *Rummage* shows how we have come to throw away so much, and what lessons might be drawn from the past to address urgent contemporary dilemmas.

D1353716

Emily Cockayne is a senior lecturer in Early Modern History at the University of East Anglia. She is author of *Hubbub: Filth, Noise & Stench in England* (2007), cited by Toni Morrison as a key source for *A Mercy*, and *Cheek by Jowl: A History of Neighbours* (2012). Her own collection of reused objects is ever-growing, in range as well as size: from a   rivets, to a trunk   mbow-elled in the   wich.

Praise for *Rummage*:

'Smart and encyclopedic … a lively and frequently surprising history of what we reuse and what we throw away'
i Paper

'A marvellous history of the second and third lives of objects and, just as important, a timely reminder that there are ways out of a throw-away society'
Frank Trentmann, author of *Empire of Things: How We Became a World of Consumers, from the Fifteenth Century to the Twenty-First*

'Pertinent, fascinating and full of intricate, joyful detail'
Annie Gray, author of *The Greedy Queen*

'*Rummage* overflows with detail. [Cockayne] rescues wonderfully bizarre artefacts from rubbish heaps to plot Britain's changing attitudes to consumption and recycling … the way she embraces historical anecdote, social critique and personal reminiscence never seems ponderous … By breaking down the boundaries between waste and overlooked treasure, *Rummage* will make us think twice about what we throw away in future'
Spectator

'One of those rare books, a marvellous curiosity shop of fascinating historical gems, objects and insights, a feat of scholarship and a salutary book for our throw-away times'
Rebecca Stott, author of *Ghostwalk*

'Wonderful … worth re-reading'
Bookmunch

'An original and unusual history of recycling'
Times Literary Supplement

EMILY COCKAYNE

Rummage

A History of the Things We Have
Reused, Recycled and Refused to Let Go

P

PROFILE BOOKS

This paperback edition first published in 2021

First published in Great Britain in 2020 by
PROFILE BOOKS LTD
29 Cloth Fair
London ECIA 7JQ

www.profilebooks.com

10 9 8 7 6 5 4 3 2 1

Typeset in Garamond by MacGuru Ltd
Printed and bound in Great Britain by
CPI Group (UK) Ltd, Croydon, CRO 4YY

A CIP catalogue record for this book is available from the British Library.

ISBN 978 1 78125 852 1
eISBN 978 1 78283 357 4

This book is dedicated to the memory of Evelyn Bradbury
(1923–1992)

Contents

Time Up for *Master Humphrey's Clock*

I ventured into the back room of my local second-hand bookshop for the first time a few years ago, lured by the promise of a special set of books. Dusty piles of old leather-bound tomes surrounded me as a jolly woman stood on a chair to reach down a first-edition three-volume set of Charles Dickens's weekly serial *Master Humphrey's Clock*, published in 1840. I had located the volumes online and had been pleasantly surprised to see they were so close to home. I inspected them. The third volume had been repaired some time in the twentieth century; a tell-tale flash of purplish backing peeped out under the brown of the original spine. I wondered what had

1. *Master Humphrey's Clock* (1840). Author's collection, photographed by Taryn Everdeen.

happened to that volume, to render it in need of repair when the other two had not been. What lives had these volumes lived? I do not know how many people had owned them before me; there are no inscriptions. I do know what happened to another set from the same printing litter during the Second World War. That came within a whisker of being pulped for the war effort. Such books are not only relics of the past; they each contain the history of their own survival.

In 1942 James Ross, the City Librarian of Bristol, was tasked with examining and sorting books donated to the city's Salvage Drive. During the drive, books came from all over: some volumes were collected by children at school; others were gathered through house-to-house collections; some were pulled from papers donated to a paper mill. Sorting through the heaps, Ross and his team consigned volumes to one of four piles. Some were to be sent to the troops, each stamped with 'To Fighting Forces – wherever you may be. With best wishes from the citizens of Bristol, England.' Others went to re-stock war-damaged libraries. A portion were deemed valuable enough to be saved without further reasoning as to their use. The rest were pulped or used in munitions-making and packing. One fated afternoon, Ross will have flicked through *Master Humphrey's Clock*. The text is narrated by a lonely 'misshapen deformed old' Londoner who hoards 'piles of dusty papers' in his beloved clock, and believes butterflies will generate in 'some dark corner of these old walls'. Ross, with the Ministry of Supply instructions before him, put the Dickens volumes in the 'to be saved' pile with other books he deemed to be hard to obtain, rare, valuable or of 'bibliographical value'. By the end of the war, Ross's team had received over a million books. Countless editions of classics were pulped to make boxes for munitions or heating pipes for bombers. Through a combination of sheer luck and the efforts of people like James Ross, some treasures – including the set of Dickens I then clutched in the back of a bookshop – survived.[1] Recycling is not only about economic or ecological value. It involves cultural judgement, social norms, taste and memory. In an emergency, what would seem too precious or inconvenient to scrap, even in the face of existential threat?

Every future is, to some degree, a bricolage of the past's uncertain remnants.[2] History was once the study of the victorious. In the world of objects, antique furniture and paintings by the masters are victors of a kind; they gain value and become more cherished. Sometimes, as in the case of *Master Humphrey's Clock*, the fact that they came so close to destruction seems to imbue objects with a special worth. But the vanquished and the forgotten have value of their own. More recently, neglected people have finally received more attention from historians and, in parallel, things previously deemed irrelevant or commonplace have acquired new significance. In every age, complicated but everyday judgements are made about the usability of things: what to keep and what to discard. The relative value of leftovers constantly changes, influenced by the hopes, expectations and needs of human communities, whether we are looking at the stories, people or objects to carry from the past into the present. *Rummage* is about material redemption. It is a history of the reused, a history of extraordinary reinvention.

This book begins with *Master Humphrey's Clock* not only because it has survived the past, or because its narrator is a thrifty hoarder, but because the book itself is a paragon of reinvention. The majority of materials that came together to make it, like most books published before the end of the nineteenth century, were recycled. Bound in brown cloth, *Master Humphrey's Clock* has covers adorned with an embossed image of a clock and pages printed on paper made from linen rags. The binding glue was made from animal by-products, and even the gold used to emboss the clock may have come from floor sweepings. After the late nineteenth century, most books included a relatively insignificant fraction of recycled content: their paper was made from virgin wood pulp. Books often exemplify broader changes in the practices of reuse and recycling.

Rummage focuses on the material transformation of things, especially ingenious repurposing and material remodelling.[3] That could include recycled materials, such as the linen rag used in books, and also waste products put to a first use. The latter are not, strictly speaking, examples of reuse, because they involved little in the way of processing in order to be put to use. Still, some slip into the narrative because the public and the politicians did not always discriminate. This can be seen in the way 'salvage' during the wars included the use of human food waste for pigs, and the use of old bones to manufacture glycerine. Such processes influenced broader attitudes to the often grimy and grubby work of recyclers. The story I will tell is as much about these changing attitudes as it is about material objects.

Starting with our current age – with all the washed-up, reused remnants of the past – *Rummage* digs back through the sediments of time, picking through ever more distant episodes and showing how people extended the lives of their things. There is a tipping point at which stuff becomes simply and stubbornly worthless – unusable rubbish. *Rummage* tells the story of the stuff that was saved, not dumped: the 'unwasted'. Whether something is waste or not varies over time. Sometimes waste has secret uses later on. By working backwards, down through the deposits and layers of culture, we can unearth forgotten recycling practices and stand many basic historical assumptions on their head. Highlighting features of reuse at receding stages of British history shows us how people coped with the detritus left by those populating the chapter to follow. *Rummage* makes sense of those heaped remains and suggests new uses for them.

By casting light on salvaging efforts and past markets for recycled goods, I look to the afterlife of our own material world. During the excavation it becomes clear that industrial output could not always meet demand for goods, and that, consequently, recycling and repurposing were normal domestic activities. The lives of things were eked out; extended through being remodelled, passed down, passed on, pawned, borrowed or repaired. The urge to reuse or recycle never developed in a linear fashion but was related to population pressures, financial confidence, scientific advances, material availability and military emergency.

Recycling for ecological reasons might seem a characteristically modern idea, but in 1852 the prominent Scottish chemist and MP Lyon Playfair anticipated that industrial output would be more sustainable if people shunned waste and replicated 'natural' recycling. Economical industrial chemistry could imitate nature. The argument was picked up by an American professor of chemistry:

'Nothing is really lost in nature', wrote Peter Austen in 1901; 'give the ground filth it returns a flower. Matter is in eternal circulation.'[4] Such arguments gradually distilled into modern theories of industrial ecology, which advocates a shift away from linear or 'open loop' manufacturing systems, where some portion of the material investment was always wasted, and towards a 'closed loop' system, with all wastes turned into materials usable by other industries. As recently as 2018, the prominent Green Party member Jenny Jones echoed Playfair and Austen when she expressed her opinion that our 'use of materials needs to be a closed loop [...] Nature doesn't waste anything and neither should we.'[5] These admirable ideals obviously have complications, not least in industries that create toxic or nuclear wastes. Even if industries could all be developed in closed loops, rapidly changing patterns of manufacture and consumption cannot always accommodate closed systems, and those systems rarely achieve the resilience of natural cycles developed over millions of years.[6]

Kitchen scraps were often composted or fed to livestock. In cities, where more wholesome grain and acorns were scarce, pigs were fed an eclectic mixture of wastes: decaying vegetables, animal skins, tripe, 'the offal of rendered Tallow', carrion and bits of dead other pigs. In 1683 an author worried about pigs fed on 'Mens Dung, Pigeons Dung, or Poultry Dung'.[7] Emulating such practices, agri-feed companies in the twentieth century processed meat and bone meal, which included parts of rendered cattle. Fed to cows, this matter wrought the spectre of the disease known as BSE. Taken to extremes, recycling doesn't mimic nature. There is nothing natural about cows eating cows. The recycling loops of earlier times were less visible but also tighter, since waste was often redeployed within the house, though they never closed entirely. The energy used to repurpose things degrades the loops, producing unforeseen wastes which escape them.

The recycling of some items has traditionally been so ubiquitous that little contemporary attention was paid to them. Metal objects such as horseshoes and printers' type were melted down once broken or redundant, leaving no physical record. Many practices of domestic economy are just as invisible. They may appear insignificant in isolation, but in aggregate they wrought large effects: we should distrust figures and statistics purporting to show the total volume of material recycled where such work is not accounted for. Instead, I consider what people thought about materials, how they manipulated them for reuse, why they shunned some wastes and embraced others – and how this might inform our outlook today.

Growing consumer fussiness has placed limits on British reuse and recycling since *Master Humphrey's Clock* was published in 1840. Although many late twentieth-century householders were sympathetic, in principle, to the concept of recycling, as consumers they were reluctant to buy the end-products, regarding items made from recycled sources as second-best. At the end of the century advertisers pressured newspaper editors to print on crisp, unrecycled paper, and customers were thought unlikely to cope with glass bottles made from mixed-coloured glass waste. Of course, householders are not the only consumers of waste. Industrialists have attempted to secure regular and uniform supplies of waste for the manufacture of a wide range of products. The struggle for regularity is linked to the need to obtain predictable profit margins. The struggle for uniformity is linked to the preference of consumers for homogeneous and reliable products. Machinery and systems have been refined to cater for this love of the invariable.

One of the many passionate proponents of recycling that we find in this book is Jon Vogler, an enthusiast for recycling for environmental reasons. He penned an apology on the inside cover of his pamphlet about recycling, *Muck and Brass* (1978). Vogler had investigated the possibility of printing it on recycled paper but found it too difficult and expensive to obtain. 'Had the policies recommended in this report been adopted in the past', he noted, 'we would have been spared the embarrassment of printing it on virgin paper.'[8] If Vogler, a committed environmentalist and recycling obsessive, couldn't find good recycled paper, then what hope was there for the average consumer in 1978? Finding a market for reused materials has often been a challenge, but when more people than ever before could buy 'new', it became virtually insurmountable.

Despite the focus on the 'natural' among generations of campaigners, the urge to reuse, recycle and repair was rarely motivated by strictly environmental concerns before the late twentieth century. Instead, these activities had been connected to the availability and expense of new products. When prices of consumables fell, the incentive to reuse and repair slipped.[9] People habitually redeployed materials using skills and ingenuities that are overlooked and underestimated when citizens have easy access to new items. In some contexts, especially wars, recessions and depressions, consumables were more cherished because of scarcity. In those times people made unique items from old things: a whistle made from buttons or a travelling trunk from remnants. Some mass-produced things also incorporated redeployed material: bone china cups, linoleum, dyes, perfumes, creosote, soap, glycerine and gelatine. Matter has multiple lives: gold from a Victorian tooth might form part of a mobile phone. Commodities are a point of entry into culture; recycled commodities doubly so.[10]

Before the twentieth century, much reusable matter was collected by private industry. Waste was sold or given to companies to process into new things. Local councils now control the gathering and storing of household materials for recycling. Around the turn of the twenty-first century, British householders had to learn to separate plastics from papers from glass from tin for recycling. At that time the *Daily Mail* encouraged a backlash against recycling, apparently incensed by fines for incorrect bin use and alternate weekly collections. The tabloid identified 'recycling martyrs' who had been punished for contaminating recycling loads. A table entitled 'More Complicated than Sudoku' listed the various recycling rules of eight different councils, with the implication that people would need to learn all eight, not just those pertaining to their own local authority.[11] In every era, calls for more recycling have been accompanied by the sophistic yelps of those unwilling to change. In fact, the issue of contamination and incorrect sorting has a long history. In 1881, items in wastepaper collections included jewellery and bootjacks.[12] Rag Brigade boys collecting paper waste in London also gathered up six pairs of silk stockings. Horrifyingly, 'a tiny baby was found, pressed almost flat' in a bag of rags at a paper mill in Victorian Edinburgh.[13] In 1918 contaminants included postage stamps, sealing wax, artificial flowers, biscuits, boots, celluloid, jars and vulcanite. Parchment was 'particularly troublesome' if it got into the paper mill, where 'unignited heads of matches' were a 'grave source of danger'.[14] More recent 'contraries' have ranged from 'all sorts of non-fibrous materials from bicycle frames and bottles of ink to such things as strings, transparent film, pitch-impregnated and bitumen papers'.[15] Samples of toiletries wedged into glossy magazines can be handled as part of domestic mixed batches of recycling, but they present problems when bulk leftovers from printers or shops enter recycling schemes.[16]

In recent years, charity shops have taken the place of 'slop-sellers', reselling all sorts of clothing.[17] Vintage shops sell items at a considerable mark-up, and there is a small but increasing market for garments that have been made using parts of older items. This is not new – previously items were cobbled together from parts of pre-worn ones, or 'translated' for other bodies. The repurposing of garments has a long history. What *has* changed is the market for such products. Whereas it was once the poorest who managed with reused things, now the rich buy unique, individually crafted one-off recycled pieces. Many boutique designers cater to this: Elvis & Kresse have made expensive limited-edition bags using recycled British fire-hose since 2005; there are myriad other examples along the same lines.[18] The sentiments behind such creations are admirable, but the fact is that such reuse is aimed at consumers who already over-consume. A genuinely 'environmental' approach to consumption might see a person buying just one bag made from old hoses, but many will have one recycled bag among several non-recycled ones.

The British have long been inconsistent reusers and recyclers. Regardless of the motives driving reuse and recycling, idiosyncrasies have constrained the scale of the endeavours. During the Second World War people were enthusiastic collectors of razor blades for 'salvage' but proved reluctant to let their iron railings go. You would think that the appeal to patriotism would be overwhelming, with the German army planning an invasion. Yet it was not so, and many resisted, sometimes successfully. The British have been keen to be seen to be 'doing their bit' by collecting tinfoil lids and other small items but have been reluctant to relinquish their property, even on the brink of disaster.

Buying recycled items comes with a guilt deduction on wasteful consumerism of other kinds. The market for recycled products and the market for virtue are increasingly difficult to tell apart. The recycling of clothes is not the only activity to foster inconsistencies. Philosophies are pursued imperfectly – particularly when it comes to the environment. The costs of transport are often not calculated into recycling. In the early 1990s I knew someone who drove around with glass bottles in her boot, only dropping them in the bottle bank when making her weekly supermarket trip. The fuel spent taxi-ing bottles around probably cancelled out some of the environmental benefit attached to her recycling. Just as in less affluent countries now, recycling in Britain was once an inventive activity in which frugal people engaged routinely. In modern prosperous societies, recycling is less dynamic, sometimes undertaken to salve consciences or meet social expectations. This does little to offset the inexhaustible hunger for new stuff. Although they are obviously related, buying recycled items is not the same as recycling: they each involve different intents, efforts and responsibilities.

Another acquaintance of mine – 'K' – once argued that people should buy cars for shopping because supermarket home deliveries came in unnecessary plastic bags, so owning a car would eliminate plastic waste. I countered that there was probably more plastic in a car than in all the supermarket bags a person used in a lifetime (forget the petrol consumption, and the energy used to make the car). K regards herself as an assiduous recycler. Her son took his PE kit to school in one of the few plastic bags to make it into their house, one formerly containing non-recycled toilet rolls. He was the real 'recycling martyr'. Yet K often drives short distances unnecessarily and expects to offset that with careful recycling. K is far from the worst offender: she rarely takes long-haul flights. Those identifying themselves as the keenest recyclers are often the most frequent flyers. Many committed recyclers find it difficult to apply their environmental concerns with consistency.[19] George Monbiot notes this paradox: 'Recycling licences their long-haul flights. It persuades them they've gone green, enabling them to overlook their greater impacts', which cancel 'any environmental savings a hundredfold'.[20] The fact that recycling is assumed to be effective also encourages some people to be less careful with their initial consumption of materials, such as paper.[21] In *Use Less Stuff* (1998), Robert Lilienfeld and William Rathje described recycling as 'merely an aspirin', which alleviated the 'large collective hangover' of over-consumption.[22] This is still true today, but the history of recycling reveals a more complicated and less morally clear reality. At the root of all this is the way we make calculations about our total effect on the world. Those calculations or guesses are based on assigning value to different sorts of processes and materials, as well as on assessing whether what we are doing is worth the waste it is causing.

A little ingenuity can allow householders to circumvent design weaknesses and extend the lives of household things. Here K is one of the good guys, as she often ensures things last longer by crafting workarounds. For years after the catch broke on her oven door she made it close by jamming in an old potato peeler, staving off a new purchase and reducing landfill. These saving types, who put odds and ends away in case they come in useful, are often regarded as hoarders. There is a growing stigmatisation of those who don't keep orderly and sparse domestic spaces, and this works against efforts to put aside useful items for later use. To embrace a make-do-and-mend attitude, householders may need to lower their standards of décor and not leap immediately to replace something if one part of it breaks. This requires a certain amount of stubbornness, plus a willingness to acquire forgotten skills, but small efforts to reuse and recycle, pursued continuously, lead to substantial material savings as well as personal fulfilment. Various factors govern the extent to which the work is possible: space, time and skill. Government attempts to intervene in domestic habits have proved bad at taking those constraints into account.

With this in mind, *Rummage* also considers the boundaries of reuse: the economies and hazards of reutilisation. When I was eight years old I got a second-hand, one-armed doll – a 'Tiny Tears' – which opened its eyes when picked up, and cried and urinated if filled with water and squeezed. Once, after filling her with water, I absent-mindedly left her in a cupboard. When she was rediscovered, the mechanism for opening her eyes had jammed shut with a ring of mould. Her arm socket and the base of each clump of nylon hair poking through her plastic scalp were blue. I threw her away. My dad, on seeing her in the bin, scolded me. 'Someone else could have had her!' he shouted. But Tiny Tears was toxic. Some things do just have to be thrown away. Deciding what those things are is a matter that every individual, corporation and public body must confront, and there is rarely complete consensus.

The supply of things appeared more finite and everything seemed to be fixed with gaffer tape in my childhood. The 1970s were a difficult time; a global oil crisis in 1973 (the year I was born) forced people to reconsider the national reliance on imports. Shortages made householders, businesses and institutions re-examine materials with a view to economising. Disturbingly, hospitals under the oversight of the Birmingham Regional Hospital Board reverted 'to less economic and sometimes less hygienic practices to save materials'. Disposable plastic forceps were reused – as were cotton items that were 'no longer sterile'.[23] *Rummage* is also interested in the threshold where recycling and reuse become disgusting or unhealthy. That is key to understanding not only domestic habits but also the segregation of certain sorts of inventive industry as well as changing political attitudes towards recycling.

As we look back at our long and intimate history with objects, we see that humans are consumers in the broadest sense – not just purchasers of things but skilled finders, reusers and menders. Some people have been better able to manipulate things. They had more time, space, ingenuity, inclination or need. People need to be taught skills if they are to make use of materials – this is true now, when certain practices are dying out because parents no longer pass them on, as it was in the past, when mothers taught daughters how to darn. Much reuse was conducted through trial and error, tinkering and botching at home. People able to make things are the people best placed to mend things, and they tend also to have a better eye for materials that can be recycled.[24]

The history of recycling and reusing involves millions of people

2. Cot-cum-compost-bin. Formerly in the author's
collection, photographed by Ben Webster.

throughout history. The majority of reuse was realised in the everyday endeavours of everyday folks. While recycling and reuse might
be a national concern, government policies take up a relatively negligible place in that history. They are examined here only to the
extent that they interact with attitudes, practices and experiences
on the ground. Nevertheless, people are both individual consumers
and also members of a community or nation of consumers. Individual reuse actions have an impact on communal resources. Take,
for example, my son's robust pine cot bed. Ned inherited it from
his sister, and when he grew out of it, I could have passed it on to
someone else. That would have been socially beneficial. Instead I
took it to my allotment, adapted it and filled it with matter to
decompose (**fig. 2**). In terms of my own economic situation, reusing
the cot made sense, negating the expense of new materials to make
a compost bin, but from the moment I filled it with allotment
waste, it filled me with guilt. Reusers make subtle calculations:
should an old (but still wearable) skirt be cut up for patchwork, to

avoid buying new fabric to finish a project, or ought it be passed on to someone else? How do we decide that one form of reuse is better than another?

Before this century, most people were material savers. They saved and reused habitually, often without any fraught reasoning. My sister and I always kept used ice lolly sticks for crafting, felt-tipping over the jokes. Many years later, by then a mother with a toddler, I spotted a pack of 200 multi-coloured lolly sticks, ready for crafting. My daughter is now grown up. There are sticks left from that pack. In the intervening years we have thrown away countless actual lolly sticks. My younger self condemns me for this waste, but she didn't have any money. I can afford more than my parents could, and the shops are full. If as a child I could have seen how we would live later, I would have found it difficult to understand. People first develop their relationships with things during childhood, when they are apt to make use of discarded matter, partly through a higher tolerance for used things, but mostly because it is all they can freely use. When I was young, I read, and re-read, *The Little Old Woman Who Used Her Head*, a collection of stories by Hope Newell. First published in the US during the Depression, but republished in 1973, the stories centred on a daft old woman and her making-do. Her errors caused me much reflection as a child. Needing buttons for geese jackets made from a moth-eaten blanket, the old woman eschewed new brass ones from the peddler, 'two for a penny', and instead drilled holes into pennies, making each a button (her logic being that she kept her cash).[25] *Rummage* tells the story of people who always thought materials could run out, and often found novel or counterintuitive ways of dealing with their fear.

Rummage is everywhere concerned with the cultural status of patching, botching, piecing; with the uncertain value of mysterious new mixes; with the possibility that leftovers, traces, grains, particles, detritus and waste might prove useful and should be heaped up in confidence of their future utility. Such possibilities shape and inform the texture and make-up of this history. *Rummage* is a heaving mass of miscellany, bits and bobs that come together into a new compound. It is not only a book about rummage but a

3. Evelyn Bradbury's cardigan. Author's collection,
photographed by Taryn Everdeen.

rummage itself. It is dedicated to Evelyn Bradbury, a woman who
recycled and mended all her life. Evelyn was my Nan. Born in 1923,
she died in 1992. The most cherished heirloom from Nan's life that
I own is a striped brown and orange crocheted cardigan made from
several older items unravelled and reworked (**fig. 3**). I lengthened
the arms using the spare bits of yarn that Nan had carefully saved.
My additions are knitted because I cannot crochet. Nan taught me
to knit and to patchwork. I started my first quilt, aged eight, under
her supervision, using old Christmas cards for paper templates
(part of that quilt appears as the background to Nan's cardigan, in
fig. 3 above). Patchworking is *en vogue* once more, and modern
patchworkers have myriad fabrics to choose from; some buy 'fat
quarters', prints in carefully selected tones made by dividing a
brand-new yard of fabric into four. Meanwhile, the British bundle
up a fluffy mountain of waste textiles to ship abroad. Nan would
never have bought new fabrics for her patchworking. Instead she
saved offcuts from needlework projects and salvageable parts of
careworn garments in a 'rag bag'. Buying new fabric to cut it up

misses the point of patchworking entirely. 'Fat quarters' would have filled her with horror and confusion. I learned much from Nan, and *Rummage* is full of her hoarding spirit. She learned much during the war. A lot has changed since then.

2

Stuffed Animals (1950s–1990s)

In 1983 my Nan gave me a toy lion which smelt of feet. To make Stinky Lion she rummaged around in her enormous rag bag. For the beast's body she fished out some ochre Crimplene left over from a dress she had made. From a jar she found two black buttons for eyes. The look was topped off with a majestic brown bow from a chocolate box. Nan kept all her old nylon tights in a box behind the settee, chopped up ready for stuffing creations such as this – whence the lion's idiosyncratic odour. I kept Stinky Lion on a shelf with a collection of passed down and second-hand toys, including the one-armed Tiny Tears.

My mother was a devotee of jumble sales and dress agencies – everything we wore above underwear was either made by my mother or Nan or obtained second-hand. When we were small, my sister and I embarrassed our mother as she pushed our double-buggy through the shopping precinct, on our way to redeem deposits from bottles of Corona dandelion and burdock by singing loudly: 'Our coats came from a jumble sale … and so did this pushchair!' OK, it doesn't quite scan, but it got lots of attention. My favourite jumble sale find was a plush Womble, made in 1974 from grey acrylic plush and orange Bri-Nylon. Forty years later, I was curious to see whether the manufacturer had kept the Wombling spirit of reuse at his core, so I eviscerated him (**fig. 4**). His stuffing consists of toy factory waste, with cardboard discs stiffening his feet – his manufacture did include material reutilisation.

4. Eviscerated Womble. Author's collection,
photographed by David Cockayne.

My Womble was made in the same year that the Conservative government implemented a 'three-day week' in response to industrial action by miners, in the form of an overtime ban, that started in late 1973. In 1974 businesses (but not essential services, supermarkets and newspapers) had their electricity consumption limited to three consecutive days. This had an effect on material supplies and triggered a temporary interest in recycling, which people assumed was new but was actually centuries old. Glass became the centre of recycling attention; bottle banks were trumpeted as the answer to material shortages and energy-saving. But glass is made from cheap and ubiquitous materials, and the energy spent recycling it could have been better harnessed elsewhere. Paper and metal recycling got less attention, and plastic recycling came to be seen as too pesky an endeavour to persist with. While the public were being distracted by glass recycling, other forms of reuse were continuing quietly – and some of these activities were to create public health crises as the century progressed.

Womble plastics

The Wombles, a stop-motion animation children's programme, first appeared on the BBC in 1973, adapted from books by Elisabeth Beresford. Looking like aardvarks in sheep's clothing, they made 'good use' of the things that the 'everyday folks' left behind on Wimbledon Common. Beresford was the daughter of the author John Davys Beresford. After her parents' relationship ended, she lived with her mother, Beatrice, in Brighton, where they took in lodgers. Beatrice was from a provincial middle-class background; her father had been a wholesale clothier, a dedicated parish councillor and overseer of the poor.[1] This altruistic atmosphere, plus experiences marshalling the belongings of lodgers and guests, may have sensitised Beresford to waste even before she served as a wireless telegrapher for the Wrens in the war. In an interview, she discussed the Wombles' 'fight against pollution by collecting the rubbish we humans leave on the common and hoarding it in their warrens'.[2]

Womble names imply émigré origins: Great Uncle Bulgaria, Orinoco, Bungo, Tomsk, Madame Cholet. Immigrants to Britain have often pursued activities and industries involving the reuse of materials, as we will see in later chapters. In the episode aired on the day I was born, Tobermory mends a broken rocking chair using an abandoned tyre found by Orinoco. Beresford suggested that the recycling in *The Wombles* was a whimsical novelty; she even claimed that they had 'invented recycling'.[3] *The Wombles* were instantly popular – they even inspired a novelty pop group. Although they were hijacked by the 'Keep Britain Tidy' brigade, Wombles were much more than pickers up of litter; they were salvaging gurus. Their popularity came at a moment of increased sensitivity to the value and utility of things. For a while there was renewed interest in repair, reuse and recycling. There was a boom in domestic crafts reutilising materials: the British made rag rugs, patchwork quilts and corn dollies. The Wombles were at home in this ephemeral zeitgeist, and they addressed a young and impressionable audience.

Whether or not through the influence of *The Wombles*, children became involved in many recycling schemes in the 1970s. The 'War on Waste' green paper (1974) noted the need to involve children as volunteers.[4] Through schools, and organisations such as the Scouts and Guides, waste merchants made good use of the free labour of children. Charity collections were encouraged during periods of strong demand. In 1977, 10 per cent of wastepaper for cardboard-making was gathered via charitable collections (compared with 15–20 per cent from local authority collections).[5] *Blue Peter* campaigns tapped into this juvenile eagerness, but they did not always run smoothly. In 1970 their appeal was for old cutlery, but some parents in Farnborough, Hampshire, found that their offspring had got carried away and sent off the family silver.[6] Two years later, millions of woollen items saved for the appeal had to be salvaged from a warehouse fire in Camberwell.[7]

Of course, this appetite for recycling was not without its hypocrisies; consumers are inconsistent. Just as groups of Girl Guides were discovering that approximately 5 million milk bottle tops made a ton of aluminium, individual Guides pestered their parents to buy *Womble* merchandise, much of it made from nylon or plastic.[8] During the 1970s, the use of acrylics and synthetic plastics boomed. Plastic replaced paper as the prime packaging material. Seen as hygienic, plastic was robust and non-degradable: even cuddly toy Wombles came in plastic bags. The traditional means of salvaging metals, paper and glass were relatively straightforward, if labour-intensive, but plastics posed new recycling headaches.[9] Many people had expected plastic recycling to be simple, but the sheer variety of types and forms of plastic available by the 1970s made it complex. Training was required for the separation of thermoplastics, which are often soft and flexible, from thermoset plastics, which were generally more rigid.[10] There were even obstacles to correct identification of plastic itself: 'soggy parchment can easily be mistaken for a plastics film, helically wound tubular paper containers with glossy overprint look and feel like plastic containers.'[11] Non-synthetic plastics had been developed by Victorians using recycled materials, so this later difficulty surrounding synthetic plastics was no simple irony.

Officials were aware of the problems, and plastic recycling was mostly experimental in the 1970s. A scheme to turn polystyrene cups into shoe heels, devised during the oil crisis, was never put into action.[12] In 1973 some companies toyed with the idea of installing regranulation plants to recover and recycle plastics, but few actually did, and many subsequently vanished. The cost of recycling did not create a product significantly cheaper than one made from new substances, reducing industrial inclinations to recycle.[13] By the end of the 1970s, the last *Wombles* episode was five years old and memories of the three-day week had faded. Margaret Thatcher was Prime Minister, and her government heralded a more luxuriant, environmentally naïve era. Thatcher herself was keen to eke out the life of her own possessions, but her policies encouraged a social mentality that minimised communal sharing and magnified profligacy.

No return to the returnables

While Girl Guides were interested in the bottle tops, the glass milk bottle itself became the focus of recycling debates in the 1970s and '80s. Since the 1970s, much British energy has been spent on transporting intact used glass somewhere to be crushed, re-melted and added to raw materials to make new glass bottles and jars. Things are different on the Continent, where the bottles are returned and a deposit redeemed.[14] Back in the 1970s, the gentle chink of the bottle on the doorstep and the quiet whirr of an electric milk float were still familiar sounds of reuse. Approximately 150 million milk bottles were circulating daily in Britain in 1973, and each could be reused many times, after sterilisation. The area served by a dairy determined how long the bottles might recirculate; in southern suburbia they might last fifty trips, but in 'some parts of Glasgow' they only managed six. It was also reported that 'return rates are low in tower blocks and bedsit areas where people buy milk from stores'.[15] In 1972 *The Times* confidently reported that the future of the milk bottle was safe, predictions of its demise in the 1960s had proved unfounded; the milk bottle had 'reemerged as that most fashionable of all products, the returnable container'.[16] Back in 1961, the same newspaper had gloomily found that 330 million milk bottles were lost each year, some of them jettisoned into 'the nearest hedge or ditch' by ignorant tourists.[17] By 1973, the

5. Rinse-and-return milk bottles from the 1970s. Author's collection, photographed by Taryn Everdeen.

estimated number of bottles needing replacing each year was 500 million, putting the figure for the '60s into the shade.[18]

Milk was not the only product to come in returnable vessels. The production of non-returnable beer bottles soared by nearly 600 per cent between 1950 and 1967.[19] A tipping point came in 1969, when, for the first time, the majority of bottles were non-returnable. This trend has never been reversed.[20] One reason why drinks companies preferred to move to single-use bottles takes us back to 1928, when a Glaswegian, May Donoghue, ordered a 'Scotsman's ice cream float' (a mix of ice cream and ginger beer) in a Paisley café. She got more than she bargained for: decomposing remains of a snail plopped out of the ginger beer bottle and made her ill. The snail had probably got into the bottle before it was reused by David Stevenson's drinks company. The washing process had been compromised in some way. The bottle was not see-through, but dark opaque brown, and so the café owners could not have known the snail was there. Donoghue's solicitor steered her to sue Stevenson's. The writ against that company described the plant as a place where 'snails and the slimy trails of snails were frequently

6. Allsopp & Sons (early twentieth-century) beer bottle. Clensel bottle from J. Paterson & Co. Author's collection, photographed by Taryn Everdeen.

found'. There were various appeals after the initial case, heard in the Scottish Court of Session, which found Stevenson to be negligent, and in 1932 the case was heard in the House of Lords.[21] Companies selling drinks in reused bottles needed to invest in bottle-washing apparatus. Caustic solutions and hot water sterilised the bottles, but machinery took up space and required maintenance and oversight.[22] By the 1960s many companies wanted to avoid these hassles, and found they had more control over processes involving single-use bottles – with no chance of accidental snails.

The move towards single-use bottles did not happen quietly. In 1971 the newly established British group of Friends of the Earth, a campaign network that developed from one started in San Francisco in 1969 in response to environmental concerns, reacted to the announcement that Cadbury Schweppes was to abandon returnable bottles. In one of their earliest direct-action campaigns Friends of the Earth dumped loads of bottles on the steps of Schweppes's headquarters in London.[23] The deputy chairman of Cadbury Schweppes retorted with a woolly statistic: 'About 75 per cent of our trade', he said, 'is mainly in returnable bottles.'[24] The hackles of other campaign groups were also raised, but for different reasons. The Council for the Protection of Rural England worried about bottles littering the countryside. Cadbury Schweppes, struggling to tell apart the motives for the various campaigns, issued the statement that 'Litter is not caused by manufacturers: it is caused by litterbugs', without addressing specific concerns about sustainability and energy use.[25] Friends of the Earth enjoyed a surge in membership, but Cadbury Schweppes's decision to move away from reusable vessels was not reversed, and the charity, in seeking to promote *reuse*, came instead to be regarded as a champion of *recycling*.[26]

Recycling can be organised, and it creates new products – it is sexier than reuse. Recycling also adds a layer of processing that makes consumers more comfortable with products, but it is generally less environmentally sustainable than reusing. Environmental activists and academics continued to agitate for a return to returnables, arguing that they conserved energy and represented savings for consumers, but customers slipped ever further out of the habit of returning returnable bottles to shops; many were dumped in landfill.[27] R. White and Sons Ltd noted a drop in the return rate for their 4p deposit bottles. A spokesman speculated that 'the throw-away bottle campaign has lulled people into believing all bottles are now non-returnable'. A glass shortage in the early 1970s caused panic in the industry. An overtime ban at a Cheshire factory that produced soda ash for bottle-making saw drinks manufacturers run out of bottles. Makers of vermouth had particular difficulties bottling their products in the run-up to Christmas of 1973.[28] Some – even a few members of Friends of the Earth – agitated for plastic sachets as the way forward.[29] Beechams, whose products included Corona, Ribena and Lucozade, were one of the few companies that spoke positively about returnable bottles, stating they would be 'happy to abandon the limited use' they made of non-returnables, 'provided that our competitors do likewise'[30] – a sneaky caveat that proved to be the catch. In 1974 the Scottish firm AG Barr, producers of Tizer and Irn-Bru, increased the deposit on bottles in England but kept their original deposit in Scotland, since recovery rates there were 'much higher' (presumably, outside of 'some parts of Glasgow' from which the milk bottles never returned). Barr's retained a deposit scheme until 2015.[31] Tetra Paks and plastic containers came to dominate the milk packaging market during the 1980s, when consumers turned away from the milkman and towards the supermarket.

Save with the bottle bank

Glassmakers felt under threat not just from demands to keep returnables in production but also from the ever-expanding development of plastics, including robust polyethylene terephthalate (PET) bottles, which were gaining enthusiasts.[32] With mounting pressure from charities and academics to continue making reusables, the glass industry, represented by the Glass Manufacturers' Federation (GMF), responded by instead funding research into alternative uses for waste glass, such as glass-fibre and tiles.[33] Realising that the work of waste glass collection would mostly be carried out by volunteers and local councils, the industry also climbed aboard the recycling bandwagon. The GMF sponsored some of the first 'bottle banks', which were essentially skips with colour-coded sections for different types of glass. The scheme was the most visible encouragement to recycling since the wartime salvage.[34] To assist with the deployment of waste glass (known as cullet), the GMF compiled a directory of merchants and manufacturers willing to buy it.[35] In 1979, on the verge of expanding the scheme, the Glass Manufacturers' Federation reported on the progress of the bottle banks. In two years they had recovered 21 million glass containers (while 150 million milk bottles circulated). Collection was most viable for places near glassworks. Before a glass recycling plant was established in Harlow in the early 1980s, the capital's bottles were transported (somewhat incredibly) to West Yorkshire for recycling.[36]

As always, contaminated loads caused problems. Glassmakers had always used their own cullet in their manufacturing processes, but they were wary of 'foreign' supplies, of variable quality.[37] Bottles of different colours had to be kept separate as each contaminated the others. The 'chemicals in green bottles produce the

worst discolouration in clear glass', but a limited amount of contamination could be tolerated: clear glass could have up to 2 per
cent brown glass in it. Brown glass needed to be made from at least
80 per cent brown glass bottles, but the mix could stretch to 15 per
cent green glass. But the real menace in contamination was metal:
one aluminium bottle top in the mix of glass cullet could form a
slug of molten metal so potent as to drill a hole in a multi-million-
pound furnace.[38]

Not everyone was convinced of the merits of bottle banks. 'The
ballyhoo and publicity surrounding the recovery of glass containers
for remelting should not obscure the fact that minimal financial
returns accrue', warned Andrew Porteous, an academic with the
Open University. Once skip hire and energy costs are factored in,
the money and energy saved was 'quite small'.[39] Reporting on
bottle banks in 1980, Margaret Allen wondered whether the £2
million spent by glassmakers since the mid-1970s to encourage the
proliferation of non-returnable bottles was significantly greater
than 'the investment required to produce returnable bottles'. Allen
argued that, in costing energy expenditures, the GMF had overlooked the fuel spent by individuals transporting empties to skips.[40]

Jon Vogler, an engineer as well as an environmental activist, also
thought that returnable bottles with deposits were the best option,
and so lamented government inaction. He summed up the situation thus: 'Conservationists decry the industry's unwillingness to
re-use it; the industry points to the public's failure to return it [...]
shudders run through the bodies of glass industry executives at the
prospect of a "bottle bill".'[41] To Vogler bottle banks were merely a
sop to public opinion, and he highlighted the initial hesitancy of
the glass industry to embrace recycling. Why did glassmakers
bother with cumbersome collections of cullet, which offered only
negligible energy savings? Unlike other industries, their product
was made from abundant and cheap raw materials. Although glass
recycling made people feel virtuous for doing their bit for the
planet, environmentally it was 'of limited value'.[42] Looking back,
the environmental historian Brian Clapp has optimistically suggested that 'bottle banks must be useful – otherwise glass
manufacturers would not promote them'.[43] But this is naïve: their

utility was to distract from the question of returnables, which threatened to damage the glass manufacturers' trade. Little effort was made to collect bottles from places where reliable and plentiful supplies would be forthcoming, such as hospitals, hotels and restaurants, because bottle recycling was all about visibility and not sustainability.

Jon Vogler, who died in 2017, was the son of Sidney and Thérèse, and born in Hackney just before the outbreak of the Second World War. Thérèse was the daughter of Cicely Ashton-Jinks, a prolific but now little-known author of historical novels, who combined writing with a business focused on fur renovation and alterations to school uniforms. Cicely lived with Vogler's family when Jon was young. Sidney was a sanitary inspector, of Polish-Russian parentage.[44] Vogler's life, works and writings put me in mind of some of the municipal engineers who devoted their energies to dealing with waste and reuse – men such as Jesse Cooper (J. C.) Dawes and John Arthur Priestley, whom we meet in later chapters. Vogler, like Priestley and Dawes, spent time abroad: he was one of the first people to do Voluntary Service Oversees, and worked as an engineer in Nigeria in the 1960s.[45] All three men, with backgrounds in engineering or sanitary inspection, were often way ahead of their times; they all raised the profile of reuse and recycling in the face of resistance.

Before he became a critic of the schemes, Vogler had set up his own glass reclamation scheme in Huddersfield – but it failed because glass firms paid low prices for cullet.[46] Reflecting on the effects of world poverty and the finite nature of most resources, by 1975 campaigners for the charity Oxfam had become involved in recycling. Oxfam and Vogler combined with the Kirklees Metropolitan Council for an experimental recycling scheme, Wastesaver, which focused on various materials.[47] Huddersfield, the largest town in Kirklees, was the focus of an enterprise that salvaged glass, paper, cardboard, tinplate and rags, with net profits being returned to the council. In addition to being able to bring oil for recycling to the Oxfam car park, householders were given multi-compartment units made of tubing (called 'Dumpies') which facilitated the separation of domestic wastes into coloured bags (newspaper; mixed wastepaper; glass, plastic and tin; fabric, books, and bric-à-brac). Eventually the Dumpy was dumped in favour of a single sack.

A couple of thousand local housewives volunteered their time. Fearful that local dustmen might grumble about infringements on their 'totting' privileges (a customary right to keep decent rubbish), they were offered a bonus. Collections were taken to a disused textiles mill. Tins were reduced to 'small flat chips' and taken to a de-tinning plant on Teesside; glass went to Redfearn Glass in York. Ultimately, the scheme was also less successful than it might have been. Vogler pointed to increasing labour and transport costs coinciding with decreasing sums realised for waste items. But at least it took in more than just glass bottles, and the Wastesaver scheme survives today, in an attenuated form.[48]

Derek Dey's old newspapers

Bottles got a lot of attention, but glass recycling has never been the most essential repurposing activity: paper has long had a more intense but less visible role to play in the story of reuse. Over the course of the twentieth century, improvements to household heating had diminished reliance on domestic fires: and so the papery content of the bins increased as less was burned in hearths.[49] Yet local authority wastepaper recovery shrank steadily between the late 1950s and 1980, from 40 per cent to 10 per cent.[50] Few local councils collected wastepaper from households in the early 1970s. In some places, such as Solihull, small schemes remained as a holdover from wartime salvage. Why did collection fall at a time when waste was on the rise?

One key hindrance was fluctuating demand.[51] The bottom fell out of the wastepaper market when large quantities of wood pulp started flowing into the UK. Pulp was favoured because it made better-quality paper. Wastepaper was a last resort when pulp was in short supply, or prohibitively expensive. Householders and volunteers who collected wastepaper with zeal in 1973 and 1974 became disillusioned when over-supplied mills started refusing to take it in. By the summer of 1975 supplies of old newspapers that had been collected by some keen councils started to pile up, causing storage headaches.[52] During fire service strikes in 1977, warnings about the

7. Thames Board Mills bookmark, 1974. Author's collection.

risks of hoarding paper put another nail in the recycling coffin.[53] Four tons of twelve-year-old Derek Dey's collection of old newspapers, gathered to pay for a kidney dialysis machine, were destroyed in a fire near Aberdeen in 1979 while awaiting sale to a paper merchant.[54] Derek himself wasn't harmed.

Paper-mill managers were reluctant to invest in new plants with higher recycling capacity without being sure they could regularly draw in enough wastepaper. Makers using wastepaper needed to be mindful, inevitably, of contaminants, which were more numerous than the occasional stray bottle top in glass. They were listed in 1977: self-adhesive envelope seals, rubber bands, chewing gum, cigarette packet foil, chip fat and 'plastic spots used to correct errors on computer tabulating cards', which were described as 'the very devil to remove'.[55] A few paper mills did focus on using wastepaper; some used it to make the 'flutes' comprising the central bulk of corrugated cardboard.[56] The Mugiemoss paper mill collected wastepaper from households in Aberdeen each Wednesday in 1978, processing it into paper bags for bakeries and lining papers for plasterboard – the low-quality stuff. It was to Mugiemoss that Derek Dey's papers had been destined.[57]

This bookmark (**fig.** 7), made from over 60 per cent recycled wastepaper, was produced in 1974 by Thames Board Mills in Purfleet, Essex, along with similarly be-sloganed leaflets, posters and bin wagon banners. Note the order of the benefits to be gained from saving wastepaper listed on this bookmark: saving money and reducing imports take priority, with environmental concerns coming last. Promotional materials were distributed to local authorities and voluntary groups in order to stimulate local collections.[58] By 1972 the company was utilising over half of the wastepaper collected by local authorities, some of it being made into Fiberite, a tough cardboard packaging material.[59] Recycled paper was regarded as inferior, reducing the potential market to such an extent that few companies were willing to invest in it. By 1980 only four companies produced paper that was 100 per cent recycled; two of them were based in Scotland.[60]

In local authority and media discussions, the problem of the depletion of natural resources was barely mentioned. Economic viability and profitability mattered most; immediate pragmatic issues were at the fore, not medium-term conservation. Journalists who noted that paper production minus waste input meant more

8. Reused envelope from 1982. Author's collection.

trees would be felled did so in afterthoughts.[61] An Oxfam pamphlet reminded readers that 'private profitability' was not 'always an adequate criterion for decisions about the reclamation of Britain's waste resources', citing fair global resource management, employment opportunities and environmental concerns.[62] Friends of the Earth kept their message up: the world was at stake, so recycling did not need to turn a profit, just to constrain human impact on the Earth's systems. Again pursuing reuse rather than recycling, the charity created envelope reuse labels. Despite their best efforts, environmental concerns made little impact on paper recycling rates.

Housewives and houses

In the 1970s governments were criticised for not amending the law to favour waste reusers and for not doing more to stimulate demand and change attitudes: for example, by using recycled paper for ballot papers.[63] Going on the apparent or assumed unwillingness of their constituents to engage with, or fund, recycling efforts, some councils backed landfill-tipping schemes in the 1970s.[64] Recycling was seen as unlikely to break even or would make too small a profit to make the effort and upheaval worthwhile.

Women, the most enthusiastic adult recyclers, bore the brunt of the blame for the delinquency of authorities and businesses. Housewife apathy, if not actually a reason for a lack of progress in the decade, was cited by local councils to justify the rejection of collection proposals.[65] The success or failure of a glass recycling scheme in 1974 was said to hinge on whether the housewives of York could 'be bothered to clean out the jars and bottles properly'. This uncertainty was said to destroy 'the economies of the recycling operation'.[66] 'War on Waste', a green paper issued in September 1974 under Harold Wilson's Labour government, carried a realistic notion that there would be 'no easy answers' to the situation. The document massively underplayed the anticipated roles of charities and housewives, with charities referred to twice and housewives just once (compared with dozens of references to the role to be played by trade unions and local authorities). The 'householder' referred to on the fifth page is a 'he'.[67]

Women argued back. At a time when women's role in the home was coming under radical scrutiny, with some calling for wages to be paid for housework, campaigning women were divided. Left-wing feminist activists did not identify themselves as housewives – that was antithetical to the movement – but a group of house-wives formed who defined themselves as 'not Women's Libbers'. The National Housewives' Association claimed to fight inflation, and to teach housewives how to mount effective consumer com-plaints. Women at both ends of the political spectrum campaigned for more recycling. In 1974 Sandra Brookes, chair of the National Housewives' Association, led a delegation to Gordon Oakes, the Under-Secretary at the Department of the Environment, to encour-age him to consider installing reclamation centres throughout the country. 'At a time of economic crisis like this', he was told, 'we cannot afford to waste anything.'[68] Oakes pledged support, but government reaction was hesitant and sluggish; it was left to private industry to develop solutions.

Not everybody underestimated the recycling abilities of house-holders. In 1972 Graham Caine, an ambitious architectural student, built an eco-house which functioned by reusing human waste, harvesting rainwater and harnessing solar heat. Built from scavenged materials, it was constructed on fields belonging to Thames Polytechnic; a banner on the exterior read 'From Here We Grow'. The project was experimental and forward-thinking. Caine, part of Street Farmers, an anarchist group, planned the house at the start of the 1970s. It was not built in response to the oil crisis, but it did put forward and test potential solutions to energy short-ages. The Street Farmers understood self-sufficiency as a statement against consumerism; eco-houses presented opportunities to live 'off the grid'. Caine and his family lived in the house. Their excre-ment was converted into methane for cooking, and they grew fruit and vegetables in a hydroponic greenhouse. When an emergency forced them to be away for some weeks, an architectural student was trained to house-sit. Unfortunately, antibiotics taken for an illness worked through the sewage system, killing algae that pro-cessed the sewage and halting the production of methane. This time women couldn't be blamed. The systems in the house were derailed: 'antibiotics killed the house!' In 1975 the structure was demolished.[69]

Recycling animals

While the media was getting to grips with the term 'recycling', and local and national governments were causing mutual frustrations that scuppered meaningful recycling at the domestic and civic levels, farmers continued to recycle, as they had for centuries – but also in novel ways. Two developments, both involving recycling, led to food scares in the later twentieth century. The first was the use of antibiotics in animal breeding. In 1949 nutrition researchers in New York noticed that animals fattened when their feed was supplemented with vitamin B_{12}. Thomas Jukes and his team eyed up the vats of grain mash which had been used elsewhere in the laboratory complex to cultivate mould from which the antibiotic chlortetracycline was extracted for human use. They recognised the mash as a potential source of vitamin B_{12}. Chicks fed the dried mash gained additional weight. This was an in-house re-employment of a fermentation by-product. When experiments were repeated on pigs and cattle, it was discovered that vitamin B_{12} alone could not have produced such accelerated growth. Residual chlortetracycline in the mash had caused the startling improvements. So began the use of growth-promoting antibiotics in livestock. Henceforth animals could be raised in more confined spaces and would yield more meat while being fed the same amount of food.[70] This use of chlortetracycline was banned in 1969, but other antibiotics have since been used as animal food supplements, with serious implications for the effectiveness of antibiotics in treating human diseases.[71]

Another food scare hit the headlines in 1986, when it was reported that a cow had died the previous year of Bovine Spongiform Encephalopathy (BSE). Also known as 'Mad Cow Disease', BSE is a fatal neurodegenerative cattle disease that infected over 180,000 animals in the UK.[72] Uncertainty remains about the exact cause of BSE, but most experts think the likeliest explanation is connected to feed, specifically the protein-rich meat and bone meal (MBM) that formed an increasingly large part of British cattle's diet. Farmers had been feeding MBM to cattle for decades, turning naturally vegetarian animals into meat-eaters. Companies were often vague or secretive about the composition of their meal.[73] MBM had been advertised as ideal livestock feed since the 1880s.[74] In 1953 Sheffield Council erected a by-products plant to deal with condemned confiscated meat, which was rendered and ground into MBM to be fed back into the system.[75] An article from the same year asked, somewhat presciently, 'Do dogs get cleaner meat than family?'[76]

Other animal feeds also incorporated wastes. In 1981 microbiologists announced concern about the consequences of using sludgy animal wastes in feeds. 'Wastelage' was produced by mixing fermenting cattle slurry with molasses and straw. Apparently beef cows could be 'successfully maintained on wastelage alone'.[77] The 'War on Waste' green paper noted that wastes could be deployed as animal feeds as long as precautions were taken 'to avoid health hazards'.[78] And yet by late 1987 MBM was identified as the likely carrier of the pathogen causing BSE in cattle, suggesting that the disease scrapie, from sheep carcasses, was to blame. Animal tissue had been in cattle feed for a long time by the 1980s, so experts wondered if a relaxation of rendering regulations in 1980, effected by the Conservative government, might have allowed the MBM to become contaminated. Previously batches had been heated and then treated with solvents and steam: the hydrocarbon extraction method. After 1980 rendering companies could skip the solvent extraction. Most did. Only two firms in the UK retained the full process: both were Scottish.[79]

By 1985 nearly half of all animal waste in the UK was administered by the Prosper de Mulder Group (PDM). They sold MBM, blood, animal fat, tallow, hide and skin to distributors of livestock and poultry feed and pet food. The group was investigated by the Monopoly and Mergers Commission, which drew attention to anti-competitive market tactics.[80] Other companies were sucked up piecemeal by PDM. A report by the Secretary of State for Trade and Industry concluded that a merger with Croda International in 1991 was not contrary to the public interest. It was noted that the 'rendering industry has undergone difficult trading conditions since 1985 because of a collapse in the prices of its products and the effect of health scares concerning salmonella and BSE'. The report stated that 'there are important issues other than competition in this case. The merger is likely to improve PDM's efficiency and bring wider public health and environmental benefits.'[81] PDM employed a Conservative MP, John Whitfield, as a solicitor. Whitfield arranged lobbying meetings with John Selwyn Gummer, the Minister for Agriculture. PDM donated money to the Conservative Party.[82] When the use of MBM for enhancing the diets of ruminants, including cattle, had been prohibited in 1988, piles of unusable MBM were shovelled into warehouses across the country. By 1990 there was a mountain of MBM, estimated at 400,000 tonnes, destined for landfill.[83] During the late 1980s and up to 1996 the Conservative government repeatedly denied that there was a link between beef consumption and a human disease: a transmissible spongiform encephalopathy – variant Creutzfeldt-Jakob Disease (vCJD). Gummer infamously fed a burger to his daughter Cordelia to reassure the public about the safety of beef in 1990. Concerns mounted about the safety of mechanically recovered meat, which included spinal cords, leading to a ban on British meat exports.

Recycling and reusing, as we will see, are things that many businesses, including farming, have habitually undertaken for centuries. The late twentieth century witnessed some idiosyncratic reuse developments. While the public was being encouraged to see recycling as a new industry, and not an activity that had always been carried out by householders, industries and businesses, some agricultural recycling was taken to an extreme and unchecked level. At no point did the government seem to have a grip on any situation: either the encouragement of domestic recycling or the oversight of agricultural reuse.

🗑 🗑 🗑

Before 1974 recycling had fallen under the remit of the Department of Sport and Recreation. While it may have been exercising, British housewives could confirm that recycling was no recreation. The government had a lacklustre approach and was accused of not incentivising widespread recycling.[84] Under Gordon Oakes, recycling moved to the purview of the Department of the Environment; Oakes was dubbed 'the Minister of Waste'. Despite this shift, successive governments underfunded research and curbed long-term local government spending, meaning that decisions about recycling needed to prioritise short-term viability, and were unable to attract the significant capital investment that was required in the long term. Legislative action was weak: local councils and industrialists blamed national government, which in turn blamed the local authorities and businesses. Local authorities required industry to guarantee prices before they could commit to collections. Industry wanted local authorities to guarantee supplies before they would agree to buy collected waste. Neither could provide necessary guarantees because neither had full control of the situation. The local authorities needed to work with householders, whose goodwill was thought to be fickle. Industrialists needed to be able to switch to alternative raw supplies when they were cheaper, superior or more plentiful. The result was stalemate. Only national government could take steps to stabilise the market, but this did not happen.

Weak national leadership, fettered local control, price fluctuations and waning enthusiasm saw little genuine recycling in the 1970s, beyond the traditional reuse of by-products in industry. The entire decade was taken up in inaction and frustration. Some people lamented the lack of economic urgency at both local and national levels. In July 1973 a Mr C. Pollard wrote to *The Times* to express his bewilderment that, in these days of 'supposed scarcity', 'many hundreds of thousands of tons of basic valuable materials' were just dumped in tips, like the one he worked at in Bushey in Hertfordshire. British governments did far less than other European countries, such as Germany and Sweden.[85]

Virtually nothing in Britain was printed on recycled paper, not even the 'War on Waste' green paper of 1974. A BBC *Horizon* programme titled 'What a Waste!', shown on 24 May 1973, highlighted the throwaway culture. Henceforth, viewers would still chuck out reusable wrappings, but guiltily, rather than in ignorance.[86] Britain joined the European Common Market in 1974, bringing some

changes to rules controlling pricing for scrap metal and other commodities, and also introducing some environmental regulations,
but the British lackadaisical approach continued.[87] There is no evidence of widespread concern for the environment. In 1978 a
journalist wrote words that still summarise many people's concep-
tions of recycling: 'recycling is the type of activity which many of
us think of as happening elsewhere.'[88] *Plus ça change*.

Again and again the Scots were shown to be the nation keenest
to reuse and recycle properly (Glaswegian milk-bottle wasters
aside). Perhaps this was down to their own economic history or
maybe it was the result of a deeper, religious abhorrence of waste or
a moral preference for thrift. Following the relative success of the
Scottish Nationalist Party in the October 1974 General Election, a
new Womble arrived in the second television series, aired in 1975:
Cairngorm MacWomble the Terrible, with bagpipes and tartan,
channelling Scottish chic. In stuffing toy animals with whatever
waste was easily to hand, my Nan followed an ongoing tradition.
Vintage bears had been stuffed with wood shavings; my Pedigree
Womble was padded out with factory waste. Real animals did not
fare much better. In these strange times some people started driving
bottles to car parks and stuffing them into skips. Beyond that, few
recycled habitually. Much material reuse remained unseen: inside
toys, inside cardboard, inside cows.

3

Nellie Dark's Bundles (1930s–1950s)

On the afternoon of 1 March 1940, Nellie Dark stood in the small hallway of her flat, flanked by unsightly bundles of papers, food waste, bones and rags. A jagged tin snagged a stocking, so she put it aside for darning. Just like thousands of other women, Nellie had been eager to obey the order to avoid waste and not to throw into the dustbin anything that might be of material use. But nobody had come to collect her hoard – not even the Boy Scouts. The Darks lived in Bedford Court Mansions, off London's Bedford Square, Holborn. Nellie's husband, Sidney, was the editor of the *Church Times*. The Dark family were no strangers to salvage. Sidney was related to John Dark, a dust contractor whose evidence had been taken for a government-commissioned report on sanitation by Edwin Chadwick in 1842. There was a family rumour that John was the original Noddy Boffin, the 'Golden Dustman' from *Our Mutual Friend*, 'having been known to Charles Dickens'.[1]

Sidney would be back soon, and Nellie didn't want him to clatter into one of the heaps as he shed his overcoat. Had the Darks lived in Finsbury or Tottenham, the authorities would have been more prompt, but, Nellie concluded, 'Holborn cares for none of these things'.[2] What else would the daughter of a journalist and the wife of an author do but pen a letter to *The Times*? In response, the paper contacted Joseph Parr, the borough engineer and surveyor for Holborn, who argued that his department had been strenuous in their endeavours to collect salvage, giving the matter publicity and

9. Excerpt from a Ministry of Supply leaflet. Author's collection.

appointing a 'woman canvasser, to call on [...] people, inviting them to salve all they can'. Parr arranged for this woman to visit Nellie.[3]

'Every woman can help to win this war'

It is easy to think that the spirit of salvage was born in the Second World War – but those being encouraged to recycle in the 1940s often found it no easier than those struggling with colour-coded bottle systems in the 1970s. Just before the Second World War, British bins were mostly filled with a combination of cinders and ash, food scraps, bones, paper and tin, along with broken pots and glass. As the war progressed, the government requested that various types of waste be separated from the rest of the rubbish. Limitations on time or space scuppered many salvage opportunities. Nellie Dark's plight stemmed in part from the arrogance of official-dom and the assumptions they made about householders' circumstances. Officials often appeared to lack local knowledge or an awareness of how parts of the community actually lived: for instance, not every citizen inhabited a separate house, and so not everyone had their own bin, or their own gate to hang salvage on. Many had serious space limitations.[4] Parr suggested that Nellie put her bundles of salvage outside her block, by the dustbins, but surveys found that most 'people will not carry stuff to outside dumps and bins, and often haven't room outside their backdoors to put it out for collection'.[5]

Nellie, like many city flat-dwellers, as well as those who occupied back-to-back properties, would have found the instructions difficult to comply with. A Hampstead dustman noticed that houses of multiple occupation 'produced practically nothing, in all classes' for salvage, in contrast to separate family houses. A Kilburn landlady took umbrage when a Women's Voluntary Service (WVS) woman 'just waved at the hall and said, "There's plenty of room there"!', after she queried where salvaged materials should be stockpiled. 'I can't have paper piled up in the hall', she said, 'I have to think of how things are going to look.' A mother refused to keep three separate salvage pails in case her baby fell into them.[6]

Uncollected salvage created litter, discouraging householders from putting more out. People who worked during the day missed ad hoc collections of salvage. To advertise their presence, some salvage gatherers used loudspeaker vans. One called out: 'Now then ladies, the salvage van is in your street now; will you please bring all your salvage down to your front gate to save them wasting their time; we will take it up as we pass.' In Hampstead it was assumed that some housewives – and it was inevitably housewives – would serve as centres for collection, but 'official enthusiasm' was not 'wholly echoed by the housewives themselves'. One complained that she had no room, and no time to answer the door 'all day to people with bits of paper'. Householders in Willesden made the not unreasonable request that the frequency of salvage collections be increased, from fortnightly to weekly, and for a fixed collection day. Complaining about the poor organisation of wastepaper collection in Kilburn, one resident snorted 'that is ridiculous' when instructed to hang a sack for salvage by her gate: 'with all the rain and the wet it would all be pulp before they came round for it [...] I thought it was nonsense frankly.'[7]

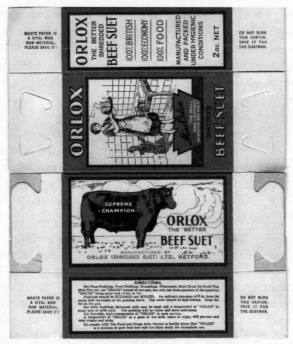

10. Orlox beef suet box, 'Do not burn this carton. Save it for the dustman.' Author's collection.

A Salvage and Waste Materials Order of 1942 mandated fines of £500 and two years' imprisonment for wasting paper, mixing paper with other refuse or littering. The contents of bins shrank.[8] During the war many wastes were put to new and ingenious uses: ashy remains could level bomb sites or fill sandbags; sewage grease was turned into candles, camouflage paint and dubbin; cereal boxes became cut-outs for target practice; and abattoir blood was reused in miners' safety helmets.[9] Slogans such as 'Shirts off to Berlin!' helped secure cotton and woollen rags, worn out clothing, tattered dusters, undarnable socks, ancient curtains and old carpets and sacking to be turned into uniforms, blankets or padding for tank seats. Even oily rags were in demand: 'wipers can be de-oiled, cleaned and used again and again; even the recovered oil can be used again.' Rope, string, twine and rags were made into charts and maps, insulation and camouflage netting.[10]

In 1941 salvage collectors in London found mouth organs and a dead dog in loads of wastepaper.[11] These particular misplaced items couldn't quite be put down to the complexity of sorting materials, which was nevertheless a recurring wartime grumble – what went where? Were cans ferrous or non-ferrous metals? 'Should old mackintoshes be put out as rags or rubber salvage?' asked *The Times* in 1942. In fact, the newspaper had answered this question six months previously: rubber-proofed mackintoshes contained 'too small a proportion of rubber to make reclamation worthwhile'. Old macs were 'rags'. Such quandaries led a commentator to look for the silver lining:

> So the disposal of rubbish, which had seemed the simplest of duties, becomes the most complex and recondite of tasks. We are, as a nation, inclined to casualness in our thinking, looseness in our definitions, and it may well be that [...] these constant demands for precision and nicety are good for us.[12]

That demand for precision and nicety wasn't necessarily heeded by waste collectors. The careful separation of items was not always maintained: wastes were often jumbled into one cart together, which made many a householder reflect on the time they had wasted.[13] A London diarist wrote that 'to my horror I observe the dustmen tipping all the carefully sorted paper, bones, tins etc. into their cart together. Really this sort of thing is the limit. The maids – whom I've ticked off several times for not saving paper properly, etc. – were absolutely furious.' She would have written 'a raging letter to the Borough Council' – had she had time.[14]

While the content of waste bins was under scrutiny, the marshalling and reuse of items inside the house was an equally contentious matter for the home front. Housewives were exhorted to 'Make do and Mend' by darning socks, turning collars and cuffs, patching and unpicking. Socks for a soldier's kitbag might be knitted from an unravelled swimsuit. Official leaflets issued by the Board of Trade instructed householders how to perform various mending tasks. Suggestions for reinforcing children's clothes and other tasks appeared in leaflets featuring 'Mrs Sew-and-Sew' (a demented doll figure, wielding a sewing needle).[15] A homiletic tone is evident: missives recycled ancient advice about stitches in time saving nine. The Ministry of Supply and local authorities aimed their salvage instructions at housewives with good reason. Women sorted the domestic salvage and helped to arrange local collections, often as part of voluntary societies such as the Women's Institute (WI) and the WVS. Much of the organisation and promotion of salvage in Gloucester was undertaken by Ina Skinner, known as Judy (later Baroness de Turckheim). Skinner became the 'Assistant Honorary Salvage Adviser' for the council in March 1941. The local WVS organiser recognised Skinner's enthusiasm, remarking that her rousing talks on salvage managed to turn 'a very dull subject into an interesting one'.[16] One leaflet issued in 1942 made its own attempt to add excitement to the dull affair of taking out the bins, with the eye-catching title 'Bombs & Bullets from Your Dustbin', commencing the text with 'Every woman can help to win this war'.[17] The only appeals geared towards men concerned oil or rubber items, because men were more likely to be motorists and sportsmen, with spare tyres and rubber balls to donate.[18]

Wartime appeals for salvageable material tested the values of the nation. Just a week into Britain's involvement in the war, Dr Bernard Stark, an Austrian legal academic living in Hampstead, wrote to *The Times* suggesting that a National Defence Fund be established, using gold donated by citizens. Dr Stark offered his last gold ring as an initial contribution, having lost the rest of his property to the Nazis. Lady Lucy Amos, the wife of another legal academic, asked the same the following year, noting that Italian women had donated their rings, and wondering whether 'some under-employed factory might make for us in exchange rings of any base metal that can be spared'.[19] The idea was not novel: Prussians sacrificed gold wedding rings for iron replacements during the Napoleonic Wars in 1813. Items with sentimental attachment were exchanged for those with purportedly higher virtues, iron treasures of a fatherland that offered Iron Crosses to heroes. In 1916 iron jewellery inscribed with 'Gold for Defence, Iron for Honour' was given to Germans who gave up their gold heirlooms, as a mark of patriotism.[20] In March 1940 the Finns were reported to be collecting gold from households to buy aircraft, but 'iron for gold' never caught on in Britain.[21]

Other personal sacrifices do peep through the humdrum tales of hundredweights of wastepaper and tons of scrap metal: in July 1940 an aluminium bracelet made from a First World War relic, given to a sweetheart, became one of Gateshead's 'gifts-with-a-story-attached'.[22] The press loved these. Several reports detailed the giving of old love letters to salvage, some of them received during the First World War. The stories served a patriotic end. Hundreds of letters were donated by one Ulster woman, while a faded bundle sat 'unopened' in the Amersham salvage depot; their donor had 'not the heart to destroy the letters personally, but wanted them to help the war effort'. The Cheltenham Corporation gave reassurances that love missives would not be read, and advised that they be neither shredded nor crumpled.[23] Faced with such tales of painful sacrifice, how could the reader neglect to hand in old newspapers and out-of-date ledgers?

Nellie Dark correctly identified Tottenham as one of the keener districts for salvage. In February 1940 the Minister of Supply and fifty of his colleagues had sat down for lunch in Tottenham Town Hall after touring the nearby salvage works, considered a paragon of good practice. Their menu cards, made from reused envelopes, bore the slogan 'We can't scrap without scrap'.[24] Although plenty of citizens embraced the ethos of material salvage during the war, some did not even try to engage with salvage schemes, especially if they were inconvenient or time-consuming. Others got involved if there was an element of competition or gimmick (often involving collecting more than their counterparts in a neighbouring town), but then lost enthusiasm.[25] In some places salvage non-compliance was rife.[26] A middle-aged woman living on Holloway Road let on that she burned her paper: 'it's wasteful not to.' She burned her lodger's papers too, fearing that collected papers would breed vermin.[27] Even after fines were set for wasting, a west London housewife declared boldly, 'I'll put it in my dustbin and chance it'. A flat-dweller in Willesden said that she and her neighbours had detected no organisation of paper salvage and saw 'an awful lot that goes to waste […] we don't bother any more'.[28]

Driving salvage

At lunchtime on 10 July 1940, Lord Beaverbrook, who as well as being publisher of the *Daily Express* and the *Evening Standard* was also the Minister for Aircraft Production, launched a nationwide campaign for aluminium salvage. Domestic items would be turned into aircraft, urgently needed to repel Luftwaffe bombers on their nightly raids: 'We will turn your pots and pans into Spitfires and Hurricanes, Blenheims and Wellingtons.' Lady Reading, as head of the WVS, broadcast an appeal on Beaverbrook's behalf, asking 'the women of Great Britain for everything made of aluminium […] the things that you are using, everyday, anything and everything new and old, sound and broken'. France had capitulated, and the threat of invasion had become real; British citizens took the campaign to heart. Before the broadcast ended kitchenwares were piled up on the streets. Women were buying aluminium products from hardware stores and having them sent direct to the WVS headquarters in Tothill Street. Aluminium legs were donated: in South Shields a man gave his spare, and two were taken from a dead man in Newcastle.[29] Egg whisks, colanders and curlers joined part of an old Zeppelin and rods from incendiary bombs in tottering piles.[30] One woman gathered an 'immense amount', declaring that her collection was not junk 'but it isn't really indispensable either' – a heap of spare things and items 'not used often enough to justify keeping them, 5 saucepan lids, ditto ditto, Woolworth's fish slices etc.'.[31] The princesses donated miniature teapots once given to them by the people of Wales. In Whitechapel, eleven-year-old Gilda Cohen did the same, donating a set she had won as a school prize. An old lady offered a solid silver tea set. When told it was not aluminium, she replied, 'but you can sell it and buy aluminium'.[32]

Caught up in the fervour to collect, the under-secretary at the Treasury agreed to authorise expenditure of £100,000 to purchase brand new aluminium pots and pans from shops to melt down for aircraft production. The request for the money came from a temporary civil servant, who later admitted that he had 'somewhat untruthfully' implied that the request had come directly from Lord Beaverbrook.[33] Others also came to regret their generosity. Poorer families found it hard to cope without their fish kettles and pans. Cooking was difficult enough already, with the effects of rationing, and shops were not always able to supply shoppers with substitute enamel kitchenwares, because stocks were low (although they often had new aluminium pans in stock). A shop in Davenport had 'immense quantities of aluminium knick-knacks, not labelled as aluminium, but [which] obviously are so'.[34] Grumblers thought that the collection was 'a dirty swindle', 'taking the poor old dames' pots and pans [...] They've got dump yards full of the stuff.' Firms were accused of profiteering from aluminium hoards while the government sent collectors to 'loot the poor people's kitchens'. Seeing wastage in piles of salvage, one woman had an interesting take: 'It does seem a shame, all these good things [...] It's taken time to make them.' There were various motives for donation, ranging from simple patriotism to 'the general tendency to do what is the "done" thing, even if grudgingly'.[35] 'I can't get any tea, let alone teapots', complained one woman in Stepney. 'If I had teapots to give away I wouldn't be living in this sort of house.'[36] Only the comfortably off could assume that everybody had spare household paraphernalia to give away.

Many aluminium dealers actually saw the 'Saucepans for Spitfires' boost their trade. Some eventually reaped the benefits of a mass abandonment of goods:

> We can sell aluminium things all right. I don't say today or tomorrow, not while the ladies are all thinking about the aeroplanes (*laughs*). But next week they'll all be wanting aluminium again [...] And it's not like a bit of beef, it'll keep.[37]

Salvaged aluminium was not always easy to reuse – some pans needed separating from non-aluminium handles before being melted down. The whole enterprise devolved into a logistical nightmare. Beaverbrook's energy might have been better directed at the aluminium suppliers who continued to keep stockpiles. An Air Raid Warden in Liverpool worried about the waste of aluminium

11. Postcard sent in 1940. Author's collection.

in the aircraft factories: 'ask men who work there, country crying out for scrap, tons being buried.' He had written to Beaverbrook, who had swept it under the carpet.[38]

Encouragement and education came via exhibitions, newspaper appeals, post marks (see above, **fig. 11**) and leafleting campaigns (and the occasional beseeching menu), but they couldn't win: exhibitions attracted the already converted, while leaflets drew criticism for wasting paper. One diarist was unimpressed by 'pictures of housewives [...] armed with lamp standards and dustbin lids, or hurling saucepans and bundles of paper at the heads of Hitler and co.'. Most householders already did their bit, and those 'who remain impervious to the example of their more virtuous neighbours are hardly likely to respond to a mere newspaper appeal'.[39] She may have been right: people ignored leaflets, complaining about samey contents, and some households in the north did not even have letterboxes to receive leaflets. The impact of leaflets was 'rather slight', especially of those sent by local authorities.[40]

In 1941 Beaverbrook (concerned for his newspaper businesses, no doubt) appealed for 100,000 tons of wastepaper. The response varied enormously between boroughs. In Willesden little attempt

12. 'Belle Vue salvage week', *The British People at War*
(Odhams Press, 1944). Bridgeman Art Library.

was made to encourage citizens to dig out their paper, besides the
pinning up of one poster. In St Pancras, thanks to the efforts of one
'energetic' woman, the response was overwhelming. She spoke at
film showings, her 'modulated voice' reaching thousands of women
each night (the Public Relations Officer for St Pancras said 'it's no
good having a Cockney voice').[41] Beaverbrook's campaign came as
the weather turned cold: much wastepaper went in hearths. The
paper shortage became dire. One housewife claimed to have given
away all her old books long ago, and was using the *Express* and
biscuit wrappings to light fires.[42] J. C. Dawes, the Deputy Controller of Salvage, complained that tons of unwanted books were not
being handed in to make munitions in 1941: 'Liberty is a priceless
quality, but it can give rise to dangerous indolence even in time of
national emergency.'[43] Dawes had much pertinent experience,
having garnered much praise for his work in waste salvage during
the First World War.[44]

In Bristol, where the volumes of *Master Humphrey's Clock* were
saved in 1942, book salvage drives were organised with enthusiasm.
Plans for the first started in earnest in July 1941, at a meeting chaired

by H. G. Judd, the National Controller of Salvage. The previous month Judd had started to ramp up the rhetoric, claiming that the need for wastepaper had never been so great, and calling for 'those dust-covered old books'. Making a bizarre comment about the public holding on sentimentally to books given as wedding presents, he ordered: 'You will never read them again. Hand them over for salvage.' On Judd's suggestion, the Thames Board Mills were approached to provide paper leaflets and cardboard exhibits to advertise the scheme. Adverts were also shown in cinemas, including *Salvage with a Smile*. In this case, the leafleting and advertising seemed to work: in 1942 the Council exceeded their aim for 750,000 books, most of which were re-pulped. Two years later an ambitious drive to gather in another 500,000 volumes was thwarted, gathering only 268,011. The committee channelled John Milton to put a positive spin on things: 'Although the total amount of material collected fell below the target figure [...] the leaves were thicker than the "Autumnal Leaves that strow the Brooks in Vallombrosa".' Although it was claimed that nothing of real value had been handed in, various seventeenth- and eighteenth-century works were submitted, including the sixth edition of Arthur Warwick's *Spare Minutes* (1637), Robert Venables's *The Experience'd Angler* (1676) and eighteenth-century editions of works by Isaac Watts, Ann Radcliffe and Joseph Priestley.[45] Of the books that were rescued by James Ross and his team in Bristol, one was a 1676 copy of Paolo Sarpi's *History of the Council of Trent*. A quick perusal revealed the probable interest in the book's fate shown by the William & Mary College in Williamsburg, Virginia. In 1704 it had been the 'gift of Captain Nicholas Humfrys Commander of the Ship Hartwell' to the library. Ross, the conscientiousness civil servant, later returned it. In a letter of thanks, the librarian speculated that the book may never have been delivered to the college, or that it had been 'picked up and carried away' in the confusion of a fire in 1705. Later fires had seen the whole library destroyed – the Sarpi volume could be the sole remainder of the early eighteenth-century library.[46] These books were saved, but other treasures were surely lost – not every book salvage was scrutinised as closely as by James Ross's team in Bristol.

Beveridge enveloped by El Greco's jacket

The household was not the only source of gatherable salvage. Companies and firms handed in unwanted files and paperwork, and dug out more reusable stuff. A Southend solicitor's office cleared out fifteen years' worth of old newspapers. Railway companies rummaged around in their offices, engine sheds and yards. One reconditioned 55 million sheets of paper, a million envelopes and 750 lb. of pins.[47] There were suggestions that retailers should take more responsibility for their packaging, creating less matter to be wasted. From 1941 the laundry powder Persil was temporarily packed in 'thinner cardboard from waste paper'.[48] Haberdashers were identified as packaging offenders: buttons and press studs came with card backings.[49] Cigarettes and chocolate were still marketed in 'unnecessarily elaborate wrappings of tin foil, cardboard, and paper'.[50]

A street sweeper who cleaned the Cambridge Road area of Kilburn drew attention to the masses of greasy litter left by patrons of the three fish and chip shops in the vicinity. He mused that 'people should be made to bring plates'. A shop customer trying to refuse receipts from shops was amused when one 'was pushed in my shopping basket – and when I opened it, it read "Save Paper"!'[51] In 1942 J. C. Dawes was probed about a decision by the Bank of England to save paper by keeping banknotes in circulation for longer. Leaflets had expressed 'the hope that people would not object to using dirty notes', but calculations suggested that 'ten times more paper was used in making the announcement than would have been used in the new bank notes'.[52] One disgruntled citizen chuntered: 'it's penalties this and penalties that. It's all a lot of talk. It's the Government itself who wastes all the paper [...] It's all a lot of twaddle.'[53] The author and editor John Collings Squire criticised government priorities in 1942, when the paper shortage was affecting 'the more civilised sort of books and periodicals'. He identified one of the most egregious offenders as Herbert Morrison's 'Go To It' campaign for the Ministry of Supply, which had been 'plastered, on very large sheets, on every hoarding in the country between advertisements of drinks and cigarettes which needed no publicity and for which the demand often exceeded the supply'.[54]

Commercial businesses needed to keep churning out wares to stay profitable, but consumption seemed counter to efforts to marshal resources. Many companies made compromises or created workarounds. While householders were carting their old books off for salvage, publishers were still sending out new books, but they struggled to find supplies.[55] Paper for dust jackets ate into the rations. A doctor wrote a letter to *The Times* in December 1942 imploring publishers to abandon dust jackets entirely and to stop including on them any illustration referred to in the text, which made the jacket an 'integral part of the volume'. The publishers George Allen & Unwin responded, explaining that 'we (and doubtless others) are to a large extent using the reverse side of pre-war jackets of books of which the sheets were destroyed by enemy action'. Jackets protected the cloth covers. Had they been abolished, it was argued, 'paper to prevent rubbing would have [had] to be slipped between separate books when they were dispatched in bulk'.[56] A George Allen & Unwin edition of William Beveridge's report on *Full Employment in a Free Society* (1944) re-employed a wrapper designed to clothe a book about El Greco published in 1938 (**fig. 13**).[57] Oxford University Press opted for very simple wrappers, reutilising covers from their back catalogue. On the inside back flap of C. K. Allen's *Democracy and the Individual* (1945) is the terse statement: 'An *Economy* Jacket. Please disregard the matter on the inside of this wrapper.'

This was a practice that didn't stop with the end of the war. In the post-war period, George Allen & Unwin reused war maps for jackets. Karel Čapek's *Three Novels* (1948) has a jacket printed on the reverse of a map of part of Holland. The paper is so thin that a red line on the new front bleeds through to the other side. Bertrand Russell's *History of Western Philosophy* (1946–47) reused coloured maps, including a map of the Alps published by the War Office in 1944.[58] These wartime maps were surplus to requirements in peacetime. A collection of wartime recipes called *Off the Beeton Track* (1946) used as a dust jacket the reverse of a US Army map of part of Germany from 1944. The frugality of the recipes and the jacket jar oddly with the bright lithograph prints used to illustrate the book. Recipes include 'Soup with left-over meat', offal-based dishes and uses for dried eggs.[59]

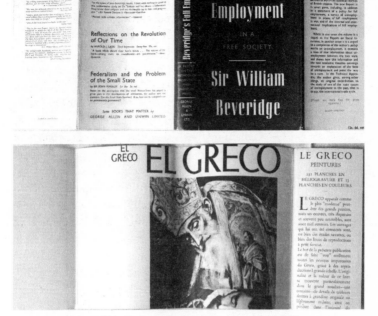

13 (a and b). Both sides of the dust jacket for William Beveridge's *Full Employment in a Free Society*. Author's collection, photographed by Taryn Everdeen.

Letters were also enveloped in reused matter, and most letters were sent by businesses. This envelope (**fig. 14**), used to contain a letter sent to a French Catholic church in London's Soho in 1946, was made using a redundant map of part of France. The stamp used on the envelope also appears to be reused, taken from another mailing, with writing on the reverse.

During the First World War, Fred Ridout of Whitstable in Kent had devised an adhesive label to enable the reuse of envelopes, little different from those later issued by Friends of the Earth.[60] The Stationery Office had printed millions of labels by the end of that war.[61] Labels were used again during the Second World War, but

14 (a and b). Map reused as an envelope in 1946,
showing front and back. Author's collection.

on an even grander scale.[62] A report for the *Nottingham Evening Post* from September pointed out a snag: the additional thickness of envelopes holding successive labels might fatten them, taking them overweight and requiring extra postage.[63] *The Times* celebrated these corpulent packages, which bore 'records of their previous journeys' if 'peeled off layer by layer' by dexterous fingers.[64]

At the end of the war businesses took advantage of a surplus of specific types of paper waste. Envelopes were redeployed and maps were reused. War had made people discuss things we still haven't resolved today: ought retailers to take more responsibility for waste packaging? Should customers take their own crockery to the chip shop? How can we guard against inconsistencies – the sort that consumed more paper in memos about banknotes not being reissued than in reissuing new banknotes?

Captain Wills's railings

Aluminium was not the only metal in demand during the war, and, like old books, old metal often raised questions of historical as well as material value. A hundred tons of horseshoes recovered from a railway horse depot were handed in along with fireside fenders from waiting rooms, old rails, sleepers and turntables.[65] Faulty gas meters, nearly one thousand per week, were sent to a Manchester smelting works. Residual gas clung to the insides, so they were pierced with holes before being shovelled into a huge furnace. One box-piercer said: 'Wish this were Goebbels, he's just another gas bag.'[66] Dozens of guns being used as street posts on Tower Hill in London were sent off for scrap. It was assumed they were Napoleonic, but the Master of the Armouries spied a Tudor rose on a few of them, so they were saved from the salvage. Tudor guns were rare, as a consequence of William Pitt's campaign to scrap old guns to make new ones for the battles of Trafalgar and Waterloo in the early nineteenth century.[67] Museums were lucky recipients of at least some scrap sent for salvage.[68] A barrel from Nelson's HMS *Victory* sent to scrap in Keswick was returned to the ship in Portsmouth, with the agreement of the Ministry of Supply. The gun's wheels were not saved, being judged to be later additions.[69] In early 1941 Gloucester Borough Council made steps to grub up old tram rails to send to salvage, but were prohibited from doing so by the Ministry of War Transport because the bother of resurfacing the roads negated the value of the metal in the rails.[70] Other rails could be taken more easily.

Back in April 1938, Göring had ordered German councils to grub up railings for scrap. At that time, Clough Williams-Ellis had wondered if Britain might follow suit – after all, it would make streets more welcoming, and remove dangerous spikes which impaled children.[71] This suggestion was not surprising – Williams-Ellis was an architect with an eye for reuse, having designed the Welsh village of Portmeirion to include fragments of demolished structures in many of its buildings. The 'Cloughed-up' village includes a Town Hall built in the 1930s as an (in the architect's own words) 'unabashed pastiche of venerable Jacobean bits and pieces', with panelling, windows and a ceiling depicting the Labours of Hercules reclaimed from Emral Hall in Flintshire. The Town Hall is topped with a cupola made from an upturned and upcycled pig-boiler.[72]

Whether or not they were eventually convinced by accusations of child impaling, in 1940 the Ministry of Supply did requisition unnecessary railings. They began to be removed from city and royal parks.[73] Some were melted on-site, with celebratory pomp and luncheons, part of municipal efforts to encourage householders to offer up their own railings more willingly. By June 1943, 580,000 tons of railings were reported to have been collected.[74] Ironically, in January 1943 an appeal was made on behalf of the owners of Gayton House in Northamptonshire to be permitted to retain their gates on the grounds that they had been designed by Clough Williams-Ellis. The appeal was successful: the gates were deemed to be 'not of historic value, but of exceptionally good original design and good craftsmanship'.[75] Williams-Ellis's opinion on the matter is unknown.

Right across the country, what constituted 'superfluous' was up for debate. Railings often formed boundaries between public and private property, and could also demarcate civic, state, sacred and secular spaces. A variety of arguments about competing priorities were had around railings. Local authorities surveyed the railings of institutions and private dwellings in their districts, but their decisions could be overruled by the Ministry of Works.[76] Notices were posted identifying those railings destined for salvage, but these could be challenged by owners.[77] Appeals were cumbersome. Early in 1942 a land agent employed by Captain Arnold Stancomb Wills wrote to the Clerk of the Brixworth Rural District Council to argue that Wills's entrance gates at Thornby Hall in Northamptonshire should not be requisitioned. Wills had already been informed that the 'unnecessary' railings would be taken as scrap unless he could prove they were required for reasons of safety or animal corralling, or if they had 'special artistic merit or historical interest'. His agent argued instead that the railings were necessary for the protection of the property. This proved to be an unacceptable argument. The appeal was rejected because 'the individual interest must be subservient to the need of the nation'. The agent switched tack, asking that their architectural value be assessed by a panel of architects.

The Council consented. Their correspondence reveals the tortuousness of some requisitioning negotiations. It also reveals a certain degree of inattention on the side of Wills and his agent; he had not, for example, taken the opportunity to claim that the railings were required to protect cattle. Wills bred dairy shorthorns and might presumably have shuffled some stock into place – it was a tactic that others had used to great effect in the battle for railings.[78]

Alongside dubious cattle excuses, numerous reasons were given by owners of grand houses as to why they should get to keep their railings. Lady Wimborne, hoping to save the gates of the Manor in Ashby St Ledgers, Northamptonshire, pulled in their designer, Sir Edwin Lutyens, to appeal on her behalf in 1941, and also penned her own letter, pointing out that other railings from the estate had been freely given over. She also claimed that the gates had great sentimental value, having been her last gift to Lord Wimborne before he died. The gates were reprieved.[79] F. Brodie Lodge, owner of Flore House in the same county, successfully argued for the retention of his ornamental gates from the reign of William IV, claiming that 'their value to the nation is obviously infinitely greater in their present form than it would be in that of pig iron, even in time of war'.[80] Mina Allfrey, at Chacombe House, near Banbury, appealed against the decision to remove her wrought-iron gates in 1943: 'I do not think that the gates in question are of any value for your purpose.' They allegedly comprised only thin bands of metal. In an attempt to copper-bottom her case, she claimed they were necessary to 'prevent stock from straying on the road', and she also offered 'a bigger amount of old iron free of charge' for an exemption. The old cattle trick didn't work this time, and nor did her iron bribe. Her case was rejected: the gates were not necessary or exceptional.[81]

While the landowning gentry were busy penning defences of their ornamental gates, city-dwellers were no less keen to keep their railings. In cities across the country, humble railings from terraced houses were being taken up. Their owners would rarely be able to argue artistic or historic exemption, let alone plead the corralling of cattle, so arguments for retention took other angles. On Wellington Street in Gloucester, a homeowner claimed that the railings were necessary to prevent people falling into the front cellar pit during black-outs. Reginald Bevan lived opposite the Empire Cinema and appealed, unsuccessfully, in January 1942 on grounds of privacy. He thought the loss of his railings would make a 'currently intolerable' situation worse, 'with the nuisance of the boys' coming out of the cinema. Removing the railings would 'make the house almost uninhabitable' (the occupants have coped without the railings since 1942).[82] In late 1942 a Mrs Malcher asked the Northampton Rural Council 'why necessary railings are taken and unnecessary ones around graves in churchyards are left'. Some of these had been installed to discourage body-snatchers in the nineteenth century, but this was less of a pressing issue in the twentieth. She also wondered about 'that ornament on the Market Square' in Northampton, a large fountain of cast iron with bronze attachments, which was eventually removed in 1962, presumably for reasons of aesthetics rather than necessity. Mrs Malcher's own railings protected a fruit and vegetable garden, but her appeal was rejected.[83] Dismayed by the impending loss of her railings, Mary Wetherman of Clifton in Bristol complained, observing, with passive aggressiveness, that she let the forces use her property and claiming that the expense of removing her railings would exceed their value. Vegetables grew among her roses and she didn't want them destroyed by dogs. Like Mrs Malcher, Mrs Wetherman ended her correspondence with a comment about alternative civic reserves: 'Has that appalling lump of metal in front of the Victoria Rooms called a Fountain been used, it has always been an eyesore? Believe me.' Whether or not the Lord Mayor's office believed her, the fountain stayed (fig. 15) and Mrs Wetherman's railings went.[84]

Disquiet spread along Grove Road, Bushey, Hertfordshire, in May 1942, when rumours circulated that its railings were liable to be taken. One old man was 'very het up' with worry that children and dogs would gambol and frolic over his gardens. 'We don't know what's done with them', he chuntered about the proposed removals.[85] Others wondered whether the government would indemnify property owners against unrailed accidents or injuries caused by jagged bits of metal left in masonry once railings had been prised off.[86] Diary entries, newspaper reports and surviving

15. The fountain, Victoria Rooms, Bristol.

material evidence all suggest that railing removal was effected 'rather brutally' and that a mess was left behind.[87] In Daventry, Northamptonshire, railings had been removed 'by brute force', leaving walls tumbling as a result of railings being 'lugged out'.[88] A council report concerning Gloucester's Wellington Street (of the deep cellar pits), plus streets near by, found that the removal of railings had left much structural damage. The smaller bars had been 'cut with shears' and the larger ones 'loosened with a sledge hammer', which must have left quite a few jagged edges.[89] Among others, Mabel Gransmore felt 'quite ashamed that our old City should stoop to imitate the "Gestapo of Germany"'.[90] Gordon Hall Caine, the MP for East Dorset, thought the whole practice was 'a barbaric piece of socialism'.[91] Opposition to railing removal cut across party lines.

Some local authorities were thought to have obstructed the railing salvage process. At the end of 1941 only 600 of 1,500 authorities had completed their scheduling of railings collection.[92] In Cirencester there was no particular hurry to comply with the mandated railing sacrifice. George Winstone, the chairman of the Salvage Committee, stated that if the country really 'could not win the war without the conversion of railings into war material, then the Council would not stand in the way'. Winstone sensed this was not the case, however, and wanted a guarantee that the town's railings would not simply disappear into the thousands of tons of uncollected scrap. Councillor Ernest Newcombe, a builder by trade, suggested that his colleagues resist a sense of urgency, lambasting the 'sheer panic' that appeared to motivate the scheme.[93] A three-month delay in salvaging railings in Manchester was blamed on parts of the Corporation: railings there had been scheduled for removal, but appeals came in from the Parks Department, the Education Department and even from the Cleansing Department – the very team responsible for Manchester's salvage drive.[94] Discussions were held in Dumfries, where many members of the council resisted the removal of the railings around Dumfries Academy. Some argued that making an exception would smack of hypocrisy. Others voiced concerns about the security of unrailed buildings, or for children, who might tumble onto the street and break legs or necks without the railings to guard them.[95] It is possible that the removal of the railings from the entrance to Bethnal Green Underground Station contributed to the shelter disaster on the stairwell that killed 173 people in 1943, when a crush of people formed after a woman and child fell in the dark as they were descending the stairs and others, also responding to an air-raid siren, fell around them.[96]

In the end, the old man in Hertfordshire and George Winstone in Cirencester were right to be sceptical. Much of the collected cast iron was found to be too impure for its intended use. Rumour had it that much of it was dumped in the Thames estuary, off the coast at Sheerness. In 1978 a group of Canning Town dockers claimed that so much wartime scrap metal had been dumped that it confused ships' compasses.[97] Were the boundaries of parks and churches razed for nothing? Historians have disagreed about whether the railings were wasted or not, since no conclusive evidence has been provided. Recently, Peter Thorsheim has argued that 'although no consensus is likely to emerge about whether the government ought

16. Blitz damage, used to illustrate 'Scenes during the Recent Night Raids on London', *Sphere*, 1 April 1944, p. 24. Author's collection.

to have proceeded as it did, there can no longer be any doubt that railings indeed helped to feed the war machine'.[98] But this matter depends on the degree of waste involved. It seems unlikely that the amount of reusable metal recovered made the disruptions to people's lives, the changes to urban streetscapes and the damage to individual rights worthwhile.

Salvage intensified as metal imports waned, but it outstripped Britain's capacity to make use of its fruits. Dumps of old metal, gathered from bomb sites and elsewhere, had been allowed to form in places.[99] Judy Skinner, who promoted salvage in Gloucester, noted that the psychological effect of uncollected dumps did 'great harm to the salvage campaign in general'.[100] In Bristol the decision was made to clear up old dumps before starting new campaigns 'so that members of the public may not be under the impression that waste materials they have contributed are not required'.[101]

Despite efforts to clear piles of salvage, Blitz-battered cities inevitably became cluttered with heaps of twisted metal and piles of dusty rubble. What had once been proud and erect houses were cross-sectioned – the formerly private interiors revealed for all to see. This had a dampening effect on home-front morale. Nellie and Sidney Dark were bombed out of their Bedford Court flat early in the Blitz.[102] Rescue teams and civil defence workers conducted searches in the immediate aftermath of an air raid, recovering survivors and bodies. The temptation to loot proved too great for some to resist. Citizens searched through the ruins for fuel and would load up prams and trolleys with salvaged stuff. Others sought souvenirs of the air raid – a bit of shrapnel, or a bomb casing. Young boys pounced on a Heinkel III shot down in Caterham. They were eager for a souvenir; one was bemused that a chunk of aluminium was stamped 'Made in Birmingham'. Once the rescue teams were done, and the scavengers had departed, the demolition gangs cleared and tidied, hacking through the debris to barrow away loads of earth (called 'shit'), bricks, wood and other materials. Bricks were divided into two piles – whole and hardcore.[103] Much reclaimed rubble was reused during the war, sometimes to strengthen air-raid shelters. Cleaned bricks were used for repairs. A large quantity of the demolition spoil was stored and later reused as hardcore for post-war reconstruction projects such as road-building. A significant quantity of spoil was never used. The northern end of Regent's Park is elevated several feet by wartime dumping.[104]

Major Holmes's hippo

London Zoo, located in Regent's Park, closed several times during the war – and all venomous animals were destroyed in case they worked themselves free during air raids. To save petrol, camels and llamas were used to carry food to the other animals. The public were asked, in a radio appeal, to collect acorns to supplement some of the animals' diets.[105] More food was collected for less exotic animals – pig food bins were installed on city streets across wartime Britain, forming depositories for tons of kitchen waste and recalling urban habits of an earlier time. Workers at the refuse depot in Tottenham kept a herd of one hundred pigs, to which they fed waste collected on their rounds. The piggery was nicknamed 'Adolf's Kindergarten'.[106]

As early as 1939 posters had appealed for scrap bones, from which animal glues could be made for use in aircraft. Bones were also a source of glycerine, a key compound of nitro-glycerine, and had further uses beyond glues and greases. It was calculated that if one ounce of bones were collected from everyone each week for a year there would be enough to make 5,000 tons of grease, 7,500 tons of glue, 2,500 tons of pig food and 22,500 tons of fertiliser.[107] As the war progressed, more councils placed communal bone bins on streets, often next to pig food drums.[108] All bones (except fish bones) were needed, even from rabbits and poultry, and they might be stored in a 'well-ventilated container'. 'Nothing that the dog did to it – except burying' reduced their value.[109] Some containers were fixed to lampposts, 'sufficiently high to be out of reach of dogs'.[110]

Major Frank Holmes, in Essex, rummaged around and found the skull of a hippopotamus he had shot in Abyssinia.[111] Other hauls were less substantial. Despite appeals in newspapers, backed up with exhibitions explaining the uses to which bones were put, collections were generally disappointing.[112] The Ministry of Supply estimated in early 1942 that 'less than one-fourth of the tonnage entering dwelling-houses was recovered'.[113] This figure, widely reported in newspapers, sparked no sustained increase in bone salvage. A further appeal the following year for one ounce of bones per person led with the startling figure of 50,000 tons of wasted British bones per year.[114] J. C. Dawes, the Deputy Controller of Salvage, was unimpressed, 'because everyone knows, or ought to know by now, that the importation of bones means exposing the lives of gallant men in vitally important ships', costing money and time.[115] But many citizens really did not grasp the significance of bone salvage: they could not understand how bones could yield anything vital. Bone processing was more abstract than some other forms of reuse.

There was also the matter of palatable storage. Exhibiting the arrogance of officialdom, Dawes identified two problems: a lack of containers and the presence of odours. Just get a metal box, he instructed, clean and bake the bones, then give them a 'light sprinkling of lime'. Most housewives would never have had the time for all this.[116] Glib suggestions that householders use an old saucepan

17. Wartime flour bags. Author's collection.

to contain their bones neglected the fact that many such items had already been handed in for salvage.[117] Smells, maggots and flies emanated from bone buckets during summer months, and make-shift bins were 'got at by dogs'.[118] Dog owners were reminded to refrain from throwing bones to their pets: 'dogs must cut their luxuries too'.[119] After late 1942, several newspapers carried warnings that bones could injure beloved hounds by splintering and causing internal damage, concerns no doubt fuelled by an increasingly urgent need for bones.[120] A report highlighted the discrepancy between what people said they did with bones and what actually became of them, showing an awareness of the ways that socially acceptable responses could skew questionnaire results. Bone salvaging was patriotic, so few people admitted their own failings. Respondents reported that 'other people' wasted or burned bones but were reluctant to admit to wastage themselves. Many believed that their own meagre supplies of bones would make no difference: 'I don't get enough bones to make all that fuss about', wrote one. 'In general', continued the report, 'people are aware that they *should* salvage, but apathetic and unawakened as regarded the matter's urgency.'[121]

Flour bag pants with buttons on

Beyond the structured salvage operations that took up so much of housewives' time during the war, there were many more instances of piecemeal reuse and recycling – necessity being the mother of invention. This attitude was most immediately noticeable in fabric and clothing. Material shortages and the need for utilitarian clothing for women led to considerable fabric reuse during the Second World War. Men's suits were recut for women, shortening the legs but retaining the tailoring and shoulder padding. Sections of RAF surplus parachute silk were sold in London just before Christmas in 1939, ideal for underwear.[122] Flour bags, available without ration coupons, were turned into accessories and garments, even sawdust-stuffed dolls. One such doll sported skates crafted from an old office file; another somewhat creepily donned 'real hair which once adorned a baby brother'.[123] Americans had used flour sacks to make clothes for decades, especially during the Depression of the 1930s.[124] Although quietly tolerated during the emergency years of the war, flour bag reuse was considered to have got out of hand afterwards. In 1947 shocking news broke that British forces in Egypt were being supplied with underpants 'made of flour bags with buttons on'. *Quelle horreur!* The redoubtable Elizabeth Braddock MP promised to ask the Secretary for War to hold a 'full inquiry' into the circumstances that had led to this outrage. This she did in May, to peals of parliamentary laughter. The War Services Secretary assured Braddock that he was looking into the pants. Members wondered whether the buttons were strange, but it seems that they were incidental; the insult lay in the pants' floury origins.[125]

Moments of emergency triggered ingenuity during fighting as well as on the home front. A life-saving makeshift iron lung was crafted from scraps on board a ship in Gibraltar.[126] An RAF corporal interned in a prisoner-of-war camp in Burma during the Second

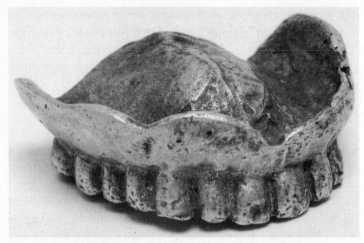

18. Upper full denture, the British Dental Association Museum.

World War found a way to create a remarkable set of dentures (**fig. 18**). Combining river sand with aluminium from a fallen plane, the dentures served as a replacement for a broken set. Mouth impressions were made using wax melted from the plane's wires. Meanwhile Auxiliary Fire Service men in Grimsby made a 'patchwork' fire engine from an old car, a lorry, football goalposts and cinema seats, 'the painting being done by a butcher'.[127] WVS volunteers made temporary kitchens using bricks, wood, drainpipes and door scrapers salvaged from bomb sites.[128] As with flour-bag pants, some ingenuity survived into peacetime. In 1946 clerks from the Ministry of Education and the Ministry of Works built a church organ out of 'old hats and carpets, picture frames, an umbrella stand, a washstand and two harmoniums'.[129] After the war, with many streets now bereft of their railings, Air Raid Protection (ARP) steel stretchers were used as replacements. Easily identifiable by the indents at each end, which had once raised stretchers off the ground in use, stretcher railings still surround many council housing estates, including the Kennings Estate in Kennington (**fig. 19**).

19. Stretcher railings in Kennington. Photograph by Toby Sleigh-Johnson.

But, despite some habits of reuse continuing, many people were keen to return swiftly to their old patterns of consumption as soon as the war was over. Without the urgencies or exigencies of war, local councils swiftly disengaged from salvage collection. The Thames Board Mills in Purfleet halted production in 1947, just two years after the war ended, because too little paper was being salvaged.[130] Thousands of tons were binned, leaving little material to make food packaging. The President of the Board of Trade noted that, if all bus and tram tickets were salvaged, there would be sufficient material to package a year's worth of margarine rationing for two million people. Uppity letter-writers alerted the newspapers to the numbers of tickets littering public transport.[131] A short-lived national campaign brought in more wastepaper, but by June 1949 a different problem was encountered – too much low-grade (already recycled) paper was being placed into the bins. The resulting paper was too coarse for increasingly discerning paper consumers. In a sign of capitulation to the times, householders were advised to burn their paper waste.[132]

Despite various positive assessments of wartime recycling, it is clear that much more could have been done to reutilise materials.[133] Salvage drives were most successful when they possessed some sort of unusual, novel or competitive aspect – or were very short-lived.[134] In 1943 the BBC covered events across London to publicise book salvage, including a troop of elephants used to move paper outside Regent's Park and a replica ship in Trafalgar Square.[135] Shop windows filled with scrap garnered much interest. One on Kilburn High Road was 'attended by little groups of shop-gazers' for a whole Saturday afternoon. An associated collection, 'without any canvassing at all, collected at least half again as much as any other in the borough'.[136]

Although cities competed to hand in the largest stash of salvage, to win national and local prizes, these prizes sometimes skewed the focus for councils. The Bristol City Engineer paid keen attention to the city salvage. In 1941 he wrote to the Lord Mayor's office worrying that salvage collection schemes organised by charities 'would clash with the municipal scheme and ultimately prejudice the total amount of waste material collected'. His concern was to have the council salvage the most, not that the most was salvaged.[137] Correspondence in Bristol suggests that a degree of foot-dragging took place in 1942: 'certain members of the Council' suspected that some Corporation departments were not putting sufficient effort into salvage. Agitating for the engagement of a dedicated officer, the mayor's attention was drawn to a national prize scheme for paper salvage, with ambitions for Bristol to win.[138] That same year, the City Engineer intervened after the Salvage Department had failed to take up an offer by a local canned goods supplier of 300 huge empty honey tins.[139]

Children were fundamental to the success of collections. Much gathered salvage was collected at schools, and in lessons pupils were taught about the reuse of waste.[140] The *Dandy* reminded young readers to give up any finished copies.[141] Children gathering waste in Northampton were described as 'excellent disciples of Syd Walker'.[142] Walker was a music-hall-style comedian whose persona of a rag-and-bone man on BBC Radio's *Band Waggon* chirped out the catchphrase 'What would you do, chums?' after posing a philosophical question concerning daily life. A young surveyor's pupil in Trowbridge, Wiltshire, managed salvage as part of his work. Spending a February morning in 1940 dealing with salvaged materials made him feel like Syd Walker.[143]

The salvage efforts of the war years have entered into British mythology, but a more contradictory picture emerges here. Some of the energies of officials such as J. C. Dawes were misdirected. The incentive of patriotism was not always sufficient to overcome the practical difficulties of a lack of time, of space and of energy. Even with a Nazi army just across the Channel planning invasion, the British would rarely intrude on the interests of large private concerns. Enormous stockpiles of vital materials could have been requisitioned from a few companies, removing the need to gather them from many households. The officialdom of local authority annoyed some people, making them reluctant to participate, especially in London, where salvage collection figures were low. WVS women in charge of the aluminium collection in the East End were 'all of a type – military-looking, school mistressy women, with white hair and medals'.[144] These officious, brusque women were skilled in the art of giving off an air of busy-ness that made the householder's time appear worthless by contrast. But when people felt virtuous about their own salvaging, their minds turned not to the salvage officials with their loud-hailers, or to J. C. Dawes and his ilk, but instead to the popular entertainer Syd Walker. 'What would you do, chums?'

4

Darn (1900s–1930s)

One evening in 1917, Lady Agnes Fox, the wife of the civil engineer Sir Francis Fox, watched as strips of paper ascended to the surface of her lily pond. They were saturated portions of blueprints, sections of the Cape-to-Cairo railway that had been separated from their backings. Fished out by Lady Fox, these fabric backings, and not the old drawings, were the real bounty. Earlier that day, her husband had rummaged through the many mounted drawings and maps in his office. Each was reinforced with 'fabrics of various kinds', including 'nainsook, butter muslin, brown Holland, linen, and the like'. All such materials had been archived once they ceased to serve any practical purpose.

Sir Francis's attention had been drawn to them because the local branch of the Red Cross Society had asked him if he could source some fabrics for them, so he had them sent to his house, Alyn Bank in Wimbledon. Stripped from the backings, the fabrics were 'sent to the laundry to be washed and sterilised', ready for use as bandaging, bedding and slings. Once dried, the paper on which the drawings had been made went as waste to the Salvation Army. The reclaimed fabrics were superior in quality to available materials: eighty-year-old backings to drawings of the London and Birmingham Railway from 1835 were especially fine. Female assistants went to work on a mass of plans, and 'soon had drawings floating in two tanks [...] some bearing the signature of such eminent men as Robert Stephenson, Sir John Fowler, W. H. Barlow of the Midland

20. Early twentieth-century darning kit. Author's
collection, photographed by Toby Sleigh-Johnson.

Railway Company, Sir Charles Fox, and many others'. When Sir
Francis Fox's own supplies dwindled, a newspaper appeal garnered
fresh drawings from government offices, industrial firms and ship-
builders. If potential donors feared that their intellectual property
might be compromised, Fox had the designs effaced. Unconvinced,
one railway company dispatched sacks of drawings cut into post-
age-stamp-size pieces. But even these were utilised, taken by Fox's
daughter, Dr Selina Fox, for use as cotton wool in her medical
mission in Bermondsey. By August 1918 Fox's scheme had recovered
over eighteen miles of fabric; a year later, thirty miles.[1]

An Army Salvage Board had been set up in 1917 under the
chairmanship of the Quartermaster-General to redeem materials
discharged on the front, but for a long time many domestic salvage
campaigns depended on the enthusiasm or ingenuity of a few
people like Sir Francis and Lady Fox. Overall, the response of the
authorities was startlingly lackadaisical in the early years of the war.
In contrast, the Germans had been stockpiling reusable wastes for
years before 1914. It was only really in 1918 that any concerted
efforts were made in Britain, and then a clash of military and

municipal cultures was evident in hurried attempts to organise efficient new national systems. The British had been caught on the
hop, and spasmodic and sporadic attempts to draw in reusable
waste were sometimes laughable. Material marshalling on the home
front can be characterised as complacent, even infuriating: Darn it!

Salvage and the enemy

Concerted efforts to collect salvage had first started in the military
camps. Posters illustrated soldiers in the act of salvage. 'PICK 'EM
UP', shouted one, showing a soldier picking up bullets from the
road. On another poster a British infantryman carries a sack of
salvage with two horseshoes in his left hand, marching above the
Shakespearean line 'use doth breed a habit in a man'. 'The salvage
of today', the viewers were told, 'will help to beat the Boche tomorrow.'[2] One poster was more direct in its appeal to the motivation
of soldiers: 'SALVE NOW & KEEP THE PRICE OF BEER
DOWN.'[3] In 1918 a manual called *Salvage* was issued by the Quartermaster-General's Branch in France, detailing the various types of
material to be repurposed on the front. Many materials were destined to be sent back to Britain, but glass (cheap to make from
easily available substances, and heavy to transport) was to be
scrapped in France. A revised edition of *Salvage*, issued just at the
end of the war, shows that the military had tightened up the marshalling of supplies during the conflict, so that 'there is practically
nothing of the debris of the battlefield that is not of value as scrap'.[4]

But the enemy was the 'real past master in the science of salvage';
elaborate German municipal systems of collection saw stockpiles of
many materials last the course of the war. This contrasted with
Britain's sluggish response in the earliest years of the war.[5] By the
end, the British were keen to learn from their enemy: an enlarged
and revised edition of Theodor Koller's *The Utilization of Waste
Products*, first translated from the German in 1902, was republished
in 1918, despite paper shortages. Koller's exhaustive study considered uses for pretty much everything except cat tails (deemed to be
'practically of no value'). Waste horn could be glued together to
make umbrella handles; the spent oxide from gasworks could
become ferrocyanides. Other examples seem more dubious but

communicate the extent of the ethos: badger skin made into knap-sack covers; fish scales made into good bonnet decorations; mother-of-pearl button-making waste become glitter; opera baskets conjured from the skin of grey monkeys.[6] It's not known if these more esoteric suggestions were followed by the British, but they certainly had lots to learn from those on the other side of the trenches. Long experience in the processing of waste materials had permitted Germany to act like 'a marine store upon a big scale'. The Germans were said to have 'ransacked the world for wastes', not only importing waste but also collecting it domestically: 'skins, rags, bones, feathers, hair, rubber-scrap, and articles too numerous to specify' from businesses and households. Farmers handed over the bodies of dead farm dogs; maids were criminalised for leaving grease on plates. Before the war some commercial products made from such waste had been sold back to the very firms from which German businesses had acquired them, 'at a considerably enhanced figure'.[7]

One significant client for Germany's canny waste strategy was Britain. Pre-war salvage had readied Germany to profit from buying waste for 'infinitesimal expense', processing it and selling its prod-ucts back to Britain. From the start of the twentieth century, the London Electron Works had offered money for 'tin cuttings'. The firm, owned by the German company Goldschmidt, approached sanitary committees across the country between 1904 and 1914, offering to buy waste tin awaiting incineration. Tins were shipped to Essen in Germany to be processed and were used to make muni-tions that would later be fired back at British soldiers and sailors.[8] That appalling irony was not lost on those looking back on pre-war efforts. Tins were notoriously difficult to recycle after the export to Germany was halted, owing to the absence of any national company with similar interests and capacity.[9] Thousands were buried, burned in 'destructors' or shot into the sea. British tin collection was still chaotic in many boroughs well into 1918, by which time a 'vast amount' of tinware could be spied languishing 'upon the disfigur-ing rubbish heaps'.[10]

Housewives and businessmen

For Britain, the military may have taken a lead in promoting salvage, but that enthusiasm did not necessarily reflect the reality of military economy. The Scottish writer Annie S. Swan was unimpressed by the waste in the army camps in 1916, where 'dustbins and streams' were 'choked with unused food which must be got out of sight at any price'. So much for hopeless Tommy, but what of Tommy's wife, left economising on the home front? Swan thought that the 'middle-class housekeeper' ought to have more say over home front economies, through a committee of British housewives.[11] This would allow women to take the initiative themselves, and allow the activity to feel more genuinely communal and transparent; insistence from higher up that British citizens should economise seemed hypocritical to people who thought their contributions were being squandered.[12]

In *The Times*, Lady Frances Balfour was scornful of the number of unmarried ministers in charge of major spending departments in the government. They could have developed 'no knowledge or understanding of the difficulties of domestic economy'. Such men patronised housewives, and that was especially galling given that

> everyone knows that the extravagance and waste of Army administration has been on an open and colossal scale. What has been wasted, destroyed, buried, and burnt in camps and at the front would have kept [...] a quarter of the families in the country in prosperity.[13]

Considerations of salvage on the home front were never far from those on the battlefield. Lady Balfour, the sister-in-law of erstwhile prime minister Arthur Balfour, was a formidable Scottish aristocrat. She had been one of the highest-ranking leaders of the women's suffrage movement before the war. She penned her letter from a fairly modest home on London's Addison Road.

From an enormous Georgian mansion in Hertfordshire came a counterblast, written by the writer Mary Ward, better known by her married name, Mrs Humphry Ward. Conceding that 'blunders are made, and will be made' because the army was so vast, Ward urged Balfour to consider the activities undertaken in the French repair depots, where 'wagon-loads' of broken objects were restored and repaired, and rags sent off for reuse.[14] Ward rarely agreed with Balfour, having been the founding president of the Women's National Anti-Suffrage League. The two women, once opponents on the issue of whether or not women should have the right to vote, now fought over whether or not soldiers wasted more than was reasonable. Writing just after the war, a journalist called Frederick Talbot argued that 'men naturally clamoured for subsistence more or less in consonance with what they had been for so long accustomed in private life. If the food did not coincide with their fancies it was promptly thrown away.' Talbot did admit that there had been some 'victualling chaos' caused by men being promoted without adequate commissariat skills, and was so good as to note that women were generally better kitchen economists.[15]

The Local Government Board eyed up these kitchen economists. In 1917 the Board charged councils and corporations with the obligation to gather from households items of glass, wool, cotton and scrap metal (especially iron, steel, lead, copper, brass and aluminium).[16] It was not such a straightforward task. Authorities were hampered by pre-existing contracts with dustmen, a lack of appropriate storage receptacles and removal vessels, clashes with private waste dealers and inchoate organisation of waste separation. A scarcity of labour had led some local corporations to urge housewives to burn their rubbish, to avoid the hassle of transporting it to municipal destructors.[17] As the war progressed, the stance on household incineration was reversed, and householders were chivvied to save and sort their waste instead.[18] Some dustmen were paid bonuses to separate waste; some local authorities implemented their own systems of collection, separation and disposal; others encouraged private dealers and voluntary groups, and even children, to organise waste separation schemes.[19]

It was not just housewives who felt the strain. Many businesses were pressed into war service, including Crittall's, a maker of steel windows, based in Essex, where they began to employ additional female workers. Francis Henry Crittall, the ironmonger who had set up the business, understood frugality. His memoirs recall cast-off clothes worn during his boyhood, 'restored, remodelled' and refit by his 'nimble-fingered mother'. He'd been bullied on account of having to wear garish garb passed down from an uncle, with 'checkered squares as large as my fist'. Perhaps it was this personal experience that prepared him for a life of recycling, remodelling and refitting. In any case, when an earthquake shook parts of Essex in April 1884, Crittall was ready to capitalise. By this point he was already making (fully recyclable) steel-framed windows, and so his trade was boosted by the need to make good the structural damage to properties. By the start of the First World War, business was booming – even the *Titanic* had been fitted out with Crittall windows.[20] Turning his premises over to munitions manufacture during the war, Crittall brought his sense of thrift to bear on that industry too. When a shortage of fuse forgings hampered production, he realised the 'importance of utilising scrap brass'. He claimed that his company was the first to melt down battlefield scrap to make the required parts. Crittall knew that a knowledge of the potential sources of supply was essential in such conditions, and implied that businessmen (and not military men) were best placed to maximise output while reutilising waste.[21] His view on housewives is not recorded. As the war developed, it was clear that many thought the wrong people were in charge of salvage, and this caused frustration: housewives were irked, householders confused and businessmen felt overlooked. Meanwhile engineers also looked on in consternation, as a council of the great and the good took over.

National salvage and the city engineers

By early 1918 steps were being taken to tackle civic waste more comprehensively.[22] The National Salvage Council (NSC) was established in March to limit waste and recycle materials efficiently, following examples set by the Controller of Army Salvage.[23] The NSC was to provide national oversight, and to increase the amount of waste recovered, rather than divert the handling of waste from the local authorities. It added a layer of oversight to the local authorities, who were encouraged to share good practice and to co-operate. A dilemma for the NSC was that the authorities wanted some guarantee that householders would supply waste sufficient to justify creating new systems, and the householders wanted assurances that the waste would be collected and really put to use. Future salvage operations would grapple again and again with this chicken-and-egg dilemma – but for the NSC it was a new, but not fully recognised, problem.

But towns and cities in the north and the Midlands had already taken the salvage lead, with Birmingham, Sheffield and Glasgow being among the most energetic collectors and reusers. Relying on female volunteers, Birmingham's citizens had collected tons of old tin each year since the start of the war.[24] An early adopter of salvage, Birmingham was regarded as a city of experimentation, with a de-tinning plant to reuse parts of cans, regular pig food collections and municipal fertiliser manufacture.[25] Some thought this was just a characteristic of Brummies, believed to be 'better imbued with the municipal spirit', and their collective success with paper recycling was seen as a 'shining example', revealing ways that business models could be brought to bear on municipal salvage.[26] Others trumpeted the energy of one man: James Jackson, the Cleansing Superinten-dent, who had been the Cleansing Superintendent in Sheffield before the war. Jackson introduced many processes designed to reuse waste profitably, including a method to extract precious metals from loads destined for incineration.[27]

Following Jackson's move to Birmingham in 1914, John Arthur Priestley had taken over in Sheffield, with hopes to rival the effectiveness of his predecessor. By March 1918 Sheffield had 'a complete and up-to-date waste utilisation plant'. Born in Vyshny Volochyok (where his father worked in a Russian cotton mill), Priestley was a sanitary inspector by training, and was one of the first to suggest that paper ought to be baled up and sent to papermakers, rather than being fed into municipal incinerators.[28] We met another of these municipal salvage heroes in the last chapter: J. C. Dawes, the Deputy Controller of Salvage during the Second World War. Dawes had been the chief sanitary inspector of Keighley in York-shire since 1911. After devoting much attention to refuse there, he came to the attention of the government, and in 1916 had been seconded by the Department of Supply to advise on salvage. These men, focused on public health before the war, morphed into resource management experts as the war progressed.[29]

The NSC was chaired by Lord Derby, the Secretary of State for War, and included representatives from the War Office, the Admiralty, the Ministry of Munitions, the Ministry of Food and the Local Government Board. In their mouthpiece, the *Municipal Engineering and Sanitary Record*, the city engineers and municipal sanitary officials recorded much puzzlement about the men filling the NSC roles, noting that 'their composition does not include men who have the necessary experience, knowledge or qualifications for dealing with the subjects referred to them'. Echoing some of Lady Balfour's concerns, the engineers calculated the expertise and experience of municipal waste shared by the military men who predominated in the NSC to be 'of an infinitesimal amount'. And yet their influence seemed disproportionately strong: representatives of municipal authorities were 'entirely ignored', and the advice of the city engineers was, seemingly, 'despised'. In turn, the municipal engineers wrote that, if the NSC wanted to keep on good terms with the local authorities, then military men should not 'be sent to borough engineers if irritation and failure are to be avoided'. Unless the NSC were not 'materially altered', they warned, it would be disregarded by the local authorities.[30] Dawes and Priestley were co-opted as advisers to the NSC and tasked with writing papers for countrywide conference tours, but their role appeared to be more one of spreading the message outwards than of informing the great and the good at the top.[31]

Meanwhile, a 'Salvage Club' was set up on Pall Mall. Cynics supposed it existed chiefly to enlarge the dining calendar. The Director-General of the NSC appeared at Salvage Club events to urge frugality after feasting on a fine lunch with other private members.[32] A magazine called *Salvage* appeared from May 1918, as the 'official organ of the Salvage Club'. It was intended that it would detail examples of material neglect and extravagance. The first edition included articles about army boot repair, reclaiming metals from tin cans and intelligence of German recycling.[33] Those who wrote for *Municipal Engineering and Sanitary Record*, and those who read it, savaged *Salvage*. Frustrated engineers wondered why *Salvage* was issued while other journals were prohibited, to save paper. A correspondent with the pseudonym 'Substitute' failed to identify a need for a 'special organ', especially one pumped with irrelevant and grating content.[34] Having received a copy 'unasked for', one surveyor queried the purpose of a journal sold for a shilling, issued by a club 'with the existence of which I was hitherto unacquainted'. 'How is it possible', he asked, for 'amateurs to start a new journal which obviously does not appeal to the public', and which was produced under 'mysterious' financial conditions. There was, he concluded, nothing useful in it, and pages of precious paper had been wasted with 'portraits of sundry officials and articles on trite and well-known topics'.[35] The resource management experts felt displaced by amateurs, just as their professional skill and experience were most sorely needed. And who could blame them?

But others were not put off – including other amateurs. Various organisations worked with the NSC, including the British Women's Patriotic League. Their Kensington branch appealed for local garage space where neatly packaged waste materials might be stored. The League gathered rags, metal and also paper to make boxboard, a timber substitute.[36] Demand for paper increased during the war: as in the Second World War, it was involved in the manufacture of munitions, ration books, war communications and food containers. Much paper evaded recapture.[37] Some went to the front in France; some lit fires; 'much suffered destruction through ignorance' or was seen to 'drift and flutter hither and thither', getting 'soiled [...] beyond redemption'.[38] Wartime limitations on import shipping meant that domestic paper mills were forced to increase production, often using rags, straw and wastepaper. Private enterprises licensed to deal in paper appealed for supplies.[39]

For a while, wastepaper from books, magazines, correspondence, businesses, packaging, old wallpaper hangings and advertising hoardings fed the mills, but these materials eventually dwindled. With evident frustration, W. M. Meredith, the President of the Publishers' Association, wrote to *The Times* in March 1918 to bemoan the initial response to wastepaper collection. 'Many thousands of tons', he stated, 'went to waste before unorganized individual effort stepped in.'[40] Lord Burnham (Edward Lawson) agreed. Speaking at a luncheon of the Salvage Club, he expressed annoyance that paper was being sucked into munitions in France while supplies to newspapers got ever tighter. Burnham, whose family owned the *Daily Telegraph*, thought that newspapers were more deserving of the increasingly precious material.[41] The *Municipal Engineering and Sanitary Record* published a dismissive account of Burnham's 'evergreen' concerns, under the headline 'Still harping on waste paper'.[42] Two months later the same journal harped on about the government wasting paper by printing a 400-page report on the Deutsche Bank and distributing a thousand copies to British banks.[43]

Salvaging the whale

Few people were more enthralled by the economics of waste than Frederick Arthur Ambrose Talbot. Born in 1880, the son of a Camden builder's clerk, Talbot wrote *Millions from Waste* in 1919 to urge his readers to absorb German habits of thrift, in order to cement victory after the First World War. He hoped to inform 'the man-in-the-street' and 'the woman at home' of the 'enormous wastage' and the 'colossal' squander occurring domestically and industrially in Britain. Talbot argued, without a hint of irony, that the subject of waste reclamation was 'romantic and fascinating'.[44] He wanted residues to gain 'accepted commercial values' but recognised that his dreams were not shared by the wider British public, who saw waste as an expense of time, space and labour.

Fat waste especially intrigued Talbot, and his hero was a man who reutilised lots of it: William Hesketh Lever (Lord Leverhulme from 1917). Lever had successfully melded several industries based on fat: soap-making, candle manufacture and margarine production.[45] He was the son of a grocer, and early in life he learned to avoid waste and keep expenses to a minimum. He started with Lever Brothers, a company that made soaps, initially in Warrington and later on the Wirral. Soap had been made from fats such as butcher's tallow and oleaginous kitchen scraps, but Lever Brothers used vegetable oils and glycerine instead. Their products included Sunlight Soap and later, after they had acquired the soap-making firm A. & F. Pears, Transparent Glycerine Soap, which could be 'used to the thinness of a sixpence'.[46]

Looking at Leverhulme's tactics, Talbot concluded that the fact that the soap-maker had diversified into the large-scale production of margarine 'merely represents one of those inexplicable coincidences of industry'. But here he was wrong. Concerned to reassure readers that Lever's strategy did not involve any ploy to get consumers eating rancid fats, Talbot failed to focus on one of the key characteristics of industries that reutilised dirty, bulky or fatty wastes: the expediency of combination.[47] It was already common to combine manufactories involving fats, and Lever was far from alone in pursuing such a strategy: Christopher Thomas & Brothers in Bristol made soaps, household disinfectants and long-lasting Arrow candles. They also distilled glycerine.[48] In 1914 the Cooperative Wholesale Society set up a margarine factory in Irlam, close to their existing soap, candle and glycerine works.[49] Ivie Hair & Co. was a well-established Glaswegian company, making soap and candles and trading in oil and lard. When their works were destroyed by fire a few days before the commencement of hostilities abroad, 'a steady stream of melted fat flowed along the gutter'.

Glycerine was a by-product of soap-making that, despite its many applications, had often been wasted.[50] It could be used topically to treat burns and skin complaints such as chilblains and eczema, and as a hair tonic. Medicinally, it served as a sugar substitute for diabetics, a cough liquid and an ingredient in laxative suppositories. It was added to foods and paints, functioned as a photographic emulsion and suspended vaccines and compass points. Householders kept it on hand to remove stains, as antifreeze or to condition leather. At the start of the war, a recipe for an economy home-made cold cream combined equal parts mutton suet and glycerine, set in empty fish paste jars that might be sold at church bazaars.[51] But the use of glycerine in toiletries was prohibited during the war, with supplies reserved for government use.[52] The recovery of glycerine was a national concern, because, combined with nitric acid, glycerine made the explosive nitro-glycerine. In August 1915 glycerine was requisitioned by the government. Three major soap-making firms agreed to process glycerine for munitions; Lever Brothers supplied the majority.[53] Eventually soap works fell under the control of the Ministry of Munitions. Getting rather carried away in 1917, *The Scotsman* described the production of glycerine for explosives as 'a process of which the inner history borders on romance'.[54] Previously soap had been the primary product, glycerine the by-product; but war turned this upside down, as it did so many other things.[55]

As a result of increased munitions stockpiling, the price of glycerine crept up before war started, and prices banked steeply with the arrival of war.[56] At the start of the conflict Britain and her allies held on to the misplaced hope that, if supplies to Germany of glycerine and waste cotton could be ended or limited by blockade, then Germany's stock of explosives would dwindle and it would be unable to prosecute a sustained war.[57] News that German supplies of glycerine had been constricted were greeted with enthusiasm in Britain.[58] In 1917 the press was hot with claims that stocks of German glycerine were enhanced using 'the bodies of their dead'. Knowledge of this practice was rumoured to have caused China to break off relations with Germany.[59] But one effect of this propaganda about dwindling German supplies – which led the British to assume the war would be quickly won – was to discourage the British from protecting their own domestic stockpiles early on. British supplies of glycerine soon thinned. Although soap-making firms supplied glycerine to the wartime government at pre-war prices, they still turned a healthy profit and were accused of profiteering. In March 1918 Leverhulme mounted a detailed and defensive account of his own business, arguing that his company had to bear loss of tonnage, plus higher wages and costs. A buoyant company, he said, could help keep the country strong.[60]

The search for glycerine sources went on in a variety of places. Army camp wastes were harvested for fats.[61] Attention next turned to householders, who were urged to save fats and bones to sell to rag-and-bone men.[62] Even parsimonious Yorkshire folk needed nudging encouragement to exchange bones for cash. The use of glycerine for medical purposes was restricted: fines were introduced to prevent doctors from over-dispensing it. When supplies became hard to obtain, companies began to advertise glycerine substitutes made from unknown substances, many 'for external use only'. A home-made substitute could be had from the run-off from turnips soaked in sugar.[63]

Relief came from unexpected sources: the sea gave up oleaginous bounties in mid-1918. First, a bottlenose dolphin swam up the Thames and became stranded at Battersea Bridge, where it died. This was an old male with worn teeth, weighing about half a ton. The carcass was taken to the Natural History Museum, where staff chopped it up, distributing some as meat steaks to the archbishop of Canterbury and the lord mayor of London. Some skin was sent off to be tanned, and blubber and waste meat were dispatched to soap-makers for experimentation.[64] Then a greater beast emerged from the North Sea: a fourteen-ton whale was beached in Bawdsey, Suffolk. A 'fatigue party at once attacked the whale with knives and hatchets, and the oil was extracted in a digester', furnishing enough glycerine for 130 18 lb. shells.[65] The bones of the beast were ground into manure. Reports of this biblical scene are instructive. The account in the *Southern Reporter* reminded readers of the enemy's 'corpse factory' where soldiers were rendered into glycerine, before trumpeting British scientific ingenuity and good fortune. There was also a reflective tone in the consideration of British waste: the whale was regarded as 'an example of the way in which the war is making us provident, even in the utilisation of cast-up whales'. Previously beached whales were monstrous calamities, left to rot and stink while people argued about whose responsibility it was to remove them. In this new era, 'whales [would] no longer be regarded as white elephants' but, rather, would be seen as vendible fortunes.[66] Talbot enthused about the chemical wizardry that allowed whale oils to harden through hydrogenation, permitting whale refuse to be turned over to margarine-making. During the early months of the war Lever Brothers had absorbed Planters' Margarine Company, apparently at the request of the government.[67] And so margarine was now made alongside far from edible substances. Margarine had been used as a cheap foodstuff since the Franco-Prussian War of the 1870s. Recipes varied – it was one of those products that could be made from many substances, including 'mammary tissues'.[68] Hydrogenation created a palatable new butter substitute to rival the hydrogen-filled German airships for ingenuity. Margarines could now be made with fish and whale oils, the fishy taste removed by hydrogen.[69] Lever Brothers snapped up relevant patents.[70] According to Talbot, hydrogenated fish oil could be an 'excellent butter substitute [...] so closely allied to the genuine article'. He couldn't believe it wasn't butter![71]

Gather nutshells in July and dog hair in vain

By the summer of 1918 increasingly wide-ranging appeals were made for things like garden hose, old boots and 'sand shoes'.[72] Industrial trials and government studies considered a vast range of products that might be recycled: bark waste, wool grease, even arsenic. Such were the mind-boggling possibilities and the new scale of the enterprise that it was eventually deemed easier to compile and circulate a report entitled 'Waste Products Considered of No Use' – the result of questionnaires sent to manufacturers.[73] Meanwhile, tatters and shreds of leather left over from the making of army boots were degreased and fashioned into discs for use in mattress-making. Eyeshades were formed from old umbrella silk, with cardboard 'retrieved from the waste-bin'.[74] The government took a close interest in the question of whether spent tealeaves might be rejuvenated, or the caffeine extracted: the army handled millions of pounds of tea refuse, but inconclusive experiments saw the scheme abandoned. Pre-war Britain imported huge quantities of potash from Germany, employing it in industries including munitions and glassmaking, as well as for fertiliser and Epsom salts. Alternative plans were drawn up to extract potash from tealeaves, cigar butts and ash, banana stalks and wash from distilleries processing molasses.[75]

During the summer of 1918, mysterious and increasingly desperate calls for nutshells and fruit stones were issued by the NSC. Lord Baden-Powell urged Boy Scouts to collect them: 'It sounds a small thing', he started, 'but, really, it is a jolly big thing.' His appeal was effective in stirring the collecting spirit in young boys. Stone and Shell Clubs were formed to help meet the demand. Boys from Victory Road School in Horsham, west Sussex, won a prize for their collection of 51 lb. of shells and fruit stones. Supplies were even gathered from Buckingham Palace. An evasive report from 23 July hinted only that they were 'for munitions purposes'.

21. Hair of the dog, ©Imperial War Museum (PST 13409).

Eventually the beans were spilled: nuts and shells produced the most absorbent charcoal and were therefore necessary for the production of effective gasmasks.[76] The Germans had long known of these special qualities and had already been processing them into charcoal for gasmasks. Britain was 'completely forestalled by the enemy': the German use of poison gas was made practicable on discovering that 'a complete antidote' for their own soldiers could be obtained from charcoal refined from the spurned nutshells of foreign countries.[77] The stones were not the only contributions towards gas masks: British photographers were asked to fish out their old negatives, to make circular 'eye pieces' for them. A layer of xylonite – an early plastic – was sandwiched between discs of glass to shatterproof it. The trimmings from the negatives were collected and sent as cullet to glassmakers; some of this was converted back into the base for more negatives.[78]

If the preservation of nutshells and fruit stones was literally a matter of life and horrible death, it could sometimes be difficult to distinguish the fateful from the frivolous in the heated atmosphere. The British Dogs' Wool Association created a poster (**fig. 21**) appealing for hair combed from dogs in Kensington, 'for the supply of comforts for the Sick and Wounded'. Miss Du Cros was a member of a family made rich by the Dunlop rubber company, whose property was attacked by suffragettes before the war.[79] She offered a prize for every pound of wool given by a dog owner. There was a national shortage of wool, despite a (rather feeble) appeal in May 1918 for the collection and dispatch to London of all stray tufts of sheep's wool found on rural hedges.[80] From temporary head-quarters in galleries of the Royal Academy the collected dog fur was spun by volunteers for the British Red Cross. Yarns would be woven or knitted into garments for wounded soldiers: white hairs made 'operation stockings and bed socks'; coloured hairs were knitted into cardigans, mufflers and waistcoats; hairs too short for spinning were stuffed into pillows and cushions.[81]

In a photograph from 1919, four women spin hair from different breeds of dog – Pekinese, chow, collies and Pomeranians – identified by labels as though they were exhibits at dog shows.[82] An exhibition was organised by the Victoria & Albert Museum that included specimens of dog hair, and the yarns that could be spun from them. A cuff made from poodle hair, first exhibited at the Great Exhibition of 1851, was on display. Members of the royal family had previously had hairs from borzois spun into yarn.[83] The owners of specific breeds of dog (including spaniels, curly retrievers, setters and Yorkshire terriers) were asked to collect clean hair in a loose muslin bag, to be soaked in soap and Jeyes Fluid for fifteen minutes before rinsing and drying. Envelopes marked 'Dogs' Wool' were to be labelled with the breed's name, and sent to the association in batches of no less than 4 oz.[84] What a palaver![85]

The scheme to use dog hair was mooted at a time when dog shows were *verboten*; some members of the Ladies Kennel Association involved themselves with dog wool collection. The *Manchester Guardian* noted that some key members of the British Dogs' Wool Association were breeders, 'a quite imposing committee, composed mainly of ladies of title'. These ladies, with time on their hands and a passionate desire to maintain the value of their dogs, now made woolly claims about the 'valuable combings' to be had from pets. Unnamed 'wool experts' were 'emphatic in their commendation', and fooled into thinking the yarn was the 'finest Angora'. Great cry and little wool. An article in *The Sphere* made explicit reference to the timing of the creation of the association, suggesting it might have the effect 'of giving pause to that section of the public which is agitating for the wholesale destruction of dogs'.[86] Only slightly tongue in cheek, the *Manchester Guardian* commented that 'on aesthetic grounds alone a Pekinese had better be broiled than shorn', and wondered when the valiant, active, short-haired breeds would get their praise for 'helping to win the war'. The paper pleaded for sanity, 'and a realisation of the fact that the dog's case for preservation does not rest on the value of his hair any more than on his skill as a pianist'. Thirty-one dog hair donors were listed in *The Tatler* a few months after the call was issued, including Lady Gertrude Penrhyn, whose two sons had been killed in action.[87] Once the ladies returned to their dog shows, their enthusiasm for exquisite yarns unravelled.[88] It may be no coincidence that the entire Kensington scheme was conceived shortly after the local MP, Lord Claud Hamilton, had suggested that Pekinese dogs might best assist the war effort in pies.[89]

Salvage continued

Towards the end of the war, hectoring rants appeared in British newspapers complaining about the late national awakening to the need to save 'numerous articles formerly discarded as useless'. That context permitted dog wool to be taken seriously by some. 'If everybody had practised in pre-war days the numberless economies the war has made compulsory', readers were advised, 'there would have been less talk of our dependence on other countries.'[90] In the final year of the war, Henry Spooner wrote *Wealth from Waste*, with a foreword by Lord Leverhulme. Spooner described a changed mindset: before the war 'everything [had been] so different; in the full enjoyment of unparalleled prosperity, extravagance and waste were rampant'; but the 'colossal requirements' of war had seen the

22. Occupational therapy 'bunnies' featured in *Quiver*, April 1919.

dawning of a new era of frugality and thrift. But it was not new and had not really dawned. It did not occur to Spooner to wonder whether there had been times (not long before) when the British had been better at this sort of thing. Like others, Spooner had been impressed by pre-war German reuse, focusing on their dye industry and the ingenuity with which coal tar was turned into a multitude of useful substances.[91] In 1918 the British finally caught on, but not up. Germans held out through the British blockades, Talbot argues, because their country had 'inaugurated a more rigid system for the compulsory collection, separation and utilization of his domestic waste'.[92] Yet Germany did not win the war.

By the time the Armistice was declared, the British had finally started to embrace collection and separation of waste. Stockpiles of salvaged objects bloomed: one salvage dump in Woolwich covered two square miles. It included brass shell cases, and periscopes – 'the mirrors of which [… were] converted into shaving glasses'.[93] Other salvage continued too. Sir Francis Fox's scheme to recover fabrics from the back of engineering and architectural drawings continued after the war ended, for some hospitals were full of casualties. In *Quiver*, a journal aimed at the middle classes, Bella Sidney Woolf

23. Lady dropping food into the pig bucket,
*c.*1917. Mary Evans Picture Library.

(Virginia Woolf's sister-in-law) outlined ways that readers (her 'army of helpers') could provide materials for use in occupational therapy for convalescing and traumatised soldiers still in hospital at the end of the war. Pieces of cloth, velvet and fur were used to make soft toys. She also appealed for old top hats, or rather just their skins, to be turned into bags: 'We had bags made from a bishop's and a general's top hats, bags of historic interest!' Gloves and colourful bags, however, were now surplus to requirements. 'It is such a satisfaction', Woolf hummed, 'to feel that we can turn out attractive toys and beat the Hun in a province which was so long his own.'[94] It is hard to imagine these toys being of comfort to anyone: they were truly horrible (see **fig. 22**).

The experience of war had revealed new ways to use waste. Materials scheduled for collection included condemned food and slaughterhouse waste in animal feed.[95] Among the pilot schemes adopted by the NSC was the speculative establishment, in Essex, of piggeries where the animals were fed on household waste.[96] Municipal piggeries were also located in Edinburgh and Birmingham and elsewhere.[97] In Liverpool old oil drums were converted into swill buckets.[98] In violation of old by-laws, pigs were even kept in crowded parts of London, including Islington and St Pancras, where a piggery was situated on an old gasworks behind Kings Cross Station. Marylebone pigs were reared on the destructor site and fed with the waste brought in.[99] Eventually, the NSC and Ministry of Food became embroiled in a public spat. The food controller in the Ministry called for all pigs to be slaughtered before Christmas 1918, saying that there was no promise of pig food beyond January. At the same time the NSC urged householders to take kitchen wastes to piggeries to overwinter city pigs. The piggies were in the middle. Reprieved by the end of the war in November, many lived to see another Christmas.[100]

Repairing

Before the war, 'VOTES FOR WOMEN' had been stamped on pennies by suffragettes, an ingenious (and illegal) repurposing that circulated their message widely.[101] Small explosive devices put together by the suffragettes in the 1910s reutilised materials including milk cans and mustard tins, and they made use of by-product materials such as candles and nitro-glycerine.[102] Suffrage banners were often custom-made by women with expertise at home or commissioned from professionals. Suffragettes wanted their skills properly recognised: exquisitely neat banners gave off a stronger message than scrappy makeshifts reusing household linens. By contrast, women campaigning for fuller liberation in the 1970s turned bed sheets into cruder banners: by then, women had become less tethered to chores of needlework, darning and mending. Thalia Campbell, who made many of the peace banners in the 1980s, some of them for the women campaigning against the storage of nuclear weapons at RAF Greenham Common, sourced her materials from charity shops and jumble sales, pouncing most readily on bridesmaid dresses in 'good quality satin and lovely colours. Many were cut up and made into banners. A bit of a feminist act too.'[103]

When the campaign for suffrage was interrupted by the First World War, women turned their hands to repair of a different kind. Female volunteers were crucial to the success of wartime salvage campaigns. By July 1918 over eight hundred women local to the

Thanet area worked in a camp processing spent munitions, returning empty shell cases for reuse. Day care and trains were laid on for the workers, who were required to 'be well shod and possess strong gloves'. Trousers were also recommended. A few months before the end of the war, the camp put out another appeal, this time for any women who could sew: 'no active woman who can sew and mend is too old for this work.'[104] Nimble-fingered women were in demand.

Hand-sewing was still a valued skill. In the early twentieth century, like centuries before, the longevity of items was extended by stitching, patching and resurfacing. Clothing items in need of complex repair work could be outsourced to a professional. Trousers were modified to make 'Bicycle Pants' by shortening them to fall at the knee, gathered in like knickerbockers to help reduce the drag when pedaling. Fur coats, stoles and muffs were remodelled and renovated.[105] Seamstresses offering alterations were found in the classified sections in local newspapers. One advert read: 'I convert old fashions, cure bad fits, clean, repair, tailoring.'[106] Mrs Bowditch in Portsea would 'neatly' repair all underwear sent her way.[107] Pre-war uncertainty, war itself and then recession in the late 1920s imposed, on some at least, an urgent need to spin out the life of their material possessions. The rise of household DIY – in the sense of redecoration and house repair – really did not start in earnest until after the Second World War, but an earlier interest in the repair of household objects is evident. Many people still sent items away to be repaired, but an increasing number of householders began gathering equipment and finding space to undertake repair tasks at home. The early twentieth century saw a flurry of books filled with helpful hints for people wanting to carry out repairs to homes and household possessions.[108]

Most textile mending has left absolutely no historical record, in this period or in any other, because most clothing repairs and alterations were undertaken in the home, by women. In her *Manual of Plain Needlework* (1930), Eleanor Griffith furnished her readers with methods to darn and patch various types and thicknesses of fabric. Darning, she taught, 'should generally be done on the wrong side'. Griffith tailored her instructions: ragged tears were to be handled differently to sharp cuts.[109] Griffith was a 'diplomee' from the National Training School of Cookery and Other Branches of Domestic Economy (NTSC) and an examiner for the Central Welsh Board for Secondary Education. She wrote her manual to give 'in as clear, concise, and workmanlike a manner as possible, all the fundamental operations underlying the practice of Plain Needlework'. Beyond the details she included in her book prefaces, little is known about Griffith's life. Between 1937 and 1945, then in middle age and as a widow, she taught cookery, laundry work and needlework at the County School, Pembroke Dock.[110]

Done well, mending is invisible, and now many of the menders are invisible too – sometimes scandalously so. Griffith's *Manual* was reprinted in 1930, then revised in 1932, and reprinted seven times by 1944, before coming out in a new edition in 1952. Clearly it was invaluable, and it remained in wide use for over two decades. Nonetheless, none of Griffith's works appears in the main catalogue of one of Britain's three deposit libraries – they appear only in supplementary material card file boxes. Her books were seen as inconsiderable by librarians, despite being published by an academic press, despite their popularity in their day and despite the

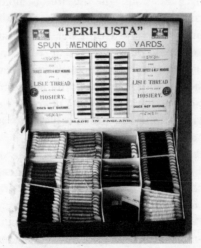

24. Peri-Lusta yarns. Author's collection, photographed by Taryn Everdeen.

25. *Illustrated War News* image of women shaking flour bags, 1916.

subcutaneous effect such works had on the nation's attitude to repair in the mid-twentieth century and beyond: mere 'women's work'. The impact of her work was still felt long after. My Nan used Griffith's books when training for her domestic science diploma. I inherited them and use them when I can find time to mend.

Mending required specific materials. This box of mending yarns (**fig. 24**) was part of leftover stock from William Stirrup Critchley's fancy goods store in Widnes. The shop was destroyed in 1929 by a fire caused by an electrical fault which ignited fireworks and celluloid items.[111] The threads came in a spectrum of colours in a durable finely spun cotton, which (according to the marketing) 'does not shrink'. This quality was important, to prevent any darned sections puckering and twisting the fabric on being washed. Peri-Lusta was a major yarn manufacturer, with a mill in Leek in Staffordshire. The wider the selection of mending yarns a darner had, the more items they could mend more neatly and less visibly. Wool yarns were used to darn woollen items, cotton yarns to darn cotton items – two-way elastic yarn was also available to repair support stockings. Using yarns like these, women darned their way silently out of history. Like Peri-Lusta yarns, the darners came in many shades and types –and like the yarns, their work was tucked away in supplementary boxes, invisible.

As the effects of recession intensified, reuse of fabrics diversified. Some housewives took to repurposing flour bags. This was not novel – references to aprons made from flour bags appear at the beginning of the century – but the practice appears to have increased, especially in Scotland, during the late 1920s and early 1930s.[112] In 1935 a competition held by the St Andrew's 'thrift class' debarred articles made from flour bags, in a bid to encourage greater ingenuity. Noteworthy entries included 'a pretty little white organdie frock [for a child] made from French knickers'.[113] As we have seen, the use of flour bags was quite concerted during the Second World War, and the subject was even raised in Parliament, but fewer households (especially outside Scotland) had bothered with them during the First World War.[114] 'Waste not, want not' was the familiar caption chosen by the *Illustrated War News* to accompany a photograph of women shaking out flour bags that had been returned from the Front to be sorted at the Hay Reserve Depot in Richmond (fig. 25). The commentary refers to the pig food made from the waste flour but doesn't allude to a use for any spent bags (although a few of the women appear to be wearing aprons that may very well have been fashioned from them).[115]

🗑 🗑 🗑

During the First World War, the language of reuse was less heartfelt than it would later become in the 1940s. Habits of reuse and salvage were shallow: this was one of the difficulties facing a government suddenly interested in pursuing organised salvage on a massive, national scale. A jaunty article in the 'Woman's Sphere' column of the *Aberdeen Press* in July 1918 claimed that the Golden Dustman in Dickens's *Our Mutual Friend* 'was not a typical Briton', and that the British have always done things 'with a magnificent disregard for trifles'. War had taught hard lessons about the value of scraps and ends, the readers were told – but they were also reassured that they need not do without. They simply had to learn the 'art of profiting by every little "wrinkle"'. 'Economy is like fishing', the column continued, needing patience and skill, but 'once acquired can be made an exhilarating sport, full of pleasant surprises'.[116] Who were they kidding? It was a tiresome and time-consuming activity, with competing petty interests and irritating egos, and mixed outcomes. Despite efforts, the British never took to it like a sport. The vanquished Germans had been the recycling pioneers. But whatever their prowess at recycling tin, neither side proved good at saving lives: those millions wasted could never be salvaged.

Wartime made some men, such as Frederick Talbot, think long and hard about the value of waste. Talbot correctly foresaw that the 'separation and collection of the residues' would be a 'perplexing' problem which would hold back the progress of recycling. He concluded that future schemes should switch away from voluntary reuse and towards paternalist coercion, even in times of peace. Compulsory measures were 'absolutely imperative, otherwise all the mickle which makes the muckle must slip through the meshes of the net, no matter how well it may be cast'. He was, however, aware that the British regarded any compulsion as an infringement of liberties. As a national outlook this can be a heroic liability.[117] Talbot himself moved to Canada after the war and was set to publicise the Grand Trunk Railway, hoping to introduce the prince of Wales to it in 1924.[118] His family had settled there a few months earlier, but Talbot died of pneumonia in October while awaiting the prince. He left a widow and three fatherless girls.[119]

Lords and ladies – people who had never understood the value of waste, or the need to eke out meagre supplies – were put in charge of British recycling during the First World War. An unfocused, ill-directed, patronising and sometimes bizarre campaign was the result. The period seems characterised by committees and discussions over lunches, rather than coherent and sustained activity, especially in the early years. The poster appealing for dog fur is emblazoned with the slogan 'Every Little Helps', used later in the Second World War (and more recently adopted by Tesco). Recycling was really just a lot of little. While the fine ladies of Kensington, bored without Kennel Club excitements, tried to make a silk sock from Pekinese ear hair, and Sir Francis Fox had Robert Stephenson's drawings trashed in his garden pond to redeem fabric, miles and miles of blankets and linens lay untouched in country house cupboards across the country. The promise of a nation 'in it together' is an ancient lie. After the war Lord Leverhulme went on a charm offensive, opening up his factories to tours for grocers. Many of the visitors came away believing that butter was made from by-products of the margarine-making process.[120] How the world had changed; how it had stayed the same.

5

Reconstructors and Destructors
(1880s–1900s)

At 2 p.m. on 3 November 1894, people queued in squally winds for a 'jumble sale of second-hand goods' held at St John's School in Werneth, an area of Oldham in what is now Greater Manchester. After paying his twopence entry fee, one man rummaged through the cast-offs and found enough hats to set himself up with headwear for life. A mother carefully judged whether fabric in a garment could be reused to make one child a frock and the other a petticoat, while eying up a perambulator missing a wheel. Lingering by birdcages, a schoolboy imagined having a tenant for one. There were 'cheap bargains for every one'. Some items had come from pawnshops, as time-expired pledges, and some had been given to servants as perks, but were found to not be useful. An old lady bought a fur muffler and an oriental vase sporting rivet repairs. Refreshments were provided: the hat enthusiast probably had a potato pie. Six years later, the newly elected MP, Winston Churchill, opened another sale at the school – a South-African-themed bazaar. The proceeds were destined to pay off debts incurred in school renovations.[1]

Bazaars and jumble sales were organised in increasing numbers at the end of the Victorian period; they quickly became 'a recognised institution in parish work'. Newspaper reports suggest that 1889 was the year 'jumble sales' were first held: one held at Wollaston in Northamptonshire in January raised money for

26. 'A jumble sale', *The Graphic*, 19 November 1892.

candlesticks and a platform for the church.[2] The following month the *St James's Gazette* inaccurately reported that 'the bazaar has had its day. The future is to the "Jumble Sale of Odds and Ends" which has just been invented by the ladies of the congregation of Holy Trinity Church, Richmond.' Some early jumble sales had auctioneers, who found the best price for 'old pots and pans, [...] rickety bedsteads, the chairs which have lost a leg, and the coats and dresses that are past wear'.[3] Ties, collars and other accessories sold well, but even old worn underwear found a market, among 'tidy but poor girls'. One of them told a kindly district lady who had helped organise a sale that she bought dilapidated articles to 'get a good pattern by which she could cut out new things for herself'. 'Philosophically speaking', *The Graphic* noted, 'the jumble sale has a fair assurance of success because it appeals to those elementary principles of popular dealing – cheapness and adaptability.'[4]

The closing decades of the reign of Queen Victoria and the early Edwardian years saw much material redeployment, including the wholesale movement of materials of heritage. At one end of the social scale, the poor bought items at jumble sales. At the other end, richer citizens purchased second-hand items on a much grander scale: oak carvings and stained-glass windows were imported from the Continent, and these lent old heritage to new builds; they reconstructed the past. Local authorities, sometimes overwhelmed with refuse, began to embrace incineration as a means both to destroy waste and to power systems of transport or lighting. Tradesmen set up businesses providing mending services for people wanting to eke out the life of their belongings. The most reputable of these established themselves in shops; the least reputable travelled around, offering itinerant services.

Just [re]do it

Many items were designed with repair and adaptability in mind: careworn pans obtained at the jumble sales could be re-tinned; shoes could be re-soled. Using techniques unchanged since the eighteenth century, pieces of broken china and glass could be riveted, giving the impression of being stapled.[5] Missing lid knobs and handles were replaced, sometimes crudely. William Larcombe priced repairs by the rivet: 'China and glass repaired 2s per dozen rivets, neat, prompt and reliable work guaranteed.'[6] Many men offering such services diversified into other fields; a significant number combined china repair with umbrella- and parasol-mending, allowing them to use tools and materials in various different ways.[7] Diversification suggests that repairers could not make a living (outside the biggest cities) if they specialised in the repair of just one thing. They could use the same equipment to mend a range of objects – the range extended when new products became available. The Barnstaple 'Umbrella Hospital' combined umbrella repairs with cutlery-grinding and glass- and china-riveting. Permanently injured in the First World War, James Richinson returned to glass- and china-riveting in Folkestone, where he also ground cutlery, recaned chairs and repaired tinwares.[8]

Itinerancy was a common part of riveting and grinding trades,

27. Rivet repairs to a glass. Author's collection,
photographed by Taryn Everdeen.

because menders did not need to keep premises – they could work
on the road. But itinerant menders were considered with suspicion:
would items be returned at all? Without a shop, how can the householder be sure? A person describing themselves as 'THE VICTIM'
wrote to the *Gloucester Citizen* in February 1887, after being swin-
dled by a gang of fake china riveters. An antique china dish had
been returned after repair, but 'on examination [...] the steel pieces
were simply stuck on with gum or resin and no single rivet put in.
The pretended work is entirely useless.'[9] The milieu encouraged
social contempt even when it was less than clearly criminal. An
Ellen Nike had been described as a 'female vagrant [...] a member
of a party of scissor grinders, now travelling the neighbourhood'
and imprisoned for sleeping rough in 1885.[10] A china-riveter
described as 'a tramp' was imprisoned for a week in 1885 for begging
in Loughborough.[11] The toughness of this lifestyle is brought home
with the unrelated sudden deaths, in 1875, of two wives who travelled with their itinerant riveting husbands, one in Leicester and
the other in Cambridge. The deaths of both of these women had
been accelerated by the severity of the winter. Jane Carter, wife of

28. Rivet repair on a jug. Author's collection,
photographed by Taryn Everdeen.

a 'travelling china rivetter', died of burns after the two coats and
shawl wrapped around her caught fire as she slept. Sarah Gray,
married to an 'itinerant glass and china rivetter', 'dropped down
dead' after she was exhausted by 'inclement weather'.[12]

Tramping botchers descended on Leamington Spa in 1912, and
the local paper railed against such 'visitors' – which were said to
include 'minclers' (tinkers and grinders), 'mush fakers' (umbrella
vendors) and 'chiny fakers' (china restorers).[13] 'Most of our readers
will no doubt have had personal experience with this fraternity',
wrote James Howorth of itinerant repairers in his late nineteenth-
century guide to riveting, 'either with their sham rivets, their bogus
secret bolting, or their alleged process of burning or fusing.'[14] The
botched repairs appeared sound initially, but quickly failed – by
which time the itinerant mender was nowhere to be seen. Localised
wariness clearly developed in Leamington Spa, as by 1923 the ladies
of the local Women's Institute had become proficient china-rivet-
ers, preferring to mend items themselves rather than send them
away.[15]

From his premises on New Bridge Street in Exeter, Josiah Nike
offered a variety of repair services. He would blow feet onto glasses,
cut decanters down into sugar bowls, burn and rivet china, and

recover and re-seam umbrellas and parasols. His advertisements from 1884 end with a curious statement: 'J. N. has no connection with any other person in the Trade.' Nike, the son of a china-riveter, had recently moved to Exeter from Plymouth. Josiah wasn't being honest, because he was the cousin of the town's notorious china-riveting John Nike. He must have known that a connection with this infamous character would be bad for business.[16] John Nike was a well-known character around town (and was also related to Ellen Nike, the itinerant scissor grinder we met earlier, who is likely to have been Josiah Nike's sister-in-law).[17] On 28 June 1884 the same newspaper which carried Josiah Nike's adverts reported that a John Nike had been charged with 'being found drunk and incapable [...] while in charge of a razor-grinding machine' on Paris Street. PC Squires tried to arrest him, but Nike, 'acting like a madman', drove his machine 'down the street to get more drink'.[18] John Nike had previous. He had smashed up Mrs Trim's lodging-house back in 1877 ('Nike and his wife skedaddled before their hostess was up') and would continue to find it hard to mend his ways. That year he was described as a 'diminutive young fellow, of sinister mien'.[19] In 1887 John Nike and his cousin were both arrested for being drunk and disorderly in Sophia Beer's pub. Nike was locked up for a week in 1899 for assaulting his wife and threatening to kill her. Again, a statement appeared in the *Western Times* to claim that Josiah Nike had no connection with John Nike.[20] In 1914 John Nike and his son, John junior, were charged with drunkenly assaulting PC Harris (Nike senior 'struck Harris a deliberate blow in the eye'). This was Nike's thirteenth conviction, his son's first. The following year John junior was killed in action.[21] Meanwhile Josiah continued in the mending trade, putting aside part of his shop to sideline in tobacco sales; he managed to maintain a superior status to his more wayward cousin John.

I like to imagine PC Squires blowing a brass whistle as he pursued John Nike and his razor-grinding machine as they veered drunkenly down that Exeter street in 1884, of the snail-shaped sort that had recently been adopted by the constabulary. A whistle (**fig. 29**) was made around that time, from two buttons. The buttons have a backmark (inside the whistle), showing their maker

29. Whistle made from buttons. Author's collection,
photographed by Toby Sleigh-Johnson.

to be James Platt & Co. of St Martin's, London. The whistle has a
home-made appearance and lacks a maker's stamp, suggesting that
it was crafted from the erstwhile wearer of the buttoned garment.[22]
While he was unlikely to have made the whistle, button-maker
James Platt was one for frugality himself, and would have approved.
In his *Order of Business* (1869) he urged his staff not to waste string
or paper: 'we keep various sizes in paper; use no larger than is necessary for covering articles to be tied up; […] line country parcels of
any size with old paper.'[23] In *Economy* (1882) Platt wrote: 'Every
thrifty person is a public benefactor; every thriftless person a public
enemy. "Waste not, want not" is a law of nature. Let no man say
that he cannot economize, *every* man can if you once get the habit
into him.'[24]

This sentiment was easy to believe at the close of the Victorian
period; items were commonly repurposed. Old garments were
modified for new bodies; old boots translated to suit new feet;
broken crockery made to last a few more meals; glassware was
raised anew, replete with metal rivets. Objects were crafted from
old parts of abandoned items, whistles cobbled from buttons. Such
was the desire to eke out the utility of things and materials that
people lined up all kinds of items for reuse: old artificial teeth
might be offloaded to dentists, for use in new mouths.[25] 'What can
be done with old sardine boxes?' asked *Leisure Hour* in 1888.

Looking to France for answers, the journal found Parisian *chiffo-niers* filling tins with clay to make bricks, a man called Drog melting solder from tins and toy-makers crafting items from reused tin to rival German trinkets. Clippings became sulphate of iron, with disinfectant qualities. In London, tin reuse was on a smaller scale: some solder was reclaimed; some waste tin was dispatched to Mid-dlesbrough as old iron.[26] Particles of tin from the scum left by other processes could be separated for use in button-making.[27] But, as we have seen, British tin re-processing was mostly in German hands by the start of the First World War; there was slippage in the adop-tion of these French recycling processes.

Mighty oaks to other places go

This extraordinary carved oak fireplace (fig. 30) was salvaged by a Bristol art collective called the Savages in the 1930s to adorn their building in the garden of the Red Lodge in Bristol where they hold gatherings.[28] It came from an extraordinary building: Canynge House, a mansion with a medieval heritage, which sat imposingly along Bristol's Redcliff Street until 1937. For most of the Victorian period Canynge House was occupied by Jefferies & Sons, wholesale stationers and bookbinders. The business also gathered and resold wastepaper, claiming to 'give a very high price for *any sort of old papers* for remanufacture, &c. into paper'.[29] In 1856 the stationers had secured lots from an auction of salvaged materials from a London stationer, Dobbs, Kidd & Co., 'now laying in the ruins' after a fire. Their haul included reams of various types of paper, some of it (according to catalogue marginalia) 'hardly touched', but 'some wet'.[30] In 1866 the company advertised a 'large lot of soiled account books, damaged writing-paper […] fit for invoices, scribbling &c'. That year the firm employed about 150 people and had just erected a new warehouse, and by the 1870s Charles Jefferies & Sons was advertising large volumes of second-hand books on the American market.[31]

In 1881 the Jefferies premises were themselves devastated by fire, destroying machinery and stock, and also a library of books (clearly an occupational hazard for stationers).[32] The ancient carved oak fire surround depicting the Judgement of Solomon survived and was

30. Carved oak fireplace. Photograph reproduced
courtesy of Bristol Savages.

depicted in an A. E. Parkman watercolour illustrating the damage:
a moat of water around a massive pile of books.[33] At the end of that
year Jefferies sold off salvaged paintings, plus damaged stationery
and other stock, claiming 'business as usual'.[34] The company also
dealt in old manuscripts: the recirculation of heritage. Like Krook
in Charles Dickens's *Bleak House* (1852), with 'so many old parch-
mentses and papers in my stock', Jefferies gathered up old legal and
estate documents. Some time after 1855 they had acquired documents relating to the family of the former MP Montagu Gore. The
documents had been ejected by his housekeeper in a fit of pique
roused by the sale of the family estate, Barrow Court. Decades later
these were purchased by the Bristol Corporation and gifted to the
Gibbs family, owners of Barrow Court. The collection is now in the
hands of the Somerset Heritage Centre, having been given up by
the Gibbs family during the Second World War.[35] These papers

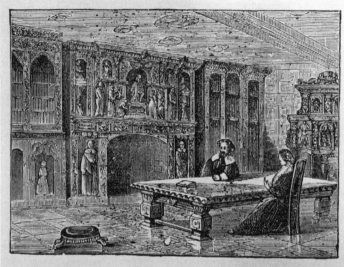

OLD PARLOUR,

31. Fireplace in the premises of Jefferies & Sons,
Bristol Archives, 37164/B/2/12.

have value because they can be used to settle estates, define territorial boundaries and preserve family stories and memories; they support and bolster heritage.

It wasn't just paper ephemera: the sale of wooden items also permitted the recycling of heritage. Some sources of reused oak had travelled long distances. Castles Ship Timber Company, established in 1838, was a firm of shipbreakers. The Castles yard became a prominent feature of Baltic Wharf in Millbank, London, where massive oak figureheads flanked both the south and the main entrances until the site was bombed in 1941 (**fig. 32**). Over the doorway to their counting house hung 'an immense wooden arm holding out a money bag', part of the figurehead from the *Commerce*.[36] Castles sold salvaged nautical booty across the country, for a variety of uses. In 1907 teak from HMS *Duke of Wellington* (broken in 1905) entered a sort of retirement as a garden bench.[37] Oaken parts could end up in higher airs: timbers from the *Duke of*

32. Castles' shipbreakers. Photograph courtesy
of Castles' Shipbreaking Archives.

Wellington and the *Caledonia* are said to inhabit the cloister of St Conan's Kirk, on Loch Awe in Scotland, renovated in the early twentieth century. The church is a curious hotchpotch, and also includes a reused window from St Mary's Church, South Leith.[38]

The Crimean War (1853–56) had revealed the vulnerability of wooden ships to shells, and iron armour was devised to clad the naval fleet. By the end of the nineteenth century more of Castles' business centred on the breaking up of armoured vessels. The market for salvage of wooden ships slackened. Some material was dumped in the Thames, along with some of the last ironclad battle-ships, such as HMS *Ajax*, launched in 1880.[39] Ironclads required much timber for their construction, and after 1860, when supplies were becoming difficult to secure, more ships were built of iron. Iron battleships lacked romance (and figureheads), but they presented massive opportunities for salvage. Parts of the modern iron hulks could be unbuttoned, de-riveted and melted down in open furnaces, then re-rolled. New materials required new tools and new skills; oxyacetylene torches cut 'through the steel plating as though it were butter'.[40] Thomas W. Ward Ltd of Sheffield established breaker's yards along the coastline, including outposts at Preston, Morecambe, Inverkeithing and Milford Haven. Founded in 1878, the company was well placed to buy ships after the First World War. Furniture and machinery would be auctioned off before the real work began. In 1921 Ward paid £36,650 (the equivalent of just over £1.5 million today) for a single ship: HMS *Dreadnought*.[41] A previous *Dreadnought* (herself formerly HMS *Caledonia*) had been used as a seamen's hospital in Greenwich, until being broken up in 1875.[42]

Deciding to establish new premises in the early 1920s, the owners of Liberty's department store in London opted for a Tudor pastiche. Teak for the external timbers and oak for the interior were inherited from two old men-of-war: HMS *Hindustan*, launched in 1841 (and sold to the scrap metal merchants J. B. Garnham & Sons in 1921), and HMS *Impregnable*, which had been the same length as Liberty's shop front on Argyll Place. Shop floors were made from ship decks. The *Sphere* congratulated Liberty's for enhancing London's street scene with an English shop which echoed 'bygone days when the ancient guilds of the craftsmen and the merchant adventurers displayed in the beautiful gabled buildings of Old London [...] the treasures for which they had sailed so far'.[43] In this case, at least part of the treasure was the ships themselves.

Fuelled by eclectic curiosity as well as bargain-hunting, a traffic in Gothic ecclesiastical woodwork and paraphernalia had developed steadily through the nineteenth century. This 'golden age' of trans-shipment occurred not only because of secularisation following the French Revolution but also through an increasing determination to re-Catholicise the Anglican church. Such materials were used by the Cockayne Cust family in the 1820s to furnish the church in Cockayne Hatley, Bedfordshire. The small church welcomed a hotchpotch of recycled medieval, ecclesiastical and

33. Carved oak choir stalls in Buxheim, near Ulm, 1883. CC

monastic items, mostly from Belgium: wooden choir stalls from an Augustan priory, a pulpit from Antwerp, altar rails and folding doors. A new west window was furnished with recycled medieval glass, this time English in origin. The church absorbed the accumulated heritage from distant places, and these items of recycled heritage had a function: to lend a more Catholic history to a church embedded in Bedfordshire, a county known for nonconformism.[44] However, to see the reuse only in these cynical terms would be to miss a more positive motivation: items rejected from elsewhere permitted the commemoration of a religious past and the creation of a new religious future, reborn in a new location.

A contemporary demand for Gothic carvings was often knowingly ill served; as in Cockayne Hatley, much of the imported material was post-medieval in origin, often Baroque.[45] Trading from 68 New Bond Street (with warehouses in Fürth in Germany and elsewhere), Julius Ichenhauser capitalised on this trend. He specialised in the sale of carved panels and ceilings for use in 'sacred and domestic edifices'. In the 1880s he imported various Baroque religious items, including 'carved oak choir stalls' from a church in Buxheim, near Ulm, that had been secularised in 1809 (**fig. 33**). He sold them to the Governor of the Bank of England, who donated them to a hospital. When the hospital was demolished in 1963, some parts were returned to their original home.[46] In 1888 Ichenhauser sold a sixteenth-century altarpiece from Brussels to the rector of Radwinter church in Essex. To inflate the sale, his catalogue claimed that Gothic imports would shortly halt, 'as the days of loot and wanton destruction are now happily over, and those religious orders which possess rarities of past ages not being likely to dispose of the same'.[47]

Through a glass darkly

The Victorian love of all things Gothic saw stained glass being imported from the Continent to adorn the windows in family chapels on country estates. Julius Ichenhauser offered up various lots of antique stained glass, including, in 1888, some that he claimed were 'given by Lord Nelson to Sir William Hamilton at Palermo in 1799'.

The trade had a long history, and Ichenhauser arrived late to it. In the early 1800s, partners John Christopher Hampp and Seth William Stevenson had acquired massive quantities of stained glass from Normandy, the Netherlands, Cologne, Aachen and Nuremberg, snapped up from secularised institutions during the periods of revolution and the Napoleonic occupation of the Rhineland. Some treasures wound up in Costessey Hall Chapel, a Perpendicular Gothic Revival construction in Norfolk that was consecrated in 1809. Old glass was reused in other private residences during the nineteenth century, but Costessey's collection was on an unprecedented scale.[48] The chapel 'constituted an extraordinary coalescence of architecture and glazing, in which medieval stained glass reemployed as medieval artefact both embodied and revitalized the spirits of its creators'. The nineteenth century was a dynamic time for British Catholics, with the abolition of laws designed to remove their civil rights. Their numbers were enhanced, first by French Catholics who had fled the Revolution and then by Irish Catholics

34. Stained glass from Costessey Hall, reused in a
Norwich house. Photograph by Frank Vickers.

driven from Ireland by famine. Various High Church Anglicans
also converted to Catholicism after the mid-century. As we will see,
much British Catholic paraphernalia had been destroyed during
the sixteenth-century Reformation, and so replacements needed to
be bought up from the Continent, if not made anew. Material
reuse lent legitimacy and heritage to an old religion trying to re-
establish itself.[49]

The tenth Baron Stafford inherited the Costessey estate in 1884
– but he was a certified lunatic, so the Lunacy Commission held on
to it until 1913. The hall passed to his nephew, who wasted no time
putting the contents, fixtures and fittings up for auction: it was
reported that 'big prices have been realised'.[50] A gold chalice found
its way to the slipper chapel in Houghton-in-the-Dale, Norfolk.[51]
Oak linen-fold panelling, once exhibited in the Great Exhibition
of 1851, was installed in the dining room at nearby Hethersett Hall,
along with a carved fireplace.[52] The buildings of the estate were
dismantled over decades: the hall was reduced to a carcass, stripped
of friezes, mouldings, chimney pieces, oak floors, ceilings, door

panels and staircases. Bricks and a hundred pine doors were auctioned off in 1920.[53] The Continental medieval stained glass from the chapel went to a dealer.[54] A parcel of this was bought and installed in a private detached house in Norwich, where it was combined with modern stained glass (fig. 34). This was a second reuse for the glass. The crucified Christ is flanked on the left side by a doting Mary Magdalene, but the glazier added jaunty Art Deco poppies to fill an equivalent space on the right side. Jesus recycled.

Glass-recycling on a more massive and less colourful scale had occurred in Sydenham, south-east England, in 1852, when workmen piled up materials taken from the Crystal Palace, the venue for the Great Exhibition, located in London's Hyde Park. *The Times* listed 'rows of prostrate columns', heaps of doors, mountains of window frames and 'an immense accumulation of glass, stowed away in a shed erected for the purpose'. Remarkably, only a small fraction of glass was broken during the transfer. A new Crystal Palace was given life from material donations incorporated alongside new additions.[55] Charles Fox, the father of Sir Francis Fox (the civil engineer who reclaimed fabrics from architectural drawings in the First World War), owned the firm tasked with re-erecting the edifice.[56] This new Crystal Palace burned down in 1936, and by 1937 the scrap merchants W. Ward & Co. had removed most of the ironwork. Tourists gawped at the ruins and left with souvenir shards of glass.[57]

Most glass reuse took on an entirely different scale and character. Bottles were commonly reused after a first use or returned to businesses for refilling. Archaeological digs at Victorian waste pits reveal a wealth of waste – mostly glass and stoneware vessels. Although such pits seem compelling evidence of squandering, glass objects skew the waste picture: the Victorian era was not the period that witnessed the start of the 'throwaway society'.[58] Glass was very cheap to make from easily available, naturally occurring substances – sand, limestone and soda ash. The weight of waste glass made it expensive to transport, rendering recycling unviable in places far from a glassmaker. People made economic calculations when assessing waste. Most of them tried to balance utility against the time and energy needed to take it to be recycled or reused. The Victorians did not just hurl everything willy-nilly into waste pits. Other materials used for packaging, such as paper and fabrics, *were* recycled.

In many places, especially cities, rag-and-bone men and 'marine store' dealers bought up used glass bottles and sold them on in bulk. Medicine bottles went 'to the dust-yards with great regularity, and with the same regularity they [found] their way back to the druggists' shops, going about the same dull round year after year'. Other bottles were melted down. A review noted an irony in 1868: '[t]he most fragile and destructible of materials when manufactured, [glass] is, perhaps, one of the most indestructible of all known substances', being repeatedly melted and reformed.[59] Such manufacture is naturally less visible than the glass that ended up in pits. A glass bottle business based in Castleford, Yorkshire, put out appeals in newspapers across England in the 1870s, seeking broken glass of any sort, casks of which might be conveyed to them by rail.[60] Thomas Greenwood, a publisher who focused on trades, described bottles that emerged from the waste heaps as a 'perfect podrida', a glittering stew of glassware. Intact bottles were bathed in acid and reused. Some glass went back into the furnace to make new bottles; some was ground up for use on sandpapers.[61] Emily Hobhouse, the British welfare campaigner, traced the afterlife of a glass broken by a 'careless servant' as it journeyed to Sweden and returned as 'neat squares of emery paper wherewith she may polish the fender'.[62]

In 1898 the *London Journal* detailed developments thought to herald a revolution in fabric manufacture, a 'most astounding and unexpected' transformation of broken glass into glass wool, which would be turned into 'soft, delicate, and durable cloth, that can be worn without the slightest discomfort next the tenderest skin'. Incombustible, strong and thick, the new fabric was touted to be the next thing in fashion for ball dresses, providing a more easily washable alternative to silk. More humdrum uses saw it fashioned into fireproof aprons and coveralls. The utility of this new manufacture would be judged, the article suggests, in relation to 'its durability and price'.[63] Clothes made from this material, somewhere between fibreglass and glass fibre, never did make it into wardrobes.

Profitability considerations

At the turn of the century, glass was given other uses. The Pilkington Brothers in St Helens processed sand, glass and iron particles from the waste thrown out in glassmaking and used them in brick manufacture. The iron content of the waste had an advantageous effect on the bricks.[64] The motivation behind these experiments was financial but also derived from the ambition to find pragmatic ways of reducing waste heaps. At the time, many companies were thinking strategically in this way. Edward Ballard, a local government medical officer, had compiled a comprehensive report about *Effluvium Nuisances* in 1882, in which he concluded that 'a considerable proportion of the nuisance trades consist in the working up and utilisation of refuse matters [...] various offensive matters which it is important to get rid of from populous places, and which are too valuable to destroy'.[65] Waste was building up and causing a public health crisis – but also encouraging new manufacturing processes. In 1894 accumulations of waste and pollution of the air and the watercourses were identified as having helped trigger 'the successful introduction of economical methods of working undreamt of a few years ago'.

Cue Pilkington Brothers and their reuse of glass waste. Greenwood also nodded to environmental issues driving development, noting aesthetic aspects and visual spoliation, not just air or water pollution. He thought that legislation helped to spur creative reuse, but saw profits as the carrot to the stick of the law.[66] By mid-century companies were encouraged to stoke engines with coke, made from 'small waste coal', which made less smoke than coal. Coke use allowed pit owners to make 'something out of a (commercial) nothing' and, in turn, removed heaps of refuse from collieries. In an extra recycling process, the dusty bits left behind by the coke makers were then bought up to make cheap artificial fuels.[67] A win for companies, and a win for public health campaigners.

The reuse of the by-products of gasification was a godsend after the 'many years' during which the wastes were 'looked upon as necessary evils, to be got rid of as quickly as possible'. Thrown into streams, gas-water and tar had 'killed the fish and poisoned the atmosphere for miles around'.[68] During the second part of the nineteenth century various industries set up next to gas works in order to make use of gasification wastes, while minimising transportation costs and increasing profits for waste producers. Although once problematic and unsaleable, profits from sales of by-products eventually defrayed the costs of gasification.[69] The functionally named Gas Products Utilising Company set up next to the Great Central Gas Company on Bow Common, producing coal tar and sal ammoniac. The company also sold alum to calico printers and dyers. The alum was extracted from coal-mine waste, which had previously covered 'acres of ground like the spelter and cinder heaps, but chemistry has found it out'.[70] Accumulations of pollutants motivated complaints and investigations, especially if they were noisome or intoxicating. Interference by inspectors of nuisance encouraged waste-producing businesses to find markets for by-products. Sanctions sometimes provoked positive and creative reactions, but motivations to utilise waste were manifold. While promoters such as MP Lyon Playfair had limited impact, profitability *was* a key consideration: self-interest, not ecological concern, was the key driving force.[71]

To increase further the profits of reuse, some businesses used the classic trick of combining trades – one consuming the waste of another. Close proximity reduced transport costs. Charles Massey's Waterloo Works in Newcastle under Lyme in Staffordshire combined paper merchanting with size and tallow sales, plus the manufacture of bone and artificial manures: this combination made the best use of bone material but also connected Massey to papermakers, to whom he could sell size. Massey claimed to make 'high-class' specially blended manures, and could supply ground steamed bones, ground raw bones, bone meal, superphosphate of lime, Peruvian guano and 'bones specially sorted and prepared for vineries'.[72]

The south Yorkshire firm of Samuel Meggitt & Sons crushed bones, made and sold chemical manure and also manufactured shirt buttons from polished bone, horn and metal. By the mid-nineteenth century business had diversified, and they ground bone into fertiliser, which they marketed as a cheaper and better substitute for guano on grass, wheat and turnips.[73] The company's understated pitch ('nothing could possibly be more satisfactory') was undermined after an increase in the price of bone-triggered experimentation to find chemical alternatives for fertiliser in 1871. The company was cagey about it, but the recipe probably included sulphuric acid, guano, bones, charcoal dust and ammonia, possibly in the form of urine, or possibly from coal gasification.[74] Whether or not they were having a positive impact on the environment, both manure and buttons were becoming increasingly difficult businesses to make a profit in, not helped by events. Meggitt's sons managed different parts of the company in different parts of the country, unified in a joint stock company in 1893. A series of misfortunes afflicted the Meggitts thereafter. In 1896 a glue-boiler died after falling into a pan of boiling water.[75] In 1902 Director and manager Joseph Bloom Meggitt went bankrupt. Liquidated, the company was split into two: one division concentrating on the button trade, the other on glues and manures.[76] Joseph's younger brother Arthur Cockayne Meggitt managed one of the company's bone mills in Mexborough, but committed suicide in 1903.[77] In July 1902, between Joseph's bankruptcy, Arthur's death and the formation of the new business, the Sheffield bone mill caught fire under 'peculiar circumstances'. A 'piece of hard substance' got in with the bones being milled, causing sparks which set the bone dust alight,[78] although the grinder had an attachment to remove metals, which were often mixed up with the bone. A horseshoe in the mix could 'emerge changed to the form of a bullet, at white heat'. Six years later the same thing happened again. It was a bad time for the Meggitt brothers.[79]

Decisions to combine industries may have made good business sense, but the noise, smell and fire hazards would have been a nightmare for those living near by. Leonard Wild was one such bad neighbour in Bolton. He supplemented the drying of waste cotton with charcoal grinding, to make blacking. This was a combustible combination which led to numerous fires, including two in one week in June 1893.[80] Combining cotton drying with waste dealing and charcoal grinding made financial sense to Wild (he would grind the charcoal spent in drying the waste), but it had unforeseen consequences. Wild's premises came to the attention of the nuisance inspectors for being so insanitary as to be 'injurious to health'.[81] In 1892 he had been at odds with his landlords, the trustees of the Independent Methodist Chapel situated next to his workshop. Wild used a machine called a 'Disintegrator' in his trades, and annoyed the neighbours with dust, 'smells' and other nuisances.[82] Wild's nuisances went further back, however, and conflicts from the 1880s give some indication of the breadth of his activities. In 1881 he clashed violently with another of the town's waste dealers, whom he accused of selling him bad quality 'skip paper'. A decade earlier he purchased cotton pickings from a dealer in Liverpool that were 'much inferior' to the sample he had examined, and were 'not dry enough to work'.[83]

Many complex businesses created massive piles of waste which blocked views or hindered passage – making their neighbours their most urgent critics. Profit was the key motivation for commercial material reuse, and although conglomerations of businesses that processed waste created nuisances, there were, nonetheless, environmental benefits for the wider community through a reduction of waste. On the whole, the neighbours had to lump it.

Fryer's clinker

We might think that the separation of reusables and non-reusables is a modern development, but by the close of the Victorian era local authorities had gradually assumed control of waste collection along such lines. Non-reusables were often incinerated onsite in a municipal 'destructor'. The destructors burned most rubbish in large furnaces, sometimes leaving hard objects made of stoneware, clay or glass only partially incinerated.[84] Alfred Fryer, aptly named, had conceived destructor technology in 1874. A sugar refiner by trade, Fryer was familiar with incinerators used to calcinate the bone used in sugar filtration, and before he moved on to creating technology to destruct all manner of objects, he had once developed a furnace to incinerate spent sugar cane.[85] We get a sense of Fryer's mindset from an unusual source, a book he wrote for his young daughter in

1880. *Vic; The Autobiography of a Pomeranian Dog* (1880), written from the perspective of a dog that busies itself hiding bones, presents a homilising canine: 'Let me entreat all young dogs', he growls, 'never to bury bones which are useless to themselves, for often such would provide a meal for some poor hungry brother.'[86] Fryer was similarly concerned with the reuse, or at least proper disposal, of waste objects. Initially destructors were thought to be clean and cheap, and the process more sanitary than sorting through rubbish and dust heaps.[87]

A fear of the transmission of communicable diseases saw the nation embrace incineration of infected matter: clothes, mattresses, wallpaper. Ernest Romney Matthews, a civil engineer with a special interest in sanitation, loved destructors: 'It is never advisable for rags to be picked out and sold, for these are likely to convey infection', he had warned.[88] All waste was shovelled in: 'ashes, dust, vegetable refuse, boots and hats in the last stages of dissolution'.[89] Animals cremated in the Blackpool destructor included, in 1907, three porpoises, four lions and a bear. The big animals probably came from Blackpool Tower Menagerie; distemper was circulating.[90]

Another key to the popularity of destructors was their supposed efficiency and cheapness as a means of waste disposal. Some produced steam used to power heating systems for swimming pools, and to drive sewage pumping engines. Preston Corporation's tram system ran on steam fed into its generating station; the steam came from furnaces fed with town refuse. Street lights in Woolwich were powered a similar way.[91] This was the clincher for the temporary success of the destructors: they appeared to return municipal value.[92] But the reputation of destructors as engines of efficient power may not have been earned. Emily Hobhouse noted in 1900 that although 'destructors intended merely to destroy are growing common, those intended also to generate steam economically on a large scale are still rare'. One significant downside to destruction was the waste produced, known as clinker. These rusty grey lumps of amalgamated chunks of burned and semi-burned matter often had bits of partially destroyed crockery and bone embedded in them.[93] Clinker accumulated in unsightly heaps. Fryer had worked

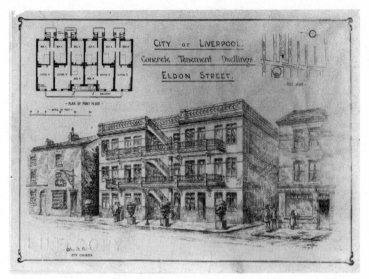

35. 'Concrete Tenement Dwellings', built from clinker, Eldon Street,
Liverpool, 1903. Liverpool Record Office, 352 ENG/2/438.

on reuses for clinker in the 1890s, but a market was slow to develop
before the close of the nineteenth century. Most was dumped,
sometimes at sea. In Poplar, east London, it was converted into
mortar and artificial paving stones, 'actually using the steam generated by the destructor for its mortar and clinker works'.[94] Elsewhere
it was ground up for use in bacteria beds at sewage works, made
into paving flags, walls or foundations for roads and paths.[95] In
Liverpool, a slab system of 'Concrete Tenement Dwellings' was
constructed on Eldon Street using clinker as the aggregate in a
concrete mix (**fig. 35**). This curious building was the brainchild of
John Alexander Brodie, the city engineer.[96]

Destructors didn't turn out to be the answer to the problem of
waste. During the First World War, engineers and sanitary inspectors urged local authorities to abandon them, suggesting that waste
cinders and ashes – at that time fed routinely into destructors –
might be reformed into fuel briquettes instead.[97] Destructors rid
towns of unsavoury and ugly waste, but to Frederick Talbot they

were a 'retrograde step in the science of economics', encouraging householders to be wasteful: 'fire [was] far too handy for the removal from sight, if not from memory, of the multitude of odds and ends incidental to our complex social and industrial existence.' Less organic material was processed into manure during the years that destructors proliferated. Talbot feared that local authorities would continue to fetishise destructors, 'strenuously declining to honour the axiom that it is often cheaper to cut the loss'.[98] J. C. Dawes, chief technical adviser to the National Salvage Council (and later, as we have seen, a prominent figure in salvage during the Second World War), hoped aloud that he would see the day 'when the destructor would itself be destroyed'.[99] Eventually, high running costs and nuisance complaints saw the closure of many of the destructors, and more limited use of the survivors. In 1900 there were approximately 300 municipal incinerators; by 2000 there were fewer than twenty.[100]

🗑 🗑 🗑

Some of the people involved in the peddling of waste and reuse were not the most savoury, although few were as wild as Leonard Wild. One of the nastiest was Julius Ichenhauser, the dealer in Continental carved oaks. From a family of bankers, he had become very wealthy through an inheritance and used that windfall to finance art dealerships in London and New York. Ichenhauser was slack, corrupt and untrustworthy. In 1884 stock abandoned in his premises on Aldersgate Street in London after Ichenhauser's lease had expired was auctioned off by the landlord. Over two hundred works of art were listed, including works by George Morland, Constable and John Crome. Works acquired on his recommendation by the Baltimore art collector Henry Walters have since been shown to have been misattributed.[101] Ichenhauser made strenuous efforts to keep his art business separate from various risky speculations he was involved in under different names, and through shell companies. Between 1883 and 1886 he traded as Schwelheimer & Co., a company managed by his brother Maurice. Julius added 'David' to his name in 1892 as part of a ruse to reinvent himself.[102] That same year Ichenhauser took the London and North Western Railway Company to court to try to recover damages of £120 he claimed to have suffered when two Oriental vases (each one half of a pair) had been broken in transit while travelling, as a consequence of employee negligence. The case caused much merriment at Queen's Bench. When asked if the vases were ugly, the counsel for the plaintiff replied, 'Fairly ugly but unique'. The Lord Chief Justice quipped that if they were old, very ugly and unique, 'they might go for any sum'. Reminded that the remaining vases were now odd halves of a pair, His Lordship responded, 'if the ones which remain are the oldest, the ugliest, and the most unique in the world, the value ought to be all the greater'. A china dealer estimated the actual value of the vases at £25 together and claimed that 'both had been broken before and joined together again'. One vase was passed to the jury for inspection. After some heavy steering, the jury sided with Ichenhauser and awarded damages of £50.[103] Ichenhauser was a freemason.[104] Various foul trading transactions were exposed in 1895, and in the same year the New York bankers the Lehman Brothers took him to court.[105]

The end of the Victorian era saw a melding of old and new: old buttons fashioned with new tin into whistles; new buildings bolstered by old oaks and medieval glass. The motivations behind reuse were numerous, but acts of commemoration and remembering were behind some of it: the recycling of heritage The reused parts reinforced more than just structures: a new building absorbed the heritage accrued to the old additions: returning a Catholic past to a new chapel, bringing the exploration of the globe to a department store. Some smaller-scale recycling also built into it the commemoration of old structures: remembering through reforming. One rosewood snuffbox claims to include inlaid pieces of oak from various original sources: Glasgow Cathedral, the House of Commons after the fire of 1834 and various ships, among them the one that carried William of Orange to England in 1688, Nelson's *Victory* and HMS *Royal George*, which sank in 1782.[106]

In the early 1900s the roof and ceiling of Glasgow Cathedral were replaced. Various items of memorabilia were crafted from the early medieval oaken parts, including snuffboxes and this cribbage

36. Cribbage box made from reclaimed oak. Author's collection.

board (**fig. 36**). Proceeds from their sale went towards renovation costs, a remaking enterprise that combined elements of bazaars and jumble sales.[107]

Various jumble sales were held in the vicinity of Exeter in the 1890s. In January 1894 a sale held in central Exeter raised funds for the Wesleyan Juvenile Foreign Missionary Society and advertised itself as having 'every kind of commodity from a bran [*sic*] new Suit of Clothes to a superannuated frying-pan'. One held at Countess Wear, a district south-east of the city, in 1895 raised money for church renovations, and was combined with a raffle for a pig. In Kennerleigh, thirteen miles from the city, a sale was held the same year, also to restore the church. At that sale there was 'a delightful assortment of things old and new, Elizabethan ruffles, Wellington boots, Grandfathers beavers and the latest most killing things in ladies hats and jackets'. A jumble sale in Ideford, eleven miles south of Exeter, in 1899 'gave a good opportunity to the poor to pick up real useful articles at ridiculously small prices'.[108] I like to think that an item of crockery once mended by a Nike, or one of Ichenhauser's ugly broken vases, featured on a stall somewhere.

6

Impostor instruments (1850s–1880s)

W. John Stonhill liked to sing at work. One day in October 1884 the singing stopped, as Stonhill found himself mesmerised by some peculiar specimens of paper. People often sent Stonhill paper samples; he was the editor of *The British & Colonial Printer & Stationer*, a weekly publication he had set up on London's Shoe Lane in 1878. But as he rifled through the sheets, he knew these samples were special. Stonhill was an expert on paper, and had just finished a long essay on the history of paper pulp for a forestry exhibition. He was familiar with the miracles that could be created using paper, from houses to carriage wheels, but these samples showed that paper 'can also be made to imitate or counterfeit almost any other substance'. Some of these paper novelties felt like fabrics from a draper's shop: one was like chintz; another felt like brocaded silk. Rummaging deeper through the samples, Stonhill's fingers danced over papers in imitation of wood, of oak, mahogany and ebony. Some even looked like parquetry. Other sheets were like a mosaic of metal: there were silvers and golds, embossed patterns that looked just like basketwork, needlework, tapestry, linoleum and tiles. Even more wondrous were marvellous imitations of crocodile skin and snakeskin, even fish skins. He played with these between his fingers and found them untearable. These were the 'most remarkable specimens of paper that we have ever encountered', Stonhill wrote in a gushing review, and they indicated the 'infinite adaptability' of the material. These novelties 'are sure, ere long, to enjoy a large sale'.[1]

The samples had been sent to him by Fritz Viëtor & Co., a firm based in the Barbican area of London. Viëtor, a German, was one of many businessmen in the late nineteenth century who spent much energy turning wastes into valuable artificial products – and papermaking was just one string to Viëtor's bow.

During the mid- to late nineteenth century, developments in recycling came thick and fast. Chemistry turned old materials into new ones, which increased the availability of stuff and permitted a democratisation of ownership; luxury items were copied for wider sales. Victorians wore colours previously too expensive for them, and filled their homes with items in imitation of Asian lacquerware made from paper and coal tar. Many of the creations were first exhibited in the International Exhibition of 1862, held in London. Giuseppe Verdi wrote a cantata for the exhibition, a *Hymn of Nations*.[2] William England's team of photographers, the London Stereoscopic Company, captured the exhibits in the International Exhibition in 350 stereo views.

The business of photography was rapidly expanding, and creating new opportunities for reuse and recycling of materials. The year after the International Exhibition, England gave a presentation suggesting how photographers might capture gold and silver from waste solutions.[3] Photographers had been making efforts to recapture silver in the waste for a few years by then; it was precipitated from the waste washings using common salt.[4] Nevertheless, by the 1870s, British photographers were not recovering as much silver waste as parsimonious German photographers. Some firms made no effort; hundreds of pounds' worth ran into sewers.[5] The *British Journal of Photography* tried to motivate readers with stories of regainable riches. Shown a 'handsome tea service' by the proprietor of one large firm, the journalist was told that this had been crafted from 'silver resulting from one year's saving'.[6] Outlays of a few shillings, combined with simple chemical and heat processes, could yield handsome returns. One firm of photographers based in Wakefield adopted all recovery processes and was able, within three years, to gather enough silver for two bars alloyed with gold, worth in excess of £44. This company was probably the Cathedral Studio, run by brothers George and John Hall of Westgate.[7] Their business was not limited to photography: they also made and repaired watches and jewellery, allowing them to put waste silver to profitable purposes in-house.[8]

Some photographic frames were made using early plastics, such as *bois durci*, invented in the second half of the nineteenth century. These new materials incorporated waste products such as rags, coal residues and sawdust. Parkesine, one of these new plastics, won a medal at the International Exhibition. They were made to imitate and replace organic materials, animal skins, meerschaum, tortoiseshell, stone, ivory and even silver and gold.[9] Lurking amid the humdrum dubiosities turned out by the magicians of reuse were things that seemed truly precious – the finest piano parts, for example. But what looked and felt like ebony or ivory was sometimes just bloody dust and nitro-cellulose.

Sweetness and bright

In the latter half of the nineteenth century many products were counterfeited by reusing wastes: even sugar. Most Victorian sugar was refined from sugar cane; sugar beet production occupied an insignificant position in Britain (although the French had a more developed beet industry).[10] But some people went further and tried to manufacture sugar from sources that were neither beet nor cane. Reports described German factories making golden syrup from sulphuric acid and starch. Rumours swirled about sugar made from sawdust.[11] While this never materialised, attempts to make artificial sweeteners were more successful, and the basis for these was quite surprising. In 1879 a Russian-born chemist working on coal-tar derivatives in Maryland discovered a delicious product made from the nasty black substance. His invention, saccharin, went into production in Germany in 1886, and by 1897 the Saccharin Corporation Limited had been established on Queen Victoria Street in London.[12] In *Saccharin: A Vindication* (1888) various British scientists clamoured to defend the new substance (the French had raised health concerns) and reminded readers that saccharin 'is now offered for sale at a considerably lower rate than cane-sugar' and had 'found ready acceptance' among distillers and jam makers, and 'in many other industries of importance'.[13]

The colour transformation involved in its manufacture perplexed and offended some. In 1888 Parliament debated the use of saccharin in brewing beer. George Goschen, the Chancellor of the Exchequer, opposed the plans. 'What a hornet's nest Mr Goschen stirred up', reported the *Nottingham Evening Post*, 'when he spoke of the beautiful white and lovely saccharine as coal tar.'[14] The Chancellor made explicit reference to the deceptively clear appearance of the artificial sweetener, stating that beer, 'like Caesar's wife, should be above suspicion'.[15] Saccharin was banned by a large majority, deemed harmful to revenues and real sugar refiners, if not harmful to health.[16]

Even the sugar made from sugar cane – true sugar – was deceptive. In a history of sugar from 1866 William Reed noted that the syrups that dripped from sugar cane were 'rich in sugar' and 'must be economised'. Reed helpfully glosses sugar jargon: '[cane] drippings, boiled, drained and cleared become pieces; the droppings of pieces, similarly treated, are bastards; and the drippings of bastards are treacle.' The sugar from which this treacle dripped, as well as the sugar syrup drippings, which were naturally brown in colour, would be decoloured to suit the Victorian taste for pale sugar. The substance used for decolorisation was 'bone-black', animal charcoal, which was made by heating bones in a vacuum and sifting away the 'pulverulent parts'. The grainy substance that remained was put into filter beds. As dark brown sugar passed through this bone-black, it became 'perfectly bright and colourless'. Reed thought this filtering process 'wonderful – almost magical'.[17] One drawback of filtering out sugar colours with bone-black was the expense: repeated cycles of filtration saturated the bone-black with impurities, reducing its efficacy. In the 1860s a new mode of 'revivifying' the bone-black using alkaline rendered the charcoal indefinitely reusable.[18]

We can see then that colour mattered to the Victorians. Not only did they care about the colour of their sugar, but new hues were forced upon a variety of objects and products. From the mid-nineteenth century on, the ingenuity of chemists who transfigured waste has rightly been trumpeted as a glory of Victorian times. The creation of synthetic dyes and other substances has been seen as a display of mastery over the natural world, 'wonders akin to the stuff of dreams', and many of the developments involved the reuse or repurposing of substances.[19] Some transformations relied on older products of reuse. Oil distilled from the process of creating bone-black, commercially known as 'Dippel's Animal Oil', was used in dye-making. The dye that imparted the colour mauve was extracted from the alkaline parts of the oily matter distilled from coal. William Perkin discovered 'aniline purple' (mauveine) by happen-stance in 1856, when he was trying to synthesise quinine, and his invention was deemed so exciting that it was showcased at the International Exhibition in 1862.[20] Others followed Perkin's suit; through experimentation the dye-makers Simpson, Maule & Nich-olson added magenta to the colour palate. Perkin threw in 'dahlia', a hue midway between mauve and magenta.[21] Judson's Simple Dyes advertised mauves and magentas in the 1860s, with these shades at the height of fashion.[22] Purples had previously been expensive dyes, available only to the rich – these developments permitted a democratisation of the colour.

Azo dyes are solids, often salts, made from organic compounds, and they were invented by a German chemist, Peter Griess, who worked for Samuel Allsopp & Sons Brewery in Burton upon Trent. Griess was an oddity. After a wayward youth he had started work in a German coal tar factory that manufactured aniline, until it burned down a year after he started. He had come to England by invitation of the Royal College of Chemistry in London, and he arrived at its gates in 1858.[23] Griess was recruited by Allsopp's to help rebut adulteration accusations, but his work at the brewery was limited to staining and identifying microbiological samples, work that did not fully utilise his skills. Bored, he made his dyes from experiments with coal tar in his spare time, paying for research help with casks of company ale. Although not thrilled to live 'in a foreign country [...] always preoccupied with beer in a town which has little to offer', Griess stayed with the brewery for the remainder of his career, but his work helped the Germans expand their syn-thetic dye industry.[24] The first brown coal-tar colours were created in the 1860s in a German re-recycling experiment, using the waste products of another dye: fuchsine. Bismarck brown was the first. Other browns followed, plus reds and yellows.[25]

In 1862 the *London Review* wrote excitedly that 'one of the most beautiful illustrations of the utilisation of waste substances, and their conversion into bodies valuable to the arts and manufacture, is afforded by the manufacture of the red and yellow prussiates of potash'. Exhibited at the same International Exhibition as aniline purple, 'no one would imagine the utter refuse from which they are prepared'. Whence followed a list of the waste substances previously considered to be beyond reuse but which were now used in this manufacture: horns and hooves of cattle, leather clippings, blood and offal, and 'the cast-off woollen garments of the Irish peasantry'. These were mixed with 'crude pearl-ash', plus old wood and old iron (including 'hoops from beer barrels' and horseshoes), all melded into a yellow concoction.[26] One stockist of this magical invention was E. Crawshaw & Co., based at the city boundary, in Fann Street, in the City of London (within spitting distance of Fritz Viëtor's firm).[27] The company advertised their aniline 'Crystal Dyes' widely from the late 1870s. An advert from around this time reveals a wide range of colours, including mauve and 'black reviver', for dying fabrics and feathers at home. They boasted an 'infinite variety of purposes – fancy, scientific and domestic'.[28]

In 1873 a journalist, Peter Lund Simmonds, himself a champion of reuse, marvelled at the 'exquisite colours' (and also the sweet perfumes) that could be obtained from repulsive products such as pitch and tar, pronouncing this to be 'a great proof of scientific progress'.[29] Others also connected dyes and aromatics with the filth whence they came. 'Nearly every article of the toilet bottle or the sachet', remarked a reviewer for *Quarterly Review* in 1868, 'is made from waste, sometimes from the most inodorous matters.'[30] Lyon Playfair, a chemistry professor and President of the Chemistry Society, commented that 'many a fair forehead is damped with the *huile de millefleurs* without knowing that its essential ingredient is derived from the drainage of a cow-house'. Such economical reuse allowed these products to reach a 'vast number of persons who enjoy them without knowing that they are only imitations of the real thing'.[31] William Crookes, a chemist and physicist, thought much the same: 'coal-tar, rags and bone, rise from the sewer and dust-heap, and are transformed by chemistry into costly luxuries, or necessities of civilization, giving employment in their transformations to a large number of our working classes.'[32] Colours and flavours were made more vivid by Victorians whose chemical ingenuity often led them to reuse, recycle and repurpose materials and substances that people left behind.

Chewing paper

The Victorians, the people who made white saccharin from black coal tar, also turned white paper into black objects. The most significant paper-repurposing trade in the Victorian period was undertaken by papier-mâché makers. Household objects were made from paper, moulded into shape, dried and then 'japanned'. Japanning was the application of a heavy lacquer, which was most commonly black.

Papier mâché and pulpware, the older cousins of plastics, were used, along with sundry other applications, in the manufacture of imitated carved wood items. Their vats accepted all papers, from high-quality white sheets to coarse brown wrappings. The trade began in Birmingham, with Henry Clay's patented process in 1772.[33] Henceforth, most of this enterprise was spent making trays and similar objects of domestic utility. Makers processed wastepaper promptly and efficiently, and their promise of total destruction made them a favoured destination for more sensitive documents:

> Bankers have sometimes tons' weight of old account books by them, which have ceased to be of use, but which they are unwilling to place in the hands of the trunk-maker or the butterman, on account of the private transactions to which the writing on the pages of such books relate [...] the banker may perchance see the relievo decorations of his own dining-room made from his own old account books.[34]

Some paper used for papier mâché was 'of a grayish colour, thick and porous; similar, but superior to the foreign paper which we find around Dutch toys and cheap magic lanterns'. Easy access to such wrapping may explain how the most impressive of all the

Victorian papier-mâché makers – Charles Frederick Bielefeld – got in on the trade.

Bielefeld's father was a wholesale toy dealer, and Charles himself started in that business. Cut-up paper was pasted onto iron moulds in layers and dried in a 'heated closet' until it had 'arrived at the proper thickness, which may vary from one-eighth to one quarter of an inch'. Then it was removed from the mould, rasped and pumiced smooth, then given multiple coats of varnish, and sometimes inlays, before being painted. A final lick of polish was applied by women. Indeed, apart from the artists and a few labourers to do heavy work, every stage of papier-mâché manufacture was undertaken by women.[35]

Predictably, the actual owners of papier-mâché factories were men. From 1839 Bielefeld's eponymous company occupied large premises on Wellington Street in London, where his family lived above the business. The company created beautiful architectural details – cornices, plinths, capitals, columns and more elaborate decorations – later expanding into a second factory in Staines, west of London. Richard Horne (working for Charles Dickens's weekly magazine *Household Words*) toured both factories in 1851. Intrigued by architectural forms being constructed for a boat for a pasha in Egypt, he described them as bronzed imitations of wood-carvings. Bielefeld even marketed portable houses built from papier-mâché pasteboards, which might be taken over to Australia and 'screwed together in a few hours'.[36] However, he was unpopular on the streets around his London premises, which tainted the neighbourhood with smoke.[37] A massive fire tore through his factory in 1854, apparently watched by a crowd that included Dickens.[38] Never recovering from this catastrophe, Bielefeld went bankrupt in 1861.[39]

Rivals lurked in Birmingham, the original home of papier mâché. Theodore Hyla Jennens and John Bettridge had bought up Clay's pulp business in 1816. Eventually they were joined by their sons, trading as Jennens and Bettridge. More famous than Bielefeld, their gaudy work was lapped up by Victorians. Whereas Bielefeld focused on architectural details, Jennens and Bettridge used papier mâché to frame painted decorations or as a base for inlaid gems and mother-of-pearl.[40] Their catalogue reveals a

37. Papier-mâché pianoforte, by A. Dimoline, Bristol. Fotolibra.

cornucopia of baubles, typified by a 'Day Dreamer Chair', for sale in 1851. Made for the Great Exhibition, this allegory on a hardened mass of paper depicted figures presenting 'joyous dreams' and 'troublesome imaginings'. Perfect for a nap. That year the company employed a whopping 160 people.[41]

The Jennens and Bettridge partnership thrived for half a century but came to a sudden halt in the late 1850s, when the partners went bust and the company was dissolved. The stock was put up for sale in a ten-day auction in 1859.[42] Bettridge continued trading alone, but his troublesome imaginings were realised the following year. A fire started in the ground-floor drying room, where there were eight stoves, and spread rapidly upwards, engulfing stock waiting to be varnished and polished.[43] But he got going again. Luxurious papier-mâché novelties floated forth from his factory: twenty-six pilasters for a steamship in 1861; fancy work for the Nizam of Hyderabad in 1863.[44] Perhaps the most extraordinary products were the elaborate working pianofortes like one made for the Paris International Exhibition in 1867.[45] Although papier-mâché products were available into the twentieth century, plastics and other composites became dominant in the production of the sorts of wares and decorative features formerly made from it, making papier mâché less commercially viable.

Plastic fantastic

Some raw materials, such as ivory and tortoiseshell and certain hardwoods, were becoming harder to acquire by the middle of the nineteenth century.[46] In response, Victorians created artificial alternatives – and one such invention was plastic. This was a turning point in material history. Plastics took elements from papier mâché and pulpware but enhanced the durability and waterproofness of both. However, their invention involved an ironic and fateful development. Many of the materials drawn together to make substances eventually classified as plastics were the products of reuse, and plastic materials owed their existence to recycling experimentation. But as these early plastics became more popular in the production of household items, and more of the materials that formed products could be custom-made, even process-dyed, this had huge implications for recycling. The woods, horns, shells, metals and leathers they replaced could often be reused or repurposed later; many of the new plastics could not. As technologies improved, plastics made from synthetic polymers (rather than modified natural polymers such as cellulose) proliferated. They in turn created the recycling nightmares of modern times.[47]

Alexander Parkes of Liverpool Street, Birmingham, invented the first proper plastic, using cotton waste or paper fibres dissolved in acids and solvents, from which the acids were removed, before camphor and oils were added to improve malleability. The best sources of cellulose were 'cotton and paper made from rags'. These were treated with acids and alkalis and water to extract impurities. The resulting product was dissolved in solvents and dyed, before being formed into blocks, sheets or rods.[48] This material was then processed into 'many articles of fancy ware', including

artificial teeth, flexible mirrors, decorations [...] spectacle frames, watch cases, buttons, pipe mouthpieces, cigar-holders, ash traps [...] brooches, hairpins, bracelets, chains, earrings, crosses, medallions [...] imitation horn, tortoise-shell, coral, ivory etc. It is also made into smooth, inlaid or stamped buttons, knife handles, prayer-book and album covers, cigarette cases, notebook covers, travelling cases, shoe eyelets, purses, pen and pencil cases, rulers, drumsticks, finger-stalls, serviette rings [...] dolls' heads and bodies, mirror and picture frames, figures for games, stick, umbrella and whip handles [...] chessmen and other figures used in games etc. For fans celluloid is somewhat expensive material, but is nevertheless used.[49]

Parkes's 'Parkesine' won a prize medal at the International Exhibition of 1862 (the exhibition that also showcased Perkin's mauveine). In America, John Wesley Hyatt experimented with Parkes's recipe, and tweaked it into a material he hoped might win him $10,000 in a competition to find a substitute for ivory billiard balls. Although no winner was formally announced, Hyatt's product was successful: he formed the Albany Billiard Ball Company in 1868 and, later, the Albany Dental Plate Company.[50] Parkes regarded himself as an artist, not a chemist. In Britain, his inventions were commercialised by Daniel Spill, a former collaborator. But neither Parkes nor Spill had much of a nose for business; others were quicker to sniff out profit from their product. Hyatt managed to make significantly more money from the new product than Parkes, despite Parkes making 'improvements' to Parkesine, to make it 'more suitable for making billiard balls and other purposes'.[51] In the mid-1870s Spill developed Parkesine into xylonite (from the Greek *xýlo*, meaning 'wood') and ivoride (made from formaldehyde and casein).[52] A company historian has described Spill as 'muddleheaded or unscrupulous'. He was probably both. Spill challenged Hyatt's patents in legal actions between 1877 and 1884, but, although the judge opined that Parkes was the true inventor of the process for making what became known as celluloid, Spill was declared to have no claim on Hyatt.[53]

Another American, Levi Parsons Merriam, developed artificial coral manufacture alongside Spill. Various similar businesses were

38. Xylonite collar. Author's collection,
photographed by Toby Sleigh-Johnson.

eventually brought together by Merriam, who along with Ernest
Leigh Bennett managed the British Xylonite Company from 1879.
The company sold xylonite billiard balls and piano keys, among
other items, in 1882.[54] Xylonite shirt collars (fig. 38) became commercially available in 1885: easily cleanable, waterproof, exchangeable
collars for men's shirts, initially made in London's East End. On the
back of the success of these collars the business expanded into a
new factory in Brantham in Suffolk, a site specifically chosen
because damage caused by any conflagration would be minimised.
For a brief period at the end of the nineteenth century, an Oxford
Street shop traded only in xylonite products.[55]

This seeming popularity hid a certain reticence. Consumers
were conservative and preoccupied with the appearance of authen-
ticity. They wanted *faux* products that looked as though they were
made from the genuine article. Artificial materials could rival, in
appearance at least, the genuine raw materials of empire. Xylonite
combs need never have looked as though they were made from
tortoiseshell, but that appearance reassured (or fooled) consumers.
Plastic developed at the same time as the morphing of black coal
tar into colourful dyes. It was not the absence of luxurious raw
materials that encouraged the production of synthetic alternatives
but the fact that the price of those natural materials prohibited
mass production. If consumers had bought *real* ivory and tortoiseshell in increased quantities, the original raw materials would have

dwindled dramatically. Replacing the substance from which combs and trinkets were made permitted wider ownership. Custom and customers dictated that robust, wipe-clean synthetic collars mimicked the delicate, stainable fabric collars, with faked linen textures and unnecessary stitching patterns.[56] The soft ivories of East African elephants made the best billiard balls and piano keys, but a rise in demand saw ivory prices hike during the nineteenth century.[57]

By the close of the century, concerns grew about the need to conserve hunted animals.[58] The trademark device adopted by the British Xylonite Company – a tortoise and an elephant walking out together – reflected the fact that early plastics were made in imitation of natural, dwindling, raw materials. While drawing attention to mimesis, the design also managed to assure any concerned consumers that more animals were now free to wander.[59] Early publicity material for Parkesine stressed that the product was an improvement on the natural substances it replaced, having 'stood exposure to the atmosphere for years without change or decomposition', and having all of the qualities of the material it replaced. Parkes's advertising bumf described Parkesine as being 'hard as ivory'.[60] 'Close inspection' was needed 'to distinguish the counterfeit from the genuine'. Although it lacked a grain, it had all of the 'strength and elasticity' of ivory, without being vulnerable to warping or discoloration.[61] Ensuring homogeneity was the key to the mass market.

Viëtorite

Throughout the Victorian period an insufficient rag supply saw papermakers struggle to keep up with demand. Paper prices soared. 'It is a curious phenomenon', remarked the journalist James Hamilton Fyfe in 1862, that rags, which symbolised beggary, 'constitute one of our chief national wants'. Fyfe put forward some reasons for the dearth, prioritising imbalances in purchasing habits: 'the consumption of paper has increased much more rapidly than the use of clothes and the consequent formation of rags.'[62] With such a shortage, sackings and rags became too valuable to be used in homely crafts such as rag-rug making, and were siphoned into papermaking.[63] Benjamin Lambert, a papermaker and chemist,

39. Postcard for Schroeder & Viëtor. Author's collection.

worried that other industries might begin to utilise rags too, divert-
ing them from the mills.[64] George Dodd, author of books about
industry, trumpeted the repurposing of a 'Hungarian shepherd's
frock or tunic-shirt, the blue shirt of a weather-beaten sailor in the
Mediterranean'; a Saxon farmer's bedding or a Hamburg burgher's
tablecloth might do too. Some paper was already well travelled
before it was written on, going through revolutions of fashion
across the Continent. As a result, quality was variable; while the
rags supplied by English housewives were clean, Italian rags were
'far otherwise'.[65]

Responding to the demand for paper and lack of rags to make
it, patents reveal a range of alternative raw and reused materials that
enterprising makers tried to turn into paper, including asbestos and
sugarcane leaves.[66] In 1839 Henry Crosley made paper from refuse
tan and spent hops, a recipe he later refined, adding old rope,
oakum pickings, cotton waste and woollen rags.[67] A German engineer called Henrik Zander experimented with well-rinsed horse

dung combined with pulped straw and linen rags.[68] Paper exhibits at the Great Exhibition included a 'sheet of coarse paper' made from hempen fibres fished up from the sunken *Royal George*.[69] In 1854 paper was made using waste products from tanneries, plus waste coconut fibre mixed with equal parts of old ropes (cordage) and rags.[70] That year *The Times* offered a £1,000 reward for a substantial alternative to rag-based paper. This prize was still unclaimed by 1861 (when the Stamp Duty repeal exacerbated shortages), despite trials of various substances, including esparto grass, nettles, hop vines, bindweed and bark. Some success was had with straw-based paper, but it was deemed too insubstantial for quality publications.[71] The efforts to find new materials from which to create paper threatened to divert waste supplies from other trades. In the 1860s attempts to make paper from the shreds and clippings of ivory meant that less ivory dust was available for doctors to make ivory jelly for patients, to help with anaemia and diabetes.[72] The paper shortage crisis was felt across Europe. Parts of catalogues for the Austrian collection at the London International Exhibition in 1862 were printed on maize paper.[73]

Several British pioneers of Victorian recycling had been born abroad, and some of these men came together as shareholders of 'Fritz Viëtor & Co. Ltd' in 1886. In spirit and substance Fritz Viëtor was a repurposing opportunist. His story and his connections with other non-nationals (mostly Germans) reveal tangled webs of reuse and imitation. Viëtor had set up in business in the 1870s in London with a man from East Friesland, the region of Germany from which the Viëtor family came. Viëtor and Ferdinand Schroeder dealt in various types of paper ('plain and embossed', 'marble, Morocco, gelatine, and printing papers'), plus lined and unlined strawboard (a material used in bookbinding and for box-making).[74] Gelatine papers were used to make luxurious cartons, delicate packaging papers and visiting cards, and were composed using a variety of reused substances, the gelatine part being made from waste animal bones and other parts.[75] The company didn't last long enough to use up their pre-paid penny postcards; the partnership was dissolved by the end of 1877.[76] Schroeder ploughed on alone until he went bankrupt in 1884, and died the following year. Theodor, his younger brother, then took on his business and married his widow. Recycling ran in the family.[77]

By 1878 Viëtor had diversified and was working alone, selling 'overstrung, iron-framed, patent check action pianos', but it was an inauspicious time for young companies.[78] The following year was a particularly bad one for entrepreneurs, with many bankruptcies and liquidations across London, but the trade in imported German pianos was more resilient than the domestic trade, and Viëtor's instruments were typically German in style.[79] An advertisement placed in November 1880 stated that pianofortes sold by 'Fritz Viëtor & Co.' were 'now' imported from Berlin.[80]

Pianos alone were not enough for Fritz. In March 1878 'Fritz Viëtor & Co. of 57 Aldersgate Street' applied for a patent for his invention of 'improvements in the manufacture of an improved compound or composition for use in the manufacture of buttons, studs, ornaments, frames and other fancy goods', and he was communicating with a Berlin engineer about the product.[81] Other men had previously made objects using pulped matter. In 1844 William Sheldon made buttons from dry matters pressed with glue into a dough; his recipe included lampblack, 'spent hops' and 'finely cut human hair'.[82] Birmingham-based japanners James Souter and James Worton punched buttons from sheets of a type of papier-mâché composition board, which they presumably then lacquered.[83] By 1880 Viëtor was selling *bois durci* ornaments that included 'music emblems' and 'heads of celebrated composers, artistically executed', and this may have been the substance to which he referred in 1878.[84]

A heavy, polishable compound of sawdust and slaughterhouse blood, hydraulically pressed into heated moulds until hard and shiny, *bois durci* could be moulded into inkstands, plaques and combs. Lacquered, gilded and glued, this substance was used as a substitute for wood and horn. It had been invented in Paris in 1855 and was the subject of a British patent a year later.[85] *Bois durci* imitated carved wood. It looked and felt like the real thing (although it would have smelt different). It pretended to be the fruit of hours of highly skilled carving labour. On the Continent there were other types of artificial wood made from different combinations of sawdust or exhausted dye-woods, plus some form of binder, pressed or poured into moulds and dried, but few other tradesmen in Britain manufactured hard substances in imitation of wood using sawdust.[86] Bound with blood, *bois durci* was distinctively dark. The blood added 'glistening specks' and increased durability; it was never an equal to resinous wood but could pass for it easily enough. A contemporary described artificial wood as 'a really firm decorative material, capable of resisting changes of temperature and damp, and in no respect inferior to expensive but fragile wood carvings'.[87]

Business seemed healthy. By 1882 Viëtor was dealing in pianos and *bois durci* from two outlets near to each other. In an advert from that year he claimed to import and manufacture pianos, and also 'Bois Durci Ornaments for Pianos'.[88] He had stalls at London furniture exhibitions. In 1881 he had shown a large assortment of *bois durci* along with 'a couple of pianofortes, [and] a handsomely carved Chinese cabinet containing collections of nature, medicines, plants and insects'. As wood-carving doesn't appear to have numbered among Viëtor's myriad skills, it is likely that this cabinet was a *bois durci* showcase.[89] Viëtor was hiding the details of his innovation: 'mouldings' is the only hint here that the ornaments were not the carved wood they mimicked. One can only imagine what Augustus Pugin, the architect famous for his revival of the Gothic, would have made of this innovation, hating, as he did, 'cheap deceptions of magnificence' which encouraged people to 'assume a semblance of decoration far beyond either their means or their station'. By Viëtor's time Pugin was turning in his hardwood gable-lidded coffin with gold-plated handles.[90]

Viëtor imported not only pianos but also 'bronze powders' from Nuremberg that he may have mixed into his *bois durci* before pressing. Used in japanning and for ornamenting wood and leather, metallic powders were first invented in Bavaria in the seventeenth century; their manufacture involved reusing 'scraps, cuttings, and fragments of [metallic] leaves'. The price of such products varied, depending on 'the demand, and the supply of the waste material of the metal leaves'.[91] Viëtor probably become interested in *bois durci* because it allowed him to build parts of his piano cases with

40. Lyre made from *bois durci*, a pressed and polished compound of blood and sawdust, probably made by Fritz Viëtor, author's own.

decorative features mimicking carved wood, in imitation of more expensive instruments. In the mid-1880s many high-end pianos were heavily ornamented with carved images. *Bois durci* offered an alternative: one could simply affix a moulded object, 'a cheap mode of truly artistic ornamentation' (according to Viëtor's own advertisement) 'suitable for pianoforte panels, &c'.[92] This ornamentation featured such delights as 'Medallions, Shells, Rosettes, Corners, Composers' Heads, Greek Heads, Lyres, Wreaths, Escutcheons, Bouquets, Griffins, Caryathides', depicting 'Fire and Water, The Seasons, Horn of Plenty, Renaissance Frames, Juvenile Musicians' and other designs.[93] Just such images adorned the fanciest pianos on the market: those of Steinway & Sons, from New York, and Bechstein, from Berlin.

Since 1880 Viëtor had also been back in the paper trade, rivalling his former partner Ferdinand Schroeder and taking as the base for his business a 'very spacious' newly erected warehouse on Fann Street.[94] Just as Viëtor settled into these premises, a few minutes' walk away there was a sale of fire-damaged paper items salvaged from a wholesale stationers after a blaze. Two auctions, each of 80 tons of paper, 'Writing, Brown, Strawboard, Surface Colour, Enamelled, Embossed, Tissue, Nonpareil, Morocco, Flints and other Paper', some of which was 'but slightly damaged', must surely have been too irresistible for him to ignore.[95] This was the perfect material with which to kick-start another business cheaply. For six years from 1880 Viëtor was included in trade directories under many headings: papermaker, pianoforte marker and 'Pianoforte Small Work Manufacturer', paper merchant and also 'Manufacturer of pianos and Bois Durci ornaments'.[96] It was while in these premises that Viëtor made the samples that impressed W. John Stonhill in 1884, and he may well have used various presses and items of machinery for both the *bois durci* and the paper items; indeed some of Stonhill's descriptions suggest that Viëtor was crafting materials that were somewhere between paper and early plastics.

In 1882 Viëtor was listed as a 'Strawboard liner' (his 'London Pasting Works' remained a key plank of his business interests until 1886), and making paste boards from straw had become a major concern for Viëtor's business by the mid-1880s. In some trade directories he is also listed as a millboard maker. Millboard was a grey stiff thick board made from cordage, old ropes from ships or sacking, milled under pressure. Strawboard, millboard and other fibrous forms used similar high-pressure technology to *bois durci* production. Strawboard could be made so dense and thick that, in the right circumstances, it could pass for timber. A Purfleet company established around this time made strawboard from farmyard manure. Ammonia 'and foreign matters' were removed from the straw before 'the usual processes for conversion into strawboard'. On one side of the factory there was 'nothing but filth'; on the other, piles of 'glossy amber-coloured board'.[97] On 13 August 1881 the *Furniture Gazette* reviewed Viëtor's stand at the Islington Furniture Exhibition. As he read this, his eyes may have flicked over to the opposite page, which featured a review of an American company that made 'solid masses or blocks of any form and size from straw'. The article described a strong and dense timber substitute. Perhaps Fritz wondered how far he might develop his own strawboard products.[98] An American exhibit at

Earl's Court in 1887 included a large 'fire-proof and waterproof' villa made entirely from straw, which no amount of huffing and puffing would blow down.[99] By then, however, Fritz's own house of straw cards had tumbled around him. What is more surprising is not that Viëtor ultimately failed, but that he had managed to apply so much capital to various trades, being non-naturalised.

Viëtorong

The beginning of the end came not through Viëtor's own business but through the many small businesses to whom he lent money. In addition to his various commercial and industrial activities, Viëtor became a creditor to small businessmen associated with the paper and board trades, as far afield as Sheffield (a box-maker) and as near as Fann Street (a book-edge gilder).[100] In 1885 he entered into an agreement with Charles Thomas Hood, who traded as Henry Jackson and was a 'shell box manufacturer'. Jackson had been described as a billiard ball maker in 1881.[101] Onno Eberhard Viëtor, Fritz's younger half-brother, entered the strawboard scene in the mid-1880s, coming over from Leer in East Friesland, as Fritz made steps to become a limited company.[102] In a new company, Onno and Fritz were each set to hold 500 shares, as would two Edwards: Tauchert and Crickmay. Lesser shareholders would be: Collan Hawke, the clerk to a billiard ball maker; Nathaniel James Whitcomb, a chartered accountant; and Maurice Spiegel.[103]

Austrian-born Maurice Spiegel was the most colourful character, having often been in trouble. Among other things, he had committed frauds involving mushrooms, wine and train tickets, and a libel against a fellow freemason.[104] Edward Crickmay was so slippery that solicitors handling the eventual de-registration of the company could not even locate him in October 1886.[105] In bankruptcy proceedings from 1888 he was described as a 'promoter of companies', but he was also involved in the sale of hard materials as a stone and marble agent.[106] Edward Tauchert was a Cheapside 'paper and fancy goods dealer' from Wongrowitz, in what is now Poland. Liabilities he had amassed by 1892, of £14,797, were attributed to the 'failure of various firms with whom he has been connected'.[107] One of these companies was E. & H. Shepherd (brothers Ebenezer and Henry), box-makers in Birmingham. In turn, the Shepherds held Tauchert partially responsible for their failure. By 1891 Tauchert owed them more than £647, which he claimed to be unable to pay. A report states that the 'transactions between them had been, to a great degree, of a cross-accommodation character'. In order to get payment from Tauchert, the Shepherds had been compelled to continue trading with him.[108] However, Tauchert's most interesting role was as the sole agent of Hyatt's Albany Billiard Ball Company', and also the manager of the Bonzoline Manufacturing Company.[109] Like celluloid, Bonzoline was developed from Parkesine. Piano keys could be crafted from it.

From 1884 legal complications blighted Viëtor.[110] Two years later, in 1886, came Fritz's *annus horribilis.* Litigation was brought against him by Julius Ichenhauser, the carved oak importer whom we have already met, operating as Schwelheimer & Co., a whole-sale haberdashery firm based in Fürth, near Nuremberg.[111] During the same year, Viëtor was also involved in a case brought by a firm of chemists, Paterson Brothers & Co., of Glasgow. Centring on a suspended payment made in May 1885, the case preoccupied Viëtor throughout late 1886.[112] William Burdett, a paper agent, is of par-ticular interest in the Paterson Brothers' case – he was one of the men who endorsed the bill and he was Fritz's most substantial debtor in 1886. Burdett was a paper and strawboard merchant on Queen Victoria Street, and was probably supplied by Viëtor.[113] His business collapsed, and Burdett petitioned for bankruptcy in May 1886.[114] This had a domino effect on Fritz's business, revealing the fragility of such cross-accommodation networks. Fritz Viëtor & Co. was registered on 17 August 1886 but was dissolved in April 1887 without issuing shares.[115] Another clue as to why Viëtor's busi-ness failed appears in the account of the collapse of box-makers E. & H. Shepherd in 1891. Their business tailed off 'in consequence of several large customers commencing to make their own boxes'.[116] Companies sourced their own strawboard: Dutch firms were key suppliers. In 1886 the debts owed to Viëtor amounted to over £2,000.[117] In early September 1886, just a couple of weeks after he registered his company and days after the conclusion of the two court cases, Fritz filed for bankruptcy, and in 1887 he sold the paper business to Onno.[118] One of Onno's major creditors was Ebenezer Shepherd of Birmingham, whose paper-box company Tauchert was later accused of scuppering.[119] Onno went bankrupt in 1888.[120]

Paterson & Sons, who had an outlet on Buchanan Street in Glasgow (just like Paterson Brothers) and described themselves as the sole Scottish importers of Steinway and Bechstein pianofortes, had published a caution in *The Scotsman* in March 1883 concerning counterfeits 'of these Celebrated Instruments bearing labels resembling the names of these Great Makers' on the market.[121] Did someone mock up fancy pianos with cheap *bois durci* decoration? Viëtor was, for certain, creating the means to puff up humble pianos more grandly. *Bois durci* was uncommon in England. It may have looked authentic to unwary buyers. Luxury pianos were supposed to be exclusive, restricted to the very few. By mimicking carved wood, *bois durci* gave the impression that many skilled man-hours had gone into the manufacture. A piano bedecked with Viëtorian knick-knackery would have looked convincing (even if it did not feel quite right), just like Bonzoline billiard balls and celluloid collars. Such products wore adopted cultural richness with arrogance.[122]

We may never know. What we can say is that men in Viëtor's business network had the know-how to make artificial ivory and carved wood. The business interests of the potential company shareholders for Fritz Viëtor & Co. in 1886 deepen speculation that Viëtor was concentrating on the manufacture of artificial hard products. Remember how his ingenuity had impressed W. John Stonhill: the exquisiteness of his Chinese cabinet, displayed at the Islington Furniture Exhibition? Viëtor's German instruments, displayed in the Furniture Trades' Exhibition of 1881, were described as having a pure tone and a fine finish to the woodwork and as being of a 'low price'.[123] There is another, less spectacular accusation one might level at a strawboard-maker in the 1880s. A Dutch manufacturer, Kuipers, placed an advertisement promising a £50 reward to 'whoever gives the information necessary for obtaining an injunction' against producers of 'inferior makes' of lined strawboard, passed off as Kuipers' Boards.[124] Ersatz artificiality was lucrative.

Whatever Viëtor's involvement in the counterfeiting of pianos, the sale of fake pianos created uncertainty in the trade in 1887. A think piece in the *Musical Opinion & Music Trade Review* included an oddly specific remark when querying how a piano buyer was to identify a 'legitimate dealer', noting that many contemporaries believed that bona fide dealers would not also sell 'stationery and paper hangings'. The Americans slapped a duty of 35 per cent on *bois durci* items.[125] Anti-German propaganda started to have an impact on the sale of German-built pianos, especially the cheaper imports, and piano fashions shed the fancy decorations and 'restless designs' that Fritz was apt to mimic.[126] Fritz, bankrupt, slips out of record and possibly headed to America. But this is not the end of this Viëtorian tale of recycling. Pianos made by Fritz Viëtor & Co. formed lots in various auctions. As early as November 1880 a Viëtor 'seven octaves' trichord piano with a 'costly ebony case, Renaissance style', was advertised for resale.[127] In June 1883 one 'superb fine toned double overstrung cottage pianoforte' was a highlight of an auction held in Leicester.[128] A few months later a 'full trichord, in fine ebonised case, by Fritz Victor [*sic*]' was auctioned, along with a Davenport writing desk, a Broadwood piano and other fineries, in Manchester.[129] Further sales took place: in Northampton in 1884 and in Sheffield and Lowestoft in 1893. A Viëtor pianoforte was auctioned off in Burnley in 1940. That may have been the same instrument, 'in walnut case, iron frame', that kept featuring in sales in the town throughout the early 1900s.[130] One of Onno Viëtor's children, Menno Heinrich, later became a hero of celluloid, that other plastic stuff, appearing in many films under the name he formally adopted: Henry Victor, star of *Freaks* (1932). In one form or another, Viëtor echoed on.

7

The Wonderland of Rework (1840s–1850s)

Without him even saying a word, Rosa Faulkner could always tell that her husband had been attending a fire when he came home, because the smell of smoke lingered on his long, forked mutton-chop beard. One night in September 1858 he came home smelling like toffee. As the head of the Liverpool Salvage Brigade, Captain Henry Cole Faulkner had been overseeing the rescue of refined sugar from a warehouse fire in Liverpool. He had led the brigade for four years, a role he held in his capacity as the Staff Officer of Army Pensions. Initially, the fire salvage corps was staffed by army pensioners, but increasingly the positions were filled by civilians. They were not firefighters, but they worked under the command of the chief officer of the fire brigade. Their job was not to put out fires but to protect wares from fire and water, and to salvage whatever they could from premises. As Faulkner's team arrived on the scene of the sugar fire at J. Leitch & Co., sited along Blackstock Street, it seemed as though half of the city was alight. The sugar warehouse was next to a charcoal works; the whole area was immensely combustible. As molten liquid sugar fuelled the flames, his men, 'with much daring', went into various premises and saved 'a considerable quantity of property'. A great deal of sugar was bailed out in leather buckets and put 'in casks for future boiling'.[1] A coping stone fell inches from one of the men helping to save the property, but the mission was a success.

Liverpool had a fire salvage corps before any other British city.

Many warehouses had gone up in flames in the early years of Victoria's reign, and a particularly nasty series of fires in 1842 had seen insurance rates rise and merchants clamouring for change. By the time Faulkner took charge in 1854 a basic uniform was issued, and under his control more horses, equipment and premises were obtained. Faulkner's men clawed out various fire-damaged materials from premises across the city, including: bottles of soda from J. Schweppe & Co. in 1862; stores of naphtha and phosphorus; and innumerable bales of cotton.[2] The corps was so effective and so central an institution that it persisted for well over a hundred years, and was only disbanded in 1984.

Other towns wanted a fire salvage corps: Thomas Erskine Austin of Luton wrote to his local paper in 1859 claiming to have witnessed every fire in the town for the last thirty years. He remarked on the damage done by overzealous volunteers, pulling items from burning houses. Austin thought Luton needed a 'Salvage Brigade' staffed by 'respectable middle-aged men' so corpulent that they could not 'run about the house adding to the general confusion'.[3] Austin did not get his wish in Luton, but after 1860, fire salvage teams were formed in Glasgow and Manchester. In 1866 London insurance companies established the 'London Salvage Corps', whose mission was to save property from fires 'whether insured or otherwise'. Their services were rendered in roughly a third of all fires attended by the Metropolitan Fire Brigade between 1866 and 1875, and included lifting or dragging valuable materials off the property, draping tarpaulins over stock to limit water damage and scattering sawdust to absorb hose water. In one of their first missions, the London Corps saved casks of oil from Crossley's paraffin manufactory on Bow Common, after fire spread from a match factory. Matches were themselves the product of reuse, using wood splints made from waste chips and phosphorus from old bones. A bone-crushing factory, one of the match company's suppliers, was also caught up in the blaze.[4]

The work of the fire salvage teams permitted the recovery and reuse of tons of objects and materials after the mid-nineteenth century. Another of the fires attended by the London Corps, in June 1873, destroyed Alexandra Palace, the construction of which had reused materials from the International Exhibition of 1862.[5] A central feature of the Great Exhibition of 1851 was a Crystal Palace, which, as we have seen, was reused in Sydenham in 1852. When this new Crystal Palace burned down in 1936, the London Corps could only salvage a few parts.[6] In many ways, however, a more interesting recycling story resides in the original, unburned, Crystal Palace, when it was filled with wondrous goods and inventions from across the globe. The Great Exhibition of 1851 energised many people and encouraged a more comprehensive recognition of the value of reuse. Mid-century developments saw businesses think in a more focused way about how they could combine trades to enhance the value of waste goods; larger-scale waste recovery industries were set up, or enlarged, and the opportunities for waste recapture increased.

Exhibitors and promoters

Another thing also happened after 1851: a plethora of British reuse popularisers emerged. People had made occasional comments about reuse previously, but after the Great Exhibition there was a more palpable sense that the public ought to be educated. Among these educators were two scientists: Lyon Playfair and William Crookes. Playfair had trained under the famous chemist Justus Liebig. He later became a Liberal politician and had been an organiser of the Great Exhibition. He pounced on almost any opportunity to promote reuse. In 1863 Crookes opened an article in *Popular Science Review* with the bold statement that chemistry had been chiefly devoted to 'the utilisation of waste substances' in the years between the Great Exhibition of 1851 and the International Exhibition of 1862: 'The visitor would be little prepared to hear that most, if not all, these important commercial products are manufactured from materials formerly thrown away, and at first sight apparently valueless and repulsive.'[7] Other educators were Henry Mayhew, the social reformer and journalist; Peter Lund Simmonds, another journalist, whose *Waste Products and Undeveloped Substances* – the first comprehensive guide to material reuse written in English – was first published in 1862; and Charles Dickens. From the mid-century these men wrote many pieces between them, focused on the potential for reusing materials that could otherwise go to waste.

In his epic study of *London Labour and the London Poor* (1851) Henry Mayhew detailed the lives of the lowest class of waste recyclers: the street finders who gathered cigar butts, cinders and other materials, and mudlarks, grubbers, sewer-hunters and dredgermen. Always keen to taxonomise, Mayhew divided these workers according to whether they collected from streets, banks or water, and included a table of master-sweeps, dustmen and contractors who also undertook 'night-work', clearing out cesspools.[8] The highest class of worker removed waste, but did not go out to find it: dustmen, nightmen, sweeps and scavengers, the majority of whom were men. According to Mayhew, the grubbers were 'with a few exceptions stupid, [and] unconscious of their degradation'. Grubbers collecting rags avoided wet days, as they could not sell dirtied and saturated pieces. Pure-finders sought dog excrement, which they took to tan-yards in Bermondsey. Some tanneries favoured 'dry limy-looking' pure (made when dogs eat much bone), but some preferred darker, moister turds. Some pure was not pure at all but was adulterated, bulked out with mortar scraped from walls.[9] Mudlarks, who were often boys but sometimes old women, salvaged stuff deposited by the tide on river banks, like wet Wombles. Such objects might have been knocking around in the river for centuries. Below the city, opportunities for sewer hunters were weighed up against the dangers – slipping, drowning, being crushed by falling sewer masonry or having hard-won finds confiscated by officials. Coins and bits of metal were sought out most keenly, but sometimes these formed massive 'conglomerates' of metals all rusted together. Coinbergs: too heavy to lift, they remained in the sewers.[10]

Mayhew shows the lengths to which some Victorians went to make a profit from stuff considered to be waste, including even the sale of waste newspaper to fishmongers at Billingsgate (although he identifies just one man at this particular trade).[11] Such people did things that others would not: their circumstances pushed them to undertake the most arduous and unrewarding tasks. *Chambers' Journal* started an 1875 article entitled 'Rubbish' with the statement that 'those who pick up and utilise the rejected trifles are benefactors to society. Let this be a comfort to the dustmen, scavengers, bone-grubbers and rag-pickers; they are not mere pariahs and dirty outcasts.'[12] That was, nonetheless, how they were treated, and depictions immerse them fully in a world of filth.[13]

Charles Dickens, a contemporary of Mayhew, was also fascinated by this world, and portrayed similar trades in his fiction. The central milieu of *Our Mutual Friend* (1865) is a yard containing 'Dust. Coal-dust, vegetable dust, bone-dust, crockery-dust, rough dust and sifted dust – all manner of Dust'. The story opens with scavengers dredging the Thames for dead bodies. The searchers include Gaffer Hexam, a waterman who robs from bodies fished from the Thames and has a home papered with handbills announcing deaths. Dickens describes the women who sifted through dust in dustyards, sorting items crudely into various elemental parts.[14] Most of this dust had previously gone to brickmakers to make bricks. However, Dickens wrote at a time when changes in brickmaking and a growing population meant that the supply of dust exceeded the demand. Profits from ash declined throughout the period, and by mid-century had become too low to make sorting and resale a viable business. By the time Dickens showered his 'Golden Dustman' (Noddy Boffin) with riches in *Our Mutual Friend*, contracting had become much less lucrative than it once was. The decline coincided with increasing concern for public health, made clear by Edwin Chadwick's *Report on Sanitary Conditions* (1842) and the Public Health Act (1848). At this point, more and more vestries stopped contracting out the removal of waste, and followed an example set by Bermondsey for municipally managed systems of waste sorting. A further Public Health Act (1875) saw the end of fixed waste receptacles and the widespread adoption of movable bins, which would be emptied weekly.[15] Dustyards continued to operate but fell increasingly under the control of municipal authorities.

The most important of all the reuse popularisers was born in Denmark, in 1814. Peter Lund Simmonds came to Britain after being adopted by a Royal Navy lieutenant. He served on board a ship when young, and was sent to the West Indies to keep books before returning to England for a career in journalism. After the Great Exhibition he devoted an increasing amount of time to educating people about the reuse of materials: curating an exhibition of waste products in Bethnal Green; editing the *Technologist* (whose contributors included Lyon Playfair) and the *Journal of Applied Science*; and authoring several editions of *Waste Products*. 'Let nothing be lost', Simmonds instructed: 're-work with profit and advantage the residues of former manufacture.'[16] Simmonds's curious books read like antiquarian rummages: archives of all possible reuses. They are absolutely extraordinary, noting every possible opportunity for reuse, eked patiently from science journals and newspapers. There was no happy end for Simmonds, and 'despite a life of constant labour he was not prosperous and became increasingly impecunious'; he died in October 1897 after being hit by an omnibus on the Gray's Inn Road.[17]

The readership of scientific journals was predominantly adult men, but, as in every era, it was women who organised most of the domestic recycling and mending efforts. Much of the waste was gathered up in homes by women: matriarchs and female servants.[18] Women were also at the forefront of industrial recycling. According to Emily Hobhouse, dust-heap sorters were mostly 'women in unwomanly dirt doing unwomanly work [...] squatting down on the unsavoury heap'.[19] Hobhouse, a welfare campaigner, described this monotonous, nasty work in 1900, depicting women picking over the heaps amid choking dust, their nimble hands cut by glass.[20] Women also collected sawdust to make firelighters; they scraped animal guts for sausage skins and violin strings; they stripped the skin from ox's feet, split them and scooped out fat to make neatsfoot oil.[21] Young women sieved through the jigged remnants of 'mundic', waste arsenical pyrites, in preparation for extracting arsenic 'soot' from it.[22] Old women collected bones from dunghills to be bruised and boiled to extract fat for soap-making.[23] Recycling was women's work.

In print as well as in life, women's contributions to recycling were undervalued and are still underappreciated. Historians have traditionally focused on publications written for an educated male readership, which underplays the importance of cheap books written by women for children.[24] Clara Matéaux, Annie Carey and Mary Anna Paull all educated children of both genders about the reuse of materials, and so made contributions to forming the habits of future generations. In her own time the books of Clarisse ('Clara') Matéaux were reviewed by men who assumed she was a man. In 1881 the *Star* published a glowing review of 'Mr' Mateaux's *The Wonderland of Work*, perfect 'for boys of a curious, inquiring turn'.[25] More recently, in *Victorian Things* (1988), Asa Briggs had no doubt Matéaux was male and renamed her 'G. L. Mateaux'.[26] *The Wonderland of Work* has an evocative chapter on 'Things that are done with', which includes passages on rags, horn, gasification wastes, bones, metals, broken china, sawdust and many other substances. Throughout the whole book Matéaux shows an exceptionally keen awareness of the potential re-applications of many materials.[27] Despite being a prolific author, the editor of a publication for children and an artist, little is known about Matéaux. The daughter of a Parisian dyer, she started her career as a wood-engraver. After winning prizes at the state-funded Female School of Art, she became a teacher there.[28]

Matéaux would have known Annie Carey, a contemporary at the Female School of Art, who was also published by Cassells. Visiting their school, Dickens lamented the cramped, ill-lit and stuffy conditions of the accommodation of the school itself, which was situated above a 'soap and sponge-shop'.[29] For her part, Carey wrote 'it-narratives', stories told from the perspective of objects, with the aim of popularising scientific ideas. In *Wonders of Common Things* (1873) a lump of iron recalls being mined from a Swedish mountain in the 1520s and melted for use more than once. The iron puts a positive slant on its own rust. The orangey powder crumbled, 'mingling with the soil' to help grow corn.[30] Carey's work was much influenced by her family. Her half-brother Eustace was a manufacturing chemist, and she was the daughter of a Baptist minister. Her writing combines her father's religious zeal with her brother's professional interests. After spending most of her life as an invalid, Carey died in 1879.[31]

Mary Anna Paull wrote dozens of books for the cheap market, mostly thinly veiled attacks on the evils of drink, and some in the style of religious 'it-narratives' similar to Carey's. Of particular interest are *Romance of a Rag* (1876) and *May's Sixpence: or, Waste*

Not, Want Not (1880). Heaving with homilies, the latter inculcated thrift and habits of reuse: 'take care of every little bit of everything', parrots one character.[32] *Romance of a Rag* is a less accomplished object narrative than Carey's and tells the tale of part of a dress worn by a schoolmistress, which comes into the possession of an alcoholic called Ned, who tears it to rags in a drunken rage. Sold to the paper mill, the rag is sallied through the sorting room by women in large aprons, shredded, washed, bashed and mushed into a pulp, moulded into paper, sized, watermarked and polished. It emerges, reformed into a 'glazed-card-board' printed with the pledge to abstain from alcohol. Pressed into Ned's hands, it preaches: 'truly existence has a perpetual sameness as well as a perpetual change, anomalous as this sounds; and we not unfrequently repeat ourselves and our circumstances and surroundings over and over again in one short life.'[33] Paull manages to convey (with some accuracy) the process of papermaking while also poking into relationships between humans, and the lives of material things.

In the Victorian period, authors of both genders tried to encourage people to think about waste. The involvement of women has since been underplayed by historians more focused on government initiatives and scientific advances than on actual domestic and commercial activity. Children's books recirculated widely, and so enjoyed a broad and lasting impact: naturally this is almost impossible to measure. Behind every grand scientific report or policy decision there were women, working quietly away. If the history of recycling was retold as an 'it narrative', the reader would encounter more women than men.

Shoddydom

Early discussions of material reuse played out in the northern press, and were concerned with the manufacture of cloth. In Yorkshire soft woollen textiles – old flannels, guernseys, stockings and similar fabrics – which were not raggable for paper, were comprehensively reformed into new cloth called shoddy, after seams and buttons had been cut away by rag sorters who were, according to Mayhew, 'entirely women'. Mayhew conceptualised shoddy as 'snatched [...] from the ruins of cloth'.[34] Samuel Cunliffe Lister, the first Baron

Masham, born in Bradford, worked for decades on finding a way to utilise silk debris, including the outer coverings of cocoons and dead silkworms. These he fashioned into a yarn to make carpets, imitation sealskins and other silken products. His project profited handsomely in the 1870s.[35]

Rags to be shoddied were shredded by machine into a dusty morass. The *Westminster Review* described the process in 1859, whereby rags were 'reduced to a filament and a greasy pulp, by mighty toothed cylinders'. 'The much-vexed fabrics', the author continued 're-enter life in the most brilliant forms – from solid pilot cloth to silky mohair and glossiest tweed'.[36] Despite this transformation, the result was sometimes less exciting. Some of the earliest shoddies fell apart if put under duress, and they were often used as paddings and linings. But improvements by the mid-century impressed Mayhew. Although he noted the inferiority of the resulting cloth, he argued that 'in some articles the remanufacture is beautiful'.[37] 'Mungo' was made in a similar way to shoddy, but was composed using finer fabrics such as broadcloths, tailors' cuttings and cloth shreds.[38] Shoddies and mungoes were fashioned into doeskins, duffels, pilot cloths, sailors' coats, stiffening pieces, cloth for institutions such as prisons and workhouses, and for export. One output was Stroud cloth, a thick felted fabric largely purchased by the government to be gifted to native North Americans 'for some privileges of a territorial nature' until the mid-nineteenth century, when the demand fell off. The bulk of the coloured blankets also went to the US, where they were mostly 'consumed in the Slave States'.[39]

The Yorkshire town of Batley became the 'shoddy metropolis'. In his *History of Shoddy*, published in 1860, Samuel Jubb described the Batley dealers and their 'visions of filthy rags being transmitted into shining gold'.[40] Alchemy, again:

> Not the least important of the manufacturing towns is Batley, the chief seat of that great latter-day staple of England – Shoddy. This is the famous rag-capital, the tatter metropolis, whither every beggar in Europe sends his cast-off gentility of moth-eaten coats, frowzy jackets, worn-out linen, offensive

cotton, and old worsted stockings – this is their last destination. […] Thus the tail-coat rejected by the Irish peasant – the
gaberdine too foul for the Polish beggar – are turned again to
shiny uses: re-appearing, it may be, in the lustrous paletot of
the sporting dandy, the delicate riding-habit of the Belgravian
belle, or the sad sleek garment of her confessor. Such, oh reader,
is 'shoddy'![41]

As suggested in this passage, some of the rags to be shoddied came
from abroad. While Irish rags, 'the most worn, the filthiest', were
often shunned, Mayhew explained that Danish rags were welcomed by tatter merchants, 'being fertile in morsels of clothing of
fair quality'. Imported rags to be shoddied were less welcome to the
general public: their sale could spark hysterical xenophobia. Sensational accounts planted the notion that shoddy might carry typhus,
cholera or even plague. When, in Hull on 30 June 1843, Mr Stamp
auctioned off 200 bales of white and coloured shoddy and mungo
from Antwerp, the *Northern Star & Leeds General Advertiser* threatened to publicise the names of bidders.[42]

In the 1840s other media fears stuck like burrs to these rough
cloths, with worries that the British trading reputation would
decline in the face of these 'shams' and 'shoddy goods'. William
Busfeild Ferrand, MP for Knaresborough in Yorkshire, attacked
shoddy traders in a whinge against manufacturing crimes. It was
suggested that some cloth destined for the American market was
stamped with Spanish names to conceal reuse.[43] The *Leeds Intelligencer*, whose readership included workers in the cloth trade, placed
the blame for 'manufacturing decay' on the 'deceit' of the shoddy
trade: 'Instead of making up cloths […] we have been making up
shams.' Exported cloths had 'dust within', and were bad wares
peddled as good, and this led to British manufacturers being
shunned by foreign customers.[44] Unwary buyers were fooled
because shoddy handled well initially, having a softer feel and finer
appearance than 'a cloth made wholly and honestly of wool', but it
was not durable, because the manufacturer had scrimped on the
new woollen content.

The northern press hoped to clarify the situation, and to

enhance the prestige of the local trade. A journalist for the *Yorkshire Post & Leeds Intelligencer* argued that there was 'nothing reprehensible in this utilising of half-worn woollen fibres' providing the commodity was not sold as 'all new wool'. Shame only attached to shoddy fabrics deliberately passed off as new, but not to shoddy wool generally. Nonetheless, some called shoddy 'devil's dust', believing that 'dirt and cheating must necessarily go together'.[45] The cloth trade generally fell into disrepute from the conversion of 'decent men's garments into sponges and riddles under the name of cloth, for the profit of the *liberal, religious,* cloth-makers of Yorkshire'.[46] 'Shoddy' became an adjective to describe behaviour in the 1860s. Imports of finished shoddied fabrics created competition for the West Riding tradesmen. By 1860 thousands of tons of shoddy fabric came through British ports, the majority of it via Hull. Much of it came from Denmark or Germany. Some factories were set up in Berlin by men from the West Riding, whose own enterprises had been thwarted by British import duties on rags.[47]

Garments made from mixed fabrics increasingly added wrinkles to the shoddy process. These 'union cloths' or 'muslin-de-laine' combined a cotton warp with a woollen weft. Processes that saved the wool destroyed the cotton, and vice versa, and yet both parts had potential reuse value. New separation schemes were developed. Frederick Oldfield Ward presented his method to the International Exhibition in 1862: union fabrics were steamed, releasing the cotton to be used for papermaking and leaving the wool charred into a powdered form called ulmate (or crenate) of ammonia. Ward was no typical recycling pioneer; he had no financial need to reuse waste. A doctor by training, and with correspondence networks including men like Charles Babbage, Ward dabbled in speculation on small patents. Rich in nitrogen, ulmate of ammonia made a good addition to manure and was especially prized by broccoli growers in Cornwall. The substance could also be used in papermaking. The Ulmate of Ammonia Company was set up next to a paper mill in Grays, Essex. In 1861 the company advertised to 'gentlemen in the manufacture of low-priced printing papers'.[48]

The potential for shoddy reuse was almost endless. Jubb was fulsome in his praise for such frugality: 'Not a single thing belonging [to] the rag and shoddy system is valueless, or useless; there are no accumulations of mountains of debris to take up room, or disfigure the landscape.' Seams and other parts of the ragged fabrics that could not be shoddied were spread on arable farming land, especially the Kentish hop fields.[49] The crenate of ammonia had another, more glamorous use. If stewed up with pearl ash, cattle horns and hooves, old iron hoops, animal blood, leather clippings and broken horseshoes, it could go towards the manufacture of an especially rich pigment – Prussian blue.[50] Valuable dyes such as cochineal were isolated and removed from the wool at the start of the shoddy process.[51] The extracted oil and grease from the shoddy process (called 'creash') was turned into a powerful artificial manure, and the suds were used to manufacture fertiliser. In 1857 the *Leeds Times* asserted that 'manufacturers are making all necessary inquiries into the matter' of waste recycling, to prevent a considerable proportion of the 'oil, soap, and natural grease of the wool' from winding up in rivers or on dunghills.[52] Mungo waste comprised fibres too small to work into cloth and was ground up and dyed to make flock wallpaper – the quintessential item of décor in the period. All elements of this paper – the paper base, the glue, the dye and the flock – originated in recycling processes.[53]

Recycling innovations of the time both responded to and affected Victorian tastes. Reuse of materials permitted a democratisation of stuff: more people could afford items that utilised waste materials; flock wallpaper made from virgin wool would have been prohibitively expensive. No doubt many of the well-to-do Victorians who papered over cracks with flock were oblivious to this reuse of rubbish. Other wastes of the wool industry fertilised the hops that helped Allsopp's to prosper, which in turn allowed Griess to develop his dyes. It is a characteristic of Yorkshire folk that they don't take waste lightly. In the history of cloth waste reclamation, most of which is located in Yorkshire, the reuse was no simple single process. Note the pride, in the descriptions of shoddying, about the ways the processes made use of everything: unusable rags, waste oil and recaptured suds were used as fertilisers; other parts went into dye-making, or wallpaper manufacture. Likewise, in Baron Masham's process for silk reclamation, waste that could not be incorporated was turned into manure. Here we inch nearer to the closed-loop processes so beloved by the modern green movement – those extra steps of reuse in the past are important ones for us to learn from. None of the Victorian processes of recycling can be described as an entirely 'closed loop', but by-products from various industries were fuelling other industrial innovations, increasing the variety of affordable goods and reducing waste to boot.[54]

Combinations of trades

A commentator in 1868 described the reutilisation of waste items as having been half-hearted until the mid-century. But rising urban populations increased waste, and increased waste gave rise to more inventive recycling practices.[55] The reuse value of pretty much everything was considered: sewers were scoured for remnants, sewage became fertiliser. Inventions and fashions provided new markets for old reuses. 'Cat-gut' (usually made from goat or lamb intestines), long used to make rope and strings for musical instruments, was marketed in 1888 to bookbinders, surgeons, racket-makers and people making bands for watches, clocks, drill bows, lathes and steam engines.[56] Gedge & Co. turned oleaginous by-products of the meat and leather industries into French polish and varnish, stains, gums and glues 'specially for the piano and music trades'.[57]

Manufactures increased and factories created wastes; secondary establishments grew up to use those as by-products, 'the one digesting the other's refuse', as innovation and demand created a virtuous circle.[58] It made good economic sense to combine trades, either to reduce movement of bulky or heavy waste from one establishment to another or to bunch all bad-smelling industries together into zones far away from posher residential areas. Combined industries became even more common in the second half of the nineteenth century. So keen were the Victorians to ensure that reusables were brought to use that industrial schemes to skim soap suds from watercourses were developed. In Bristol, where many companies combined soap- and candle-making, Matthews & Co. added to this oil-boiling and the manufacture of railway grease and black varnish.[59] In Bradford, home of the worsted wool trade, the soft soap suds that ran off from the wools were collected in tanks and treated with sulphuric acid, making the greasy matter rise to the surface in a bulk known locally as 'magma'. If distilled, this grease could be reused to lubricate cogs in mills. The harder parts made night-lights, and some of the liquid was recycled to wash wool. Some mill owners made a tidy profit from such practices.[60]

Soap manufacturers themselves had long made use of the waste products from other trades. Some soaps were made from reused fats and reused alkalis, including grease extracted from animal fat and bones, rancid butter and dripping collected from kitchens, plus potash obtained from the ashes of woody plant matter.[61] Other fats for soap were redeemed from soapy water. It was discovered that bone tallow made a good firm soap but had an unpleasant smell. Soap made from human fat was even less satisfactory: it turned yellow and dried out quickly, although you wouldn't want to think too hard about the origins of that discovery.[62] We have already seen how glycerine was a by-product of soap and candle manufactures. The better the quality of the soap, the poorer the glycerine; inferior soap left superior glycerine. Until the 1880s, much glycerine was left to run away wasted, but in a process developed by Messrs Gossage in Widnes, Cheshire, described in 1888 as 'the largest soap-works in the world', the soap lyes were boiled and distilled to isolate glycerine.[63] Some soap manufacture combined utilisation of waste fats with other waste reuse. William Valentine Wright's 'Coal Tar Soap' was marketed in the 1860s as a 'skin soap', intended to protect the body from infectious diseases. Wright himself died in 1877, having 'caught a cold in the face', which developed into a strepto-coccal skin infection.[64]

The Hammersmith Bridge Works was a huge establishment, combining soap-making, sugar-refining, charcoal-burning, gas-making and other trades.[65] In Manchester, Vickers & Sons combined bone-boiling with soap-making, plus manure and sulph-uric acid manufactures.[66] Combining trades resulted in some strange occupational titles: in 1859 George Bagshaw of Norwich described himself as a rag merchant, bone crusher, artificial manure manufacturer, poulterer and herring curer. A decade later, together with a partner, he also sold 'half-inch bones, bone dust, Peruvian guano, nitrate of soda, and other artificial manures'.[67] This reflects the fact that the trade most commonly combined with others was artificial manure-making. Bickell, Deagon & Co. sold fertilisers from their Devon manure works in South Molton and their Clappery Bone Mills in Chittlehampton. The latter produced 'genuine' bone dust, plus superphosphate and also Peruvian guano, pretty much fulfilling all local wheat-manuring requirements.[68] Guano, the solidified excrement deposited by Pacific seabirds, including 'cormorants, flamingos, [and] cranes', had rapidly became regarded as the fertiliser *par excellence*.[69] Artificial manure makers argued that their product was better than guano. Recipes included various residues. One mixed the residue from lard-making dross with street dirt and gypsum.[70] In Bermondsey, Thomas Elliott sold fertiliser made from spoiled fish and sugar scum (pressed residues left after the sugar had been filtered).[71] In their efforts to improve on nature, some tradesmen found the temptation to dilute, adulterate or cut corners too strong to resist. An analysis of artificial manures in 1878 found that some raw bone-dust samples were actually a mixture of 'raw and steamed, or glue-makers' refuse bones'.[72]

Glue-making waste was called 'scutch' or 'scotch', and this was the residue left in the pans after boiling. In 1840 Justus Liebig pondered a use for scutch since 'many hundreds of tons' were discarded annually. He wondered if it might be used as a fertiliser.[73] Shortly afterwards it was. Manufacturers of scutch-based manure did not endear themselves to their unfortunate neighbours, any more than ordinary manure-makers.[74] In 1857 a Bermondsey bone-boiling and manure-making establishment owned by William Burgess caused such a nuisance that it was taken to court. Much of the questioning in the case, heard at the Old Bailey, steered witnesses for the prosecution to testify that smells from Burgess's works were identifiable amid a particularly stenchworthy smellscape, produced by emanations from a varnish works, tanneries, a fellmonger and another manure manufacturer. Neighbours fell sick and were dizzied, developed insomnia and headaches, became depressed and watched plants wither and die. Sarah Feaver described the stench as being so disgusting it was worse than tallow-burning, itself notoriously stinky. One medical officer described breathing problems brought on by an 'excessively offensive' vapour, and a metallic taste was detectable on approaching the site. Witnesses saw all sorts of things taken onto the premises: pieces for making glue (skin, ears, noses, tails and other animal parts); several tons of decomposed wet bones; piles of dry bones; plus 'a great heap of gas finings'. Fat was spied in the bottom of a boiler in which bones had been boiled: this was identified as the source of an unpleasant 'fatty warmish smell'. The place was awash with sulphuric acid. Oblique references were made to previous manure manufacture which involved fish offal (a notoriously foul-smelling manure), and one neighbour claimed to have seen nine horse legs piled up one day, bought in from marine store dealers.

Witnesses for the defence (many were part of the workforce) underplayed the nuisance, claiming that the dry bones did not smell, that wet bones were processed quickly and that glue pieces arrived ready limed, and 'sweet'. Contradictions appear amid the statements: the company foreman claimed that 'we never made glue – Mr Burgess styles himself a glue and manure manufacturer, but we have never made glue on those premises'. Margaret Cannon, a former employee and a site neighbour, implied that some witnesses had been bribed to give false statements, and claimed that glue *was* manufactured on the premises. Burgess was found guilty on one count of nuisance, and ordered to ensure that his extractors were 'perfectly air-tight'.[75]

Despite denials, bones were probably boiled on the premises. Edward Ballard's report identified bone-boiling as 'often the cause of considerable annoyance to the neighbours'.[76] Manure, or 'superphosphate of lime', was certainly made on-site, combining coprolites (fossilised dung which was ground on site), ground and calcinated bones (bought in from sugar refiners and other sources) and the debris from glue vats (scotch).[77] It is likely that hundreds of tons of gas finings were mixed into the artificial manures, despite the improbable claim that 'nothing is done with it at all, it merely lies there'. If Cannon was correct, one of the nobbled witnesses may have been James Powell, a neighbouring 'manufacturing and analytical chemist', who tried to take the blame for the nuisance himself. Powell claimed that a one-off unspecified and vaguely recalled 'accident' on his premises, involving the production of artificial manure, created all the stink. Powell made manure from bones plus night soil (human excrement) 'and other offensive matter'. Oddly, he also boasted about his own deodorising skills from the stand: 'I am a practical chemist', he trumpeted, claiming to have fooled a hapless policeman into taking powdered night soil as snuff.[78]

As well as his controversial Bermondsey establishment, William Burgess also had various other reuse interests across the country, including in his home town of Chelmsford, where in 1856 the town council rejected his proposal to process domestic sewage into manure.[79] In his *Muck Manual*, Frederick Falkner described crude raw night soil as being too potent to apply directly to the soil. Mixing into it charred or dried peat, or finely sifted coal-ashes, rendered it 'inoffensive'.[80] Questions around the health implications of sewerage, together with the cost of desiccation and transportation, made the use of human waste unviable in cities (although not in the countryside). The development of less execrable alternatives and also a widening network of sewage facilities during the nineteenth century triggered a decline in the use of human waste.[81] Liebig argued that each 'town and farm might thus supply itself with the manure'. He even envisioned that desiccated human waste could be transported over 'great distances'.[82] But this did not happen.

Night soil was increasingly seen as a burden to those contracted to sort the waste.[83] By the time Edwin Chadwick gathered evidence for his inquiry into the sanitary condition of Great Britain in 1842, it could be stated that 'no refuse in London pays half the expense of removal by cartage', except ashes, leys and 'a few other inconsiderable exceptions'. John Darke, a London contractor, stated: 'I have given away thousands of loads of night-soil; as we have no means of disposing it, we know not what to do with it.' (Sidney Dark, a descendant of Darke's, whom we met in Chapter 3, said John added the 'e' to his surname because he 'thought it looked better on the water carts'.) Only in the East End, where there was ease of access to the market gardens, could a profit be made on carting the waste from cesspits to fields. It was noted that there 'is no filth that now, as a general rule, will pay the expense of collection and removal by cart, except the ashes from the houses and the soap lees from the soap-boilers'. A retired contractor noted that police regulations restricting the times that cesspools could be emptied and their contents transported were a further hindrance to reuse: 'Some nightmen have paid 6d per load [...] for the liberty of depositing it.'[84] In competition with other fertilisers less bothersome to transport and process, night soil lost.

For a brief period in the mid-nineteenth century some cesspool waste was fashioned into 'poudrettes', bricks of concentrated dried manure. English farmers did not embrace poudrettes. Much dried human waste was exported for use on sugar plantations 'packed up in the returned sugar hogsheads, and sent to the West Indies', but it was also favoured by the Scots.[85] Adverts stressed the natural content and national production. At this time the quandary about what to do with the waste from the sewers was also considered. Many processes were devised – some with the aim of just getting rid of the matter, others with the hope that valuable substances within it could be separated out.[86] During the 'Great Stink' of 1858, sewerage was described as 'a villainous hell broth'.[87] Lamenting the waste of sewage waste, Crookes likened it to leaving three million loaves to float down the Thames every day.[88] Many assumed that a neat solution would be devised – with human ingenuity, cycles of nature would absorb the waste in a closed loop – and that processes might provide employment and yield profits.[89] Human waste could be filtered by mechanical or chemical purification, or it could be used to irrigate land. Various schemes were thought up. Disinfectants, such as carbolic acid (phenol) made from distilled coal tar (itself a product of reuse), offered a cheap and effective method to cloak the smell, halt putrefaction and kill tapeworms, but they did not purify or cause the solid matter to precipitate. Smell remained a problem.[90] Lewis Angell, a civil engineer and sanitary inspector, acknowledged some limited successes in processing: 'to dilute two-pennyworth of manure with a ton of water, and seek to recover it again in solid form, is', he noted, 'a costly game of hide and seek.' The manure produced was less effective than guano and the process still left polluted water.[91] The problem was that the farmers were happy with guano, the use of which could be better regulated, and so they became less invested in harnessing human waste as the century progressed.

Marine store dealers

Although not all of them had nine horse legs to hand, Victorian 'marine store dealers' would buy and sell pretty much anything – but they were supposed to leave fat, grease and drippings to the 'rag and bottle' men. Marine store dealers had once obtained waste from ships' stewards, and supplied items needed on board ships, hence the name, but they had morphed during the Victorian period into general waste dealerships. In *London Labour and the London Poor* (1851) Henry Mayhew detailed items to be found in marine stores: they included 'backgammon boards, bird cages, Dutch clocks'. Quick cash could be made from these dealers, who offered an alternative to the pawnshop. One dealer claimed to have known people to 'come with such things as teapots, and old hair mattresses and flock beds'.[92]

The survival of a ledger made in the 1820s throws some light on the personality and activities of James Andrews, a Gloucester marine store dealer. Watermarks on its pages give a date for the production of the paper: 1826. Andrews first used the book to record his warehouseman accounts between 1835 and 1841. After this, a repagination indicates a new use: Andrews records the history of Gloucester's civic government, including lists of by-laws. By 1847 Andrews had turned the ledger around and started using it to make lists of items of stock for his marine store business: bones; woollen, coloured, 'seconds' and white rags; cast and heavy iron; lead; 'mettle'; junk; 'tar pauling'; horse hoofs; glass; horsehair; bottles; legs and hoofs; skins. After a few pages he turns his attention to snippets gleaned from works of local history, which he records in a beautiful hand, starting yet again with his pagination. After forty-two pages of history, seventy sheets are left blank, interrupted by a draft will from 1880 (he died in 1883). The will identifies Andrews as a man of property. More blank pages follow, before the initial accounts for the warehousing business begin again.[93]

41. Magic lantern slide showing James Andrews's premises.
Author's collection, photographed by Toby Sleigh-Johnson.

Andrews represents the respectable end of the trade. There were much less reputable dealers.

Marine store dealers waited in, passively, for people to bring things to them. James Hamilton Fyfe, who authored various works on commerce, drew a contrast between marine store dealers and the materials they collected, which were 'every day more important as elements in our national industry'. He had no confidence in the 'present provision for collecting' waste and promoted the Rag Collection Brigade (boys who collected door-to-door) as an alternative.[94] 'The mode in which rags are collected for the use of the paper maker', explained Benjamin Lambert in 1864, 'is one of the most roundabout conceivable.' The rags are collected by 'itinerant vendors' who exchange 'hearth stones, glass ornaments, crockery' and other goods for rags and bones, as well as 'soleless boots, and crownless hats'. The value of these 'miscellaneous gleanings' was calculated by some mysterious method and then conveyed to the rag shop owners, who stored them until they had sufficient bulk to sell them to paper merchants, who then supplied the paper mills.[95]

London.   The Old Curiosity Shop.

42. Postcard showing the Old Curiosity Shop under
Horace Poole's management. Author's collection.

Some waste dealers were more specialised than general marine
storekeepers. Although the name of 'Old Curiosity Shop', conspic-
uously painted onto a building on London's Portsmouth Street,
was nothing more than a promotional ruse to connect it spuriously
with a Dickens tale (first serialised in *Master Humphrey's Clock*), the
establishment did deal in books and ephemera, as well as acting as
a wastepaper merchant.[96] Taking over the establishment in the late
1870s, Horace Poole appears to have abandoned his family around
the same time, settling instead with a new partner.[97] He was making
a new start in more ways than one. While managing another establishment in 1871, Poole had been involved in a scam whereby an
employee of the Graphotyping Company had sold off current stock
of *The Children's Treasure* and *Infant's Delight* as waste. Poole claimed
that with so many editions of this work, it was difficult for him to
assess whether they had been remaindered or not, but I suspect he
knew.[98]

Poole wasn't the only waste dealer whose enterprise invited disreputable activity. In 1861 the *Devizes & Wiltshire Gazette* reported
that hundreds of pounds of stolen lead had been recovered from

John Hurd, a marine store dealer in Devizes, having been torn from roofs by youths of the town. Its author wrote: 'it is quite time that decided steps were taken to check the temptations to pilfering which the marine store nuisance offers.'[99] That same year a stolen dress was recovered from a different business in the town, having been lifted from a washing line.[100] Hamilton Fyfe wrote in 1862 that marine stores were, 'in too many cases, only disguised receiving houses for stolen property, and respectable people naturally dislike to countenance them'.[101] Marine stores, and other places storing second-hand goods for resale, came to be seen as nuisances for various reasons and were scrutinised by local authorities. Medical officers feared that rags stored on such premises might communicate contagious diseases, and vermin infestations were suspected. The trade faced increasing regulation: from the mid-nineteenth century dealers were required to obtain a licence, to register with the police, to submit to stock inspections, to paint their names above their doors and to keep records of all transactions.[102] An inspection of Newmarket marine stores in 1869 noted that skins and bones from animals rendered them 'disgusting places, and very injurious to health'.[103]

🗑 🗑 🗑

Some of the printer's ink spent informing Victorian readers about contemporary recycling was itself the product of reuse. In 1889 *The Times* detailed two forms of reuse centring on cotton waste used to clean mechanical works and railways. Obtained from waste merchants 'at a low price', it was washed, boiled and returned to the merchants. From his Bermondsey works, the engineer Charles Bastand perfected a method of reusing the grease from the dirtied cotton, turning the bulk of it into a varnish to which pigments were added to make printer's ink.[104] This ink flowed more steadily after the Great Exhibition of 1851, and commentators show that the Victorian period really was a wonderland of reworkings. Commercial and scientific enterprises multiplied. Women provided much of the initial raw work of recycling; men garnered most of the praise.

The Victorians placed a value on everything. Toplis & Harding, London auctioneers, had five tons of sugar 'slightly injured' by fire ready for viewing in late May 1859.[105] No doubt the casks of molten sugar salvaged by Captain Faulkner's team the year before had already hardened, cooled and been repurposed by then. Marine store dealers and auctioneers all relied on one thing: that people would have surplus stuff they could sell on. Both mediated between individuals and business, but auctioneers (who did not store their lots for long) were seen as much less of a threat to the health and prosperity of their neighbourhood. Marine stores became recycling centres of ill repute, hubs for collection and reuse, but also camouflaged reprobation. Many people whose work relied on the resale or reuse of materials were seen as nuisances, especially if they combined noxious trades. In Thetford, Norfolk, in June 1866, a shed located on the Croxton Road set on fire. At the time it was in the occupation of Joseph Emms, a marine store dealer, who was using it to boil bones. There was no fire salvage corps to assist, and not even the fire brigade came to extinguish the flames, so Emms's shed full of bones burned down.[106] Some marine store dealers made terrible neighbours.

8

Rebaptisms of Fire (1830s–1840s)

On 16 January 1836, accompanied by some of his men, Inspector Evans from the Phoenix Street police station in London's Somers Town climbed up a ladder and through the window of 23 Hertford Street. Picking through the filthy house, the policemen dodged shards of glass from broken bottles and stumbled over piles of scrap metal. Rats scuttled into dark corners. Eventually they found the naked body of Henry Kirkman, a notorious marine store dealer. Kirkman lay on a bed surrounded by thousands of old newspapers, his head reclined on his right hand. His left cheek and nose were gnawed away, but his pockets were full of cash. This 'most eccentric character', the son of Mary and Joseph, was in his mid-thirties. Suicide was supposed. His body was taken to the St Pancras workhouse to await the coroner. Neighbours had alerted the police after not having seen or heard Kirkman for a few days, and his doors and shutters remained closed.[1]

Kirkman was well known. His character and work – melting down ill-gotten metals – made him such a difficult neighbour that the houses either side of his had been unoccupied since 1832. The Middlesex jurors had found him guilty of ringing bells and firing guns at unreasonable hours, being drunk and disorderly, mimicking the sounds of a woman in labour, singing blasphemous and obscene songs, and exposing himself 'in a perfect state of nudity'. Mr Burden, 'a respectable solicitor' for the prosecution in one of the numerous cases against Kirkman, said he believed Kirkman was mad, and that he terrorised his neighbourhood, making the 'inhabitants of the Camden estate' determined 'to get rid of him'.[2]

Kirkman's neighbours might have had a more pleasant life had he followed in his family trade – they owned a famous company of pianoforte makers. Instead he had become a marine store dealer of the worst order. In November 1825, when he was in his early twenties and working in Chapel Street, Camden, he was charged with possession of stolen sheet lead, copper and brass. In his defence, Kirkman wheeled out 'a host of marine store people, of whom he said he had purchased various parts of the property', while a policeman claimed to have seen him steal the metal on Carnaby Market. It was recorded that 'not one of the whole set [of these marine store dealers] kept books of any sort'. Pounds of brass and copper could not be accounted for, and for this Kirkman was fined. Among the items in his possession were 'an extraordinary large dog collar, big enough to fit the neck of a calf', engraved 'Mr Stewart, Montpellier-row, Twickenham', and also breastplates from the artillery service. At the summing up there was a general admonition for all marine store dealers to beware and to attend to their book-keeping.[3] On another occasion in 1825 Kirkman could not provide a satisfactory account of why he had a 'quantity of pewter [...] about the size of a large frying-pan', which was suspected to have derived from a collection of publicans' pewter pots, recently melted down. Mr Bennett at the Globe Tavern had found himself short of seven dozen such pots.[4] The police appear to have developed a modicum of wary respect for this peculiar man, partly because he was a skilled chemist 'and by some process was enabled to remove the gold from the gilt buttons, and drove very hard bargains in purchasing metal'.[5]

In October 1826, while awaiting trial for possession of stolen metal, Kirkman killed a boy running an errand for his mother 'with a candle in his hand'. James Sullivan was knocked down by Kirkman's cart, travelling at a 'rapid pace' down East Street, Marylebone. At the murder trial the incident was chalked up as accidental death.[6] Kirkman had previously mown down a young woman and her child while carting 'wilfully and wantonly' down Broad Street, Golden Square (where his father had a music shop).[7] Later he ran over a police officer.[8] Back in court in 1830, Kirkman was convicted for 'flogging a horse most unmercifully in College Street' for ten or fifteen minutes, using a copper wire.[9] Kirkman's marine store was located on this same street. In 1832 he was in court again and fined £5 for having a large quantity of lead and other metal 'suspected to have been stolen from some East Indiaman'.[10] A couple of months later he was again fined for being unable to account for his supply of metals – this time copper, lead, brass and more pewter pots filched from taverns. On this occasion he was also 'held to bail for assaulting a policeman'.[11] Another two months on, and Kirkman was in court yet again, this time for having 'several hundred weight of old and new serviceable metal, supposed to have been stolen or unlawfully procured'. This was Kirkman's fourth conviction for possession. A search of his premises produced 'several valuable engraved copperplates, and 500lbs of brass', new hinges, printer's type, tongue-less bells, coach beading, metal cocks, copper scales, 'a military officer's epaulette', 'two cudgaree pots (used by Lascars on board ship)', a wash copper and 70 lb. of plated and brass buttons.[12]

Kirkman's Heathcliffian exploits continued, and the press began to attend to this one-man menace. With reports on each fresh court case, details of his background and demeanour emerged.[13] He was 'a small man, of sallow complexion', 'with raven locks shaking iron filings instead of ambrosia' that were seen to be 'unshorn, and falling upon his shoulders in beautiful disarray'. He was said to be 'extremely frugal in his diet, and led, in every respect, the life of an anchorite, never suffering either male or female to enter his retreat'.[14] The press never wondered why the son of Joseph Kirkman, an eminent pianoforte maker, operated as a trader of the lowest kind – but I have. Henry came from a 'very unhappy family': his father sought legal separation from his mother, Mary, when Henry was about eight years old, on account of her violent cruelty and jealous rages. Mary accused Joseph of having an affair with a servant described by the piano company foreman as 'a woman of a very disgusting appearance'; Mary verbally insulted Joseph, calling him a 'German brute'; she locked him out, bashed him with a pewter pot (unmelted), flung 'a spoonful of child's pap' at him, scratched his face and burned it with a lit tallow candle. She damaged one of her husband's grand pianos by banging hard on the keys. Henry's mother was 'not mistress of her own passions'.[15] This context created a sad beast. The coroner decided that Kirkman had died as a delayed result of head injuries he received accidentally as a result of falling out of his cart 'in a state of intoxication'.[16]

Of all the many things that Kirkman's story shows, it particularly highlights the difficulty experienced by prosecutors in trying to prove the origin of stolen metals. These could be easily liquidated, obscuring all details of ownership and provenance. Kirkman's shenanigans also reveal that it was not always easy to identify all stolen items, and give some indication as to the tendencies and social standing of the dealers. Kirkman was not the only dealer to fence stolen goods. It did not take a Sherlock Holmes to detect the crime of one 'Moriarty' in 1825, but proving it was another story. Having been observed stealing a pewter quart pot from the Fox & Peacock on Gray's Inn Lane, John Moriarty was found to have taken it to a nearby marine store, run by Henry and Morris Worms. The pub landlord saw his pot, about to be melted down: 'it had my

43. Victorian papier-mâché snuffbox. Author's collection,
photographed by Toby Sleigh-Johnson.

name and sign on it.' That was easy enough, but two further pewter pots were also missing. Near by in the store, a pan was found that contained recently melted pewter. Despite suspicions and circumstance, nothing could be proven.[17]

While Kirkman was melting down metals for his own financial gain, others looked to reuse for more patriotic purposes. The end of the Georgian period, through the seven years of William IV, the 'Sailor King' (1830–37), and into the early Victorian era, many worked on making naval vessels more secure, experimenting with coal tar and other substances made from waste products to protect their hulls. Some became interested in recycling to make layers to protect bodies from rain and bodysnatchers, to protect roads from feet and hooves, to protect ships from cannon shot or to form cleanable surfaces for houses. Many of these processes involved heating substances, to make composites, or to mould them into shape. Heat, inevitably, and especially when combined with solvents, created the hazard of fire. The story of recycling is bound up with the fear of conflagration.

Stinkle stinkle little tar

We have seen how coal tar gave colour and pure white saccharine to the Victorians. The gas that created this waste was first utilised in the cotton mills, around the time of the Battle of Trafalgar, and then for street lighting by the time of the Battle of Waterloo.[18] The waste that caused the most nuisance then – smoke – was not reused. In December 1813 William Sturt, of London's Fleet Street, bothered his neighbours with the 'noisome offensive and stinking smoke smells' issuing from coal fires making gas.[19] Neighbours of the Westminster Gas Light and Coke Company works in the 1820s complained of fumes damaging vegetation, dirtying clothes, tarnishing metal and causing illness. Gasification wastes polluted soil and groundwater, leaving a lasting legacy.[20]

Coal tar was one of three key wastes from the early nineteenth-century chemical process of coal gasification, the others being coke and ammoniacal liquor.[21] Coke could be used in blast furnaces and by metalworkers, but was found to be problematic as a railway fuel, containing too little carbon and too much sulphur, 'which is very destructive to the metal of the boiler'. Early on, it was 'completely given up on by the Liverpool Railway, notwithstanding its low price'.[22] Consequently, gas coke became a poor man's fuel. In the harsh winter of 1829 'benevolent individuals' distributed it to the poor of Bristol.[23] In the north, where coal was cheap and plentiful, coke was little favoured. In 1844 the *Kendal Mercury* sneered that cheap gas coke was a 'complete drug' to Londoners, while being 'comparatively valueless' to its own readers.[24]

Finding reuses for 'dolefully foetid and repulsive' ammoniacal liquor was more problematic. George Dodd described it as 'a most unloveable compound, which the gas-makers must get rid of, whether it has commercial value or not'.[25] Some was used in dye-works. Read Holliday set up a workshop within 'easy carting distance' of the Huddersfield Registered Gas Light Company in 1830, from whence he produced ammonia from waste ammoniacal liquor. He later used the ammonia in the production of Prussian blue dye.[26] The liquor could also be turned into salt, often known as 'ammonia salts', which were used by chemists, druggists and tinplate workers, as well as for galvanising iron and making bleach. According to Clara Matéaux, some was shipped to Russia, where it was 'used by the peasants instead of common salt'.[27]

Coal tar was derived from a destructive distillation of coal. Industry was awash with the stuff. 'There are several hundred thousand gallons of coal tar on hand at one of the gas works', wrote a London merchant. The cheapness of the product – a penny a gallon in 1824 – and the increasingly abundant supply of it created the impetus for reuse.[28] Previously it had been cast into rivers, forming 'ghastly blue patches known as "blue billy"'.[29] Coal tar arrived into a world that knew only tar derived from pine. Tar from gasification was a sludgy waste that looked like tar and acted like tar, and so it was called coal tar. People who used tar did not always distinguish the type. A similar shifting lack of precision attached to contemporary use of the words 'pitch' (tar deprived of oil through distillation) and 'naphtha' (some of which was a distillation of natural bitumen, and some of which was derived from coal tar).[30]

Driven by pragmatic and financial considerations, people found new uses for coal tar, and some supplies were harnessed by paint and varnish makers. Darker than pine tar, coal tar needed no lamp-black to make a dark coating. It went on 'smoother, with a finer skin and better gloss than common tar', which recommended it for painting coaches, for damp-proofing houses, for sign-painting and for protecting tree supports and metal tools.[31] In 1830 a Swiss chemist asserted that coal tar 'is now also used pretty extensively as paint, without any admixture, and without heat'. Rebutting claims that coal tar rotted the wood it was supposed to protect, the chemist could not resist noting the chief drawback: 'Its offensive smell'.[32] Japanned goods, such as those made by Jennens and Bettridge in imitation of black Asian lacquerware, were finished with a heat treatment using a varnish mix which could contain coal tars. Paints with a coal tar content were the cheapest and the nastiest on the market. Coal tar varnish was a rough and ready interloper, lacking the subtlety of varnishes made from linseed and turpentine. The customers who purchased japanned items were the wealthy pur-chasers of luxury; it made no business sense to draw any attention to the inclusion of coal tar as a component in the process, and the smell of the tar had worn off by the time the product reached the consumer. This makes any real assessment as to the extent of its use in japanning an impossibility.[33] In 1811 Benjamin Cook, a brass founder who made bedsteads, jewellery, toys and gas-lighting para-phernalia, was cautiously positive about coal tar, which could protect gates and fences; he made use of the tar, and, despite allud-ing to the smell of this product throughout, remarked that 'there appears to be no manner of difference' between it and a turpentine-based varnish.[34] Except there was a manner of difference: coal tar stank.

All tarred with the same brush

On 23 March 1822 the Navy Board invited a group of politicians, naval officers and scientists to watch men try to set fire to a 74-gun third-rate ship of the line on the Thameside at Millwall, Poplar. Amid the gathered throng were three men: Henry Constantine Jennings and Joseph Hume (who had requested the ship be set on fire in the first place); they were joined by a Mr Good, probably a shipbuilder, described as a 'dry-rot doctor'. The son of a shipmaster,

Hume had trained as a doctor. When he became a Fellow of the Royal Society in 1818, he was described as being an expert in chemistry.[35] Senior to his friend Jennings by about a decade, Hume was the radical MP for Aberdeen Burghs. Jennings was an 'energetic merchant' and 'practical chemist'. Nine years previously Jennings had designed a lifeboat made from calico stuffed with waste 'cork-shaving, cutting, or old cork', milled into pieces that were bigger than sawdust but smaller than peas.[36] Internal Admiralty correspondence refers to him as 'that impudent fellow Jennings'.[37] Jennings's current obsession was the use of coal tar by the navy. He was not convinced of its safety. All their expertise and knowledge was now turned to the question of whether or not a ship full of coal tar could be set alight. Each came in full expectation of watching HMS *Russell* blaze.

The Millwall experiment occurred as part of an attempt to rebuff claims that coal tar was dangerous to the fabric of ships and was likely to combust in hot weather, and its 'explosibility by red-hot shot, the flash of a pistol or the contact of flame'. The Navy had been injecting coal tar into the interstices of ships' bottoms since the end of the eighteenth century, and Sir Robert Sepping, one of the Surveyors of the Navy, who was also attending the spectacle, hoped to increase the use of coal tar within the fleet as a defence against dry rot. Meanwhile, Mr Good was trying to peddle a non-coal tar nostrum of his own making.[38] Neither candle nor pistol set the *Russell* ablaze, despite her having recently been injected with coal tar. However hard he tried, Good could not make fire 'by any fair or ordinary means'. Stubbornly unconvinced, he argued that the vapours could still be injurious to the health of seamen. Somewhat sheepishly, Jennings recalled the experiment in the *Technical Repository*: concerns had been dismissed 'by the most irrefutable and satisfactory evidence'.[39]

The experiment was merely for show and reassurance: naval weight had already been thrown fully behind coal tar as a counter to dry rot. Of 131 Royal Navy ships launched between 1815 and 1822, fifty had been injected with coal tar.[40] There was already a ship afloat that was 'most fully saturated' with it – the *Owen Glendower*. Her commander stated that it 'never caused a headache to that ship's company'.[41] In 1817 Navy Board officers investigated the purchase of coal tar in barrels similar to those in which wood tar was transported from Russia and Sweden, probably to camouflage the smellier alternative. An order for 10,000 tons was placed. However, the Navy looked to other dry rot cures after 1822, and by 1824, with the Navy no longer saturating vessels in coal tar, attempts were made to sell coal-tar-saturated rope for naval use. Described as 'having a pernicious effect on the workmen', these were rejected on account of the unpleasant odour.[42]

Mr Good's worst fears were realised in 1840, when HMS *Talavera* and HMS *Imogene* were destroyed by fires blamed on coal tar while moored in Devonport dockyard. A wooden 'bin' was located under a wooden structure which covered the *Talavera*, into which 'oakum, tallow, waste of paint, old canvass, sawdust, chips, &c.' had been disposed. This 'large mass of filth' generated 'a high degree of heat; and spontaneous combustion was the result'.[43] The fire then spread to the *Imogene*, and also to items stored in the shed: mastheads from broken-up ships; the flag under which Nelson fought at Trafalgar; and a capstan of the sunken *Royal George*, which had been recently recovered off Spithead.[44] The *Royal George* bobs tragically back into view in the next chapter.

Naval interest in coal tar products waned; the Navy was never going to relieve the gas-makers of all their wastes. Most non-naval purchasers of waste coal tar distilled it into other products, such as timber preservatives (later called creosote). In her study of gas waste consumers from the period, Mary Mills includes a study of Webster Flockton's tar distillery in Bermondsey.[45] Thomas Allen Britton, a structural surveyor, described the creosoting process perfected by John Bethell of Greenwich using oil of tar, after cruder attempts by Flockton, and more famously by John Kyan. Bethell was rumoured to have been inspired by Egyptian mummification. Smell remained a problem. Britton described creosoted wood as being more durable and less liable to 'the attack of the worms' but remarked that 'the disagreeable odour from timber so treated renders it objectionable for being used in the building of dwelling-houses'.[46]

A major repurposer of coal tar waste was John Henry Cassell, a general tar distiller and 'maker of essential oils'. An émigré whose family fled the Napoleonic Wars, Cassell was initially destined for a seafaring life like his father, but settled instead to business in Poplar.[47] During the 1820s he operated a tar works and counted Thomas Hancock among his customers, supplying him with naphtha and also coal tar for mastication experiments.[48] Cassell established a tar and varnish works at Millwall. His elder son, John Henry Cassell junior, developed a sideline, maintaining oil-based street lamps. Cassell junior was bankrupt in November 1841, one disaster precipitated by another: in July, 'in consequence of the boiling over of an immense cauldron of varnish', the works sustained 'immense' damage.[49] Around 1828 Cassell senior had commenced experiments manufacturing artificial stone that erupted into the 'Patent Lava Stone Works'.[50] Lava Stone paved Vauxhall Bridge Road and made flooring for Giblett's slaughterhouse in Bayswater.[51] Although his advertising implied that Lava Stone was made from natural bitumen, the quantity of coal tar he purchased suggests otherwise.[52] Cassell's 'hot bituminous lava' could be used to plug stone railway sleepers, make foot pavements and indoor flooring, and line sewers, canal banks and fishponds. Ready-made items of Lava Stone were available, including kerbs, drains and pipes. Riffing on a paranoid zeitgeist, Cassell foresaw a generation of dead buried in coffins sealed with hot lava, in place of leaden liners. These could be viewed in the crypt of Brunswick Chapel on London's Upper Berkeley Street, where Cassell himself ended up in 1837, presumably sealed in such a coffin.[53] Importantly, lava-sealed coffins would protect the interred from being dug up again. (The public was still alarmed by reports of the infamous Edinburgh murderers William Burke and William Hare, who had been found guilty during Christmas 1828 of selling the bodies of their victims to a doctor, to be dissected.[54]) After Cassell's death the firm passed to his partner and son Edwin Edward Cassell, who traded still as John Henry Cassell & Son until 1842, according to terms laid down in Cassell's will.[55]

Specimen Lava Stone roads had been laid by 1835, including 'on the Kennington-road, near the Horn's Tavern'.[56] John Loudon McAdam had already introduced the process of making roads using compacted crushed granite in the 1820s. The old macadamised roads had become pitted and potholed; they were noisy; and they could be alternately puddly and dusty.[57] Sweepings lifted from macadamised roads incorporated 'so much of granite' that it was 'very little use indeed' as a manure.[58] Cassell's innovation smoothed over the bumps in the road by blanketing a boiling mixture of coal tar, gravel, sand and 'road stuff' over the dry substratum.[59] The success of Richard Tappin Claridge, a tea dealer who brought over Seyssel asphalting from the Continent, increased interest in Cassell's paving.[60] In 1835 Cassell published his *Treatise on Roads* (essentially just a sales catalogue); the publisher was Dean & Munday who were based on Threadneedle Street, the first street in London to get a full surface of asphalt, applied by the Val de Travers company in 1869.[61] To go full circle, the recycling of asphalt was first recorded in 1915, and the reuse of bituminous materials increased after the 1930s.[62]

Masticating

From coating timber to lining coffins and creating roads there were yet more uses to be found for coal tar. From the 1820s distilled coal tar was combined with rubber to make waterproof coats by the Glaswegian coat manufacturer Charles Macintosh. He had purchased waste tar from the Glasgow Gas Works since 1819.[63] Initially Macintosh used coal tar (and also ammoniacal liquor, another gasification waste) to create ammonia used in the manufacture of cudbear, a violet-red dye extracted from orchil lichen.[64] Previously Macintosh (like his father before him) had used human urine as the source of ammonia.[65] Now, in extracting ammonia from gasification wastes, Macintosh turned the coal tar into pitch, and serendipitously created a volatile oil: naphtha. He used this as a solvent to dissolve rubber, so it could be rendered into 'a waterproof varnish'. A sticky mixture of liquid rubber and naphtha pressed between layers of cloth became a waterproof fabric.[66] Eventually Macintosh went into partnership with three cotton-spinning brothers called Birley in Manchester, making waterproofed canvas for ships plus sailing gear, including 'Mackintoshes'.[67]

Meanwhile, in London, Thomas Hancock was also experimenting with rubber, partly so that he could sell waterproof clothes to passengers on his family's coaches. Piles of scrap rubber expanded as orders for elastic cuffs came in.[68] Hancock found that, when strips were cut from chunks of rubber, much waste remained, and so he set about finding ways to use up the remnants.[69] To this end, he invented a masticator to shred and gnaw waste rubber into a re-mouldable form, with coal tar added as a solvent. For over a decade he insisted his workers called this machine a 'pickle' to throw possible imitators off the scent, eschewing a patent for his invention. Obtaining a patent was risky, and 'on average a successful patent would yield no profit until the seventh or eighth year'.[70] Hancock's gamble paid off. By 1821, needing room for a bigger masticator, he set up on Goswell Mews, off Old Street.[71] Macintosh merged his business with Hancock's in 1830, giving the Scotsman access to rubber that had passed through the jaws of the masticator. Previously, Macintosh's products had been blighted by inconsistencies: they were thin; they dried irregularly (sometimes never); and they had a reputation for the 'excessive odour they evolved, a consequence of the impure naphtha then which contained a considerable amount of tar creosote &c.'. Mastication permitted the use of less solvent.[72] Hancock's men worked by candlelight. The Goswell Mews factory burned down in 1834. Hancock moved some of his production up to Manchester, and the reborn London premises continued under the supervision of his nephew. Four years later, the Manchester base also burned down. Men perished; machinery was damaged beyond repair; and 'reservoirs of solutions and solvents […] of course disappeared'.[73]

Over in the US, Charles Goodyear had heated rubber with sulphur so that it was no longer affected by temperature and returned to shape after being pulled. In 1842 Hancock obtained samples of Goodyear's product via his friend William Brockedon, an enterprising inventor. By examining these he realised that sulphur and heat were required. Not a man to look a gift horse in the mouth, Hancock successfully replicated this process, effectively undermining Goodyear, whose product was likewise without patent protection. Hancock did not acknowledge Goodyear as being the creator of these essential samples in his *Personal Narrative*. Brockedon, who also took an interest, added insult to injury

MASTICATING MACHINE.

44. A masticator, from Thomas Hancock *Personal Narrative* (1857),
between pp. 84 and 85. The rubber looks like massive sticks of liquorice.

by coining a name for the process: Goodyear had developed 'vulcanisation'.[74] Brockedon experimented with rubber to make
substitutes for corks, using a core of woollen felt or cotton. For a
while these promised much, and wine merchants were excited by
the prospect of replacing 'perforated dusty' corks, but the experiments failed because the rubber imparted a bad flavour and turned
stiff and unpliable when cold.[75]

Alexander Parkes (he of Parkesine fame, one of the earliest non-
synthetic plastics) was also involved. His method of cold
vulcanisation – waterproofing fabrics using rubber and carbon
disulphide – was patented in 1846.[76] It must have been with some
embarrassment that Parkes read about his own work in a copy of
Hancock's *Personal Narrative*, given by the author himself. Parkes
is there praised for managing to effect 'a change in rubber very
closely resembling vulcanisation' without heating, using only
chemical reactions. Effectively, Hancock implied that all three men
copied Goodyear's 'diminutive' samples; he was just best at it.[77]
Parkes inscribed in the margin of his copy of Hancock's memoirs:

'He must always have on his mind Goodyear, for to him alone is the invention due.' And next to details of the process of vulcanisation he noted: 'This discovery was made first by Goodyear in America [...] Justice must be done to Goodyear.'[78] We know how ironic these laments would prove, for John Hyatt did much the same to Parkes two decades later, when he profited from his experiments with Parkesine. Parkes had a point, but Hancock remains the hero of our rubber recycling story, even if he did try to rub Goodyear out of the story of vulcanisation. In order to recycle rubber thereafter it was first necessary to identify and separate pure rubber from the vulcanised form; the latter needed to be desulphurised before it could be reformed using a masticator. The resulting product was never of top quality. Hancock complained that it was 'not very tractable, and requires a good deal of trituration'.[79] Parkes secured the first patent for reusing vulcanised rubber waste in 1846, for a process that boiled waste rubber in muriate of lime until it was devulcanised and could be made to cohere under pressure.[80] There was a mixture of talents here: Parkes and Brockedon were artists and inventors, Hancock and Goodyear engineers and self-taught chemists. Between them they applied chemical and mechanical ingenuity to make rubber both more pleasant to wear and industrially reusable.

The adventures of kamptulicon and boulinikon

Corks that had stopped wine bottles were not reused on an industrial scale. Powdered virgin cork waste could be used in the vulcanisation of rubber, and sometimes to stuff mattresses or ships' fenders – bags stuffed with shavings of cork to prevent the sides of the vessel banging into harbour walls. These limited uses did not exhaust supply.[81] Consequently, when Henry Constantine Jennings made a cork-filled calico life raft in 1813, waste cork could be got cheaply.[82] Cork dust was used in a new way from the middle of the nineteenth century. Kneaded and rolled into a rubber base, and varnished with linseed oil, it made a quiet, damp-resistant and durable elastic floor cloth called kamptulicon.[83] In 1840 John Americus Fanshawe had mixed rubber with sand, sawdust and stone particles plus cork pulverised in a coffee mill to make flooring. [84] Four years later Elijah Galloway, a civil engineer, patented a mixture of rubber and cork to make a flooring product that could also be applied as lining in warships to mitigate timber splintering under fire. [85]

The Patent Elastic Pavement Company secured a licence to make and sell the products patented by Fanshawe and Galloway. [86] Advertisements were placed in *The Builder* in 1843, by the newly established 'Patent Elastic (Caoutchouc) Pavement Company', hoping to alert 'the Nobility and Gentry' to the invention of kamptulicon, highlighting particularly its application to stable floors. In July an advertisement drew attention to 'several extensive Government & Noblemen's orders' for stabling and other purposes. [87] Kamptulicon boasted a 'turf-like feel'. It suppressed noises made by horses that were 'confirmed kickers' of their partition walls, and even cured some of this naughty habit. [88] The product's appeal extended to institutions such as prisons and asylums, granaries and breweries (it being vermin-proof) and places of mass assembly. It was also useful for 'preventing the escape of unwholesome effluviae from the vaults of churches, or other buildings exposed to the actions of noxious gases'. [89]

Experiments were conducted at Woolwich dockyard by order of the Admiralty to ascertain whether kamptulicon would 'answer as a medium for causing the adhesion of copper sheathing to the bottoms of vessels built of iron, without nails of any kind'. This would eliminate the problem of nail corrosion. The new material was found to work 'very tenaciously'. The kamptuliconed coppering protected the outside of the ships, and the crews inside might also be protected from the effects of cannon shot by the application of a kamptulicon lining, 'which catches the splinters and closes over the holes, even against the pressure of water'. [90] The new lining would close up after shot had passed through, limiting water ingress, but any benefit was undone if the shot travelled through the vessel and exited the other side, as those holes were enlarged by the presence of the lining. [91] Still, by 1844 the stables of the commissioners of Woolwich dockyard were paved with kamptulicon, as were the Admiralty courtyard and the carriage entrance to Windsor Castle. [92] It was put to more modest seafaring uses too: kamptulicon was used on the royal yacht 'to deaden the sound on decks over her Majesty's head'.[93]

In 1844, perhaps in response to a tragedy in which a company draftsman drowned after falling off Waterloo Bridge, the Elastic Pavement Company launched an oak-and-kamptulicon lifeboat designed by company manager George Walter, a lieutenant in the Royal Marines.[94] The Belgian government put in an order for three boats.[95] Kamptulicon took off, despite having a name worthy of ridicule, as in a Tom Taylor farce performed in 1849, in which one character cried: 'My brain's bewildered – Oh, what *is* Kamptulicon?'[96] It was suspected that giving a new product an 'unpronounceable, inappropriate name, is equal to seventy-five per cent of the profits of a puffed-up trade. Cosmogulpelican over the door of a devil's dust garment establishment is equal to a hundred clergyman's certificates of honesty and skill.'[97] Parts of the Houses of Parliament and the new British Museum Library (now the old Reading Room), as well as the Treasury, the Guildhall and the Bankruptcy Court, were furnished with kamptulicon flooring.[98]

Flooring manufactories were at risk of fire. In November 1849 George Walter's 'Patent Kamptulicon (India Rubber) Works' on Greenwich High Road were engulfed by a blaze 'distinctly visible from all parts of London'. A newspaper reported on the 'fearful sight': 'the whole of the factory, from the base to the roof, was one intense sheet of flame'. The great quantity of flammable rubber created such a heat that it was with astonishment that firefighters discovered 150 sheets of kamptulicon 'laying in a loft [...] almost uninjured, except darkened by smoke'. Phoenix-like, kamptulicon's reputation arose from the inferno: kamptulicon was fireproof.[99] The company was back in business by June 1850. In partnership with Ebenezer Gough, George Walter presented his company's wares at the Great Exhibition.[100] After the Great Exhibition, Gough continued alone, until he formed a brief partnership with Matthias Boyce, the two men advertising themselves as 'original patentees and manufacturers'. In 1865 they hit back at their competitors in an advertisement that insisted that 'many complaints are made of KAMPTULICON manufactured by other houses'.[101] The ever energetic Alexander Parkes got involved too, obtaining a patent for 'improvements in the manufacture of compounds in the nature of kamptulicon' in 1866.[102]

Despite all this combined ingenuity, kamptulicon suffered manufacturing glitches: it was prone to buckle and 'cockle' with wear and warmth. Charles Tayler, a surgeon who set up Messrs Tayler, Tayler & Co., realised that the addition of gutta-percha (a rubbery substance made from tree latex) to the mix would solve these. By 1869 Tayler could invest in a big factory and develop new textures and designs. But the very success of kamptulicon created limits that ultimately prevented its widespread adoption. All of the enterprise inflated the price of waste rubber and cork.[103] Kamptulicon itself had 'given a value to this reuse, which it did not possess when it was only employed for the purpose of stuffing fendoffs'.[104] This undermined the viability of the kamptulicon industry. Meanwhile, comparable products challenged its ascendancy. In 1862 linoleum was patented by Frederick Walton; it subsequently became a popular, cheaper alternative. While kamptulicon was top-coated with linseed oil, linoleum was made by mixing oxidised linseed oil with sawdust, making the oil an integral ingredient rather than just a coating. When he described his new material in a lecture of 1862, Walton's audience members included Peter Lund Simmonds, one of the Birley brothers from Macintosh & Co. and one from the Hancock family. The manufacture of linoleum was also vulnerable to flame. Only a fortnight before Walton's presentation, a 'disastrous fire' had destroyed his works, leaving him with only a small clutch of 'poor' samples to pass around.[105] Faced with cheaper competition, kamptulicon gradually fell out of production. As they struggled on, some companies were dogged by disaster: the Deptford manufactories of Tresteil, Charles & Co. and also of Tayler, Tayler & Co. were destroyed by fire: one in 1864, the other six years later.[106] A kamptulicon factory at Royal Mills in Esher, Surrey, was two-thirds gutted by fire in 1877.[107]

Around this time yet another floor covering produced from waste enjoyed a fleeting fashion. Boulinikon was composed of waste 'animal and vegetable fibres bound together by the strong tenacious skin of the buffalo [...] saturated with a solution of vegetable oxide'. It was patented in 1871 by John Bland Wood, a

Manchester surgeon and amateur botanist who had tinkered with floor cloths since the mid-1860s.[108] In addition to using up 'the rags or the worn-out fabric or the waste material of carpets [...] clothing [...] other fibrous material, of animal hair, of the hide of buffalo', the process of manufacture also shed materials that could be re-assimilated into the mix, 'the waste of bouilinikon'. This mass of mixed waste was shredded, boiled with 'inspissated' linseed oil until plastic, then rollered onto canvas.[109] Reports stressed how this flooring was 'almost indestructible by fire'. Edmund Bury, a boulinikon agent, supplied various buildings in Bolton with it in the 1870s.[110] But in this chapter almost nothing is incombustible. The manufactory of the Boulinikon Floor Cloth Company in Salford was all but destroyed by fire in March 1880.[111]

A round-up of recent improvements in the (then waning) kamptulicon trade in 1869 noted perceptively that

> in the earliest manufacture of the material its constituents were so cheap, that it is fair to suppose those persons who first realised their utilisation in this direction would have been rewarded with an ample fortune [... but] the sower is not always the reaper of the harvest.[112]

The fates of its chief proponents reflect the truth of this. John Fanshawe had been the superintendent of the Patent Elastic Pavement Company until 1845, when he attacked an employee whose complaint had led to his dismissal. A few years later Fanshawe took his former company to court for infringement of patent rights. Thereafter he moved into other rubbery areas, but never bounced back. Elijah Galloway died penniless in 1856, leaving a destitute widow and 'an extremely delicate' daughter. When the kamptulicon carpet was pulled from under Ebenezer Gough, he become a Baptist minister. This was not the end for kamptulicon itself, however: it was recyclable and might be turned into sturdy chopping boards, and so old layers were snapped up by marine store dealers.[113]

The most important instruments of national defence, the ships of the line, were enhanced and preserved using recycled materials. Waste cork and rubber gave buoyancy to lifeboats. On land, dusts were drawn back into use, lumps masticated together. In addition to his experiments with rubber, Brockedon finessed a method of pressurising graphite dust into solid sticks, thereby bringing 'into use the fragments formerly too small for pencils'.[114] Some reuse innovation maximised profits, but many of these trades were virtually subsistence industries. Reused materials became more valuable as they were put to further ends; kamptulicon's demise was hastened when the wastes used in its manufacture were transfigured. Changes to waste supplies created needs for mechanical ingenuity, and some of the recycling heroes in this chapter stepped up to meet the challenge. A motley crew were attracted to the opportunities of reuse, from the cautious and patient Hancock, who managed to turn an impressive profit, to the wild man Kirkman, who died in a vermin-ridden mess in his early thirties.

In this story of ingenious preservation, extended lives and national security, featuring naval timbers and cork floats and lava-sealed coffins, choices were guided, to some extent, by fear. Cholera haunted these years; cemeteries struggled to process the dead. Dire shortages of burial plots triggered musings about alternatives short of cremation.[115] In contrast to HMS *Russell*, the products of reuse were vulnerable to flame – there are nine infernos in this chapter alone. But bodies were not put to flame. A tongue-in-cheek article in the *Pharmaceutical Times* waded into this discussion. After setting up a sensational critique about fictitious calls for the reuse of bodies to make chemicals, it expressed horror at various potential products, 'phosphorus extracted from the brains of our grandmother, or […] a coat dyed with Prussian blue made from the flesh and blood of our father; neither should we be refreshed by smelling salts extracted from our wife or sister'. And yet, the Swiftian author mused on, perhaps this would be preferable to the current system, which saw bodies stacked up in overcrowded cemeteries.[116] At sea, corpses were committed to the depths for absorption back into Nature.

Classifications of the new recycled products were all at sea. The senses were confused. Nomenclature failed to keep up with ingenuity and failure: the first linoleums were called kamptulicon; sheets of felted or wadded insulation used on copper-bottomed ships were called paper, although they were much more like kamptulicon. Some incorporated sphagnum moss; some were dipped in 'bituminous matters'. More stenchworthy coal tar.[117] John Bland Wood suggested that boulinikon could be used to protect ships 'when armour plating is employed'.[118] Excise officers struggled to categorise sheathing paper in 1846. They were unsure if the asphalt felt used on roofs and for sheathing ships' bottoms, made using animal hair and flax refuse, came under the category of paper or not. It was a taxing question.[119] Neither ordinary people nor government experts quite knew what to make of all this newfangled stuff. The sale of mackintoshes and the use of coal tar were both hindered by offensive odours. Tar bedaubing the timbers was no longer what it had always been; it came not from pine but from dirty coal. It bore the same name and came in the same barrels, but the smell betrayed a bastard birth.

9

Dustwomen and a Straw Man (1780s–1830s)

On the morning of Wednesday, 5 December 1827, men ran towards the cries of dustwomen, who were employed to sort through rubbish on the huge dust heap in London's Spa Fields. Houses used to have dust holes, into which ashes, cinders and other wastes were tossed.[1] This material was collected by scavengers or dustmen and taken to dustyards to be sorted, for the waste to be separated into types of particle. From the dustyard, the papermaker was supplied with rags. Chunky items such as broken chinaware, shells and rubble were used for hardcore and rubble in construction and road building. Pieces of wood and charcoal could be taken as perks by the sorters. Old iron was returned to the forge, and 'the hartshorn and ivory black manufacturer gets supplied with bone'.[2] In their digging, the workers had undermined the heap, and caused a huge dusty avalanche. People lay buried while others dug with hands and shovels. A bruised elderly man was dragged from the 'superincumbent heap' with a broken leg; bone poked through his stocking. Groans were heard elsewhere, and a woman was unearthed with dislocated hips and facial laceration. Both victims were transported to St Bartholomew's Hospital on makeshift stretchers made from shutters.[3]

Sometimes the sorters were spied wearing accessories found on the heaps: begrimed kid gloves used to protect their hands, straw hats to protect their eyes.[4] During the first half of the nineteenth century, waste removed from London's houses was lucrative; contractors paid considerably for the privilege of taking it away. This astonished elite contemporaries: journalists marvelled at the peculiar ways in which muck could yield brass.[5] Some of the dust contractors could amass 'immense fortunes'.[6] Unsurprisingly, their employees remained dirt poor. Many of them were women, and many were Irish. It was the dust itself brought in the profits of the yard, the bulk of which were obtained from coal ash (called 'soil'). The resale of ash represented a huge part of industrial recycling. Soil, combined with breeze (coal fragments) and cinders (burned coal), was purchased by brickmakers. Soil was even exported – some of it to Moscow – to make bricks for new buildings following the fire of 1812.[7] Jane Dove, a dustwoman who lived on London's dingy White Lion Passage in the mid-nineteenth century, was married to William, a brickmaker. I fancy they met on the heap.[8] Their industries were very much conjoined.

The brickmakers sifted the booty brought to them from the dust heap. In London the finest ash particles were added to the brick mix along with sand; the coarser parts were used to fire the bricks. The ash gave London bricks a distinctive colour and a hard durability.[9] In 1845 *The Builder* detailed how 'most brickmakers prefer breeze or domestic ashes for their work'. If denied access to these, they would resort to 'small broken refuse coke, and sweepings from the gas-houses'. These preferences were determined by the way each substance released sulphur during the brick firing.[10] The mix of breeze and ash was called 'Spanish'. Its presence in the brick mix would reduce shrinkage and limit cracks. Inevitably, this material was contaminated with stuff not sifted out by the collectors – shells, chalk, iron fragments, bones and tobacco pipe fragments: 'London stock bricks contain holes where the organic matter has burned away, and the occasional pieces of garbage.'[11] This could be problematic: contemporaries warned that bricks which were neither heated nor limed sufficiently to destroy animal and vegetable matter might cause occupants of houses to 'breathe noxious and infectious gasses, given off by the damp, decomposing matter of the transformed dust-heap'. Decorators also had difficulty persuading walls built of these bricks to take and hold nails and plaster.[12]

Dust, soot and other particles

The time of early industrialism between the start of the French Revolutionary Wars (1792–1802) and the end of the Georgian era in the 1830s saw much recycling action, but in fits and starts. One peak of activity was, perhaps unsurprisingly, around the Napoleonic Wars (1803–1815), when imports of materials such as rags were reduced, and so domestic resources were more carefully managed. In 1803 households were called to account for their waste of rags, and in the 1820s the government was made to address a departmental waste of paper. Another period of increased interest in reuse came earlier, in 1795, when bad harvests caused grain shortages, leading to food riots later dubbed 'The Revolt of the Housewives'.[13] Inflation rocketed. Despite the hard times, by no means all of the waste spewing forth from Britain's 'dark satanic mills' was reused. It took a while to harness the wastes to come from the new gas-lighting industry. Several important recycling processes failed to establish themselves on account of the financial difficulties of their inventors. Much of the reuse ingenuity at this time was spent trying to bring small particles together: dust and ash into bricks; bits of straw into paper. Everything, absolutely everything, was eyed up as fertiliser.

Some dust-heap dust went to bulk out manures. The best manure was farmyard dung and urine (well rotted with straw and hay bedding), but arable farmers and market gardeners needed to look to other sources for the fertilisation of crops. A list of substitute manures made in 1806 includes many ingredients, such as: night soil, bones, furriers' clippings, woollen rags, tanners' bark, boiled hops and other brewhouse refuse, sugar scum, street sweepings and 'scrapings of roads'.[14] A more in-depth and critical discussion of manures by the newly established Board of Agriculture in 1795 had suggested that greater quantities of matter could be reused. The report drew attention to the underutilisation of urine, slaughterhouse refuse, soap leys, dross left in the manufacture of mustard and the scrapings of the pelts of sheep after being dressed (known as 'fellmongers' poake'). Manuals and reports proposed various recipes to combine and improve substances destined to be fertiliser, and identified the types of soil that would most benefit. Soap ashes, if 'beat small', made a rich compost and could be used as a top dressing for young crops – they were observed to 'destroy slugs and vermin'. Land dressed with fellmongers' poake was reported to be free from turnip fly. Soapers' ashes had an 'extraordinary effect' on 'stiff yellow wealdon clay'.[15] Substances used to make fertiliser were often in demand elsewhere. Horn waste shavings created in manufactures such as comb-making were also prized by toymakers, especially the thinnest shavings, from the lantern-makers, 'for being hygrometric, they curl up when placed on the palm of a warm hand': early fortune-teller fish.[16]

Chimney sweeps found a market for sacks of soot. Applying soot as a top dressing for wheat countered yellowing and improved the luxuriance of foliage. Soot from Newcastle coal was superior, thought to gain potency from 'the large quantity of ammonia it contains'.[17] One batch of soot sold to a farmer in 1811 suggests that sometimes hidden preparation occurred. Rainfall after the soot was spread revealed that it had been cut with sand and sawdust.[18] Frederick Falkner warned users of his *Muck Manual* that soot was notorious for being 'very much adulterated'. Charles Wilkinson, a sweep from Louth, had a 'large quantity of soot manure ready for sale' by the cauldron in 1827.[19] 'Soot manure' was simply soot. Even sweeps were not averse to a touch of product puffing.

A combination of less disgusting alternatives and expansions of the sewerage network during the nineteenth century saw a decline in the already limited use of human waste, despite mid-century calls to improve and develop the practice.[20] Care was needed to avoid becoming over-reliant on one type of fertiliser. Liebig identified a vineyard where an 'exclusive use of horn shavings' produced 'great luxuriance' for many years, before the vines failed completely, 'probably owing to the complete exhaustion of the potash'.[21] Variety was the spice of muck.

Bushels of human and unhuman bones

In his *Cyclopædia*, published in 1814 and based on researches from the end of the eighteenth century, Abraham Rees detailed multi-purpose horn and bone waste, the chunkier parts of which would be sent to the cutlers and turners to make toys, hafts, handles, dice

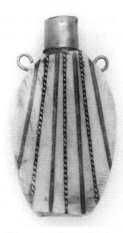

45. Small bone bottle. Author's collection,
photographed by Toby Sleigh-Johnson.

and chessmen and other items.[22] Bones collected from dustyards, butchers, marine store dealers, rag-and-bone men and slaughter-houses were processed in bone calcinatories to make bone charcoal and bone oil, and by establishments that made glue or size. Bones, and other animal parts such as ears and skin, went to various centres of reuse. The simplest of these turned out greases, fats and gela-tines. Cooks used the 'scrapings, shavings, and sawdust of bones' to make jelly: a 'well-disposed cow or sheep would not be niggardly in the bestowal of these gelatinous treasures – skin, membrane, tendon, ligament, bone, hoof, feet – all would yield gelatine'.[23] Some animal matter went to soap-makers, and some was 'manufac-tured into tempting jujubes and lozenges'.[24]

The boiling of semi-putrid fresh bones caused a nuisance.[25] Baldwin's Gardens, off Holborn, in central London, was not as idyllic as the name implies: in 1832 John Barber was convicted of nuisance after boiling 'Bones Flesh Offal Grease and other offen-sive matter' there.[26] John Michael Bray, a bone merchant and comb maker, of Norwich, was accused of committing a public nuisance in 1835. Bray's premises were located near another boneyard, the premises of a tallow chandler, a dye house and a yarn factory powered by a steam engine, all there before Bray arrived. During his trial it was asserted that the 'smell from the boiling, and the stench when the liquor was emptied into the river was worse than all'. Mr Harmer, his erstwhile adjoining neighbour, had been 'obliged to keep the windows and door shut' on account of the horrid stench which came from the bones. Mrs Harmer was made sick by it for two years before the couple moved away. Other neighbours also departed. A watchman was forced to abandon his post. For the prosecution, a surgeon argued that the smell was 'likely to produce affection in the bowels'. The neighbour whose complaint instigated the case claimed that 'bones of various animals [...] were heaped together in the yard with portions of flesh on them, until they became putrid, when they were boiled for the purpose of extracting tallow and fat'. Details of the operation emerge in the trial records. Fats extracted from the process appear to have been processed on site: there was a 'soap office' and a 'chandling office'. The defence successfully argued that Bray remanufactured 'worthless hoofs' into combs, after being cut and boiled. The court was reminded that the business secured employment for fifteen local people. Some neighbours testified (somewhat implausibly) that they liked the smell from the boneyard, the superintendent of the gasworks among them. A surgeon called in by the defence stated that the ammonia was beneficial to health, and that makers of glue, cat-gut and buckram, butchers, tanners and others working with similar substances had all been free from cholera in the latest outbreak.[27]

Size-making used up bones, but also 'clippings of hides, hoofs, horns, and feet, and the refuse from the skins of horses, dogs, and cats: and the shreddings of parchment, vellum, and white leather', and these were cleaned, boiled, skimmed, strained and cooled.[28] Size-making establishments were particularly unpleasant neighbours.[29] In Emley, near Wakefield, west Yorkshire, the Townend family had a 'sizing house' which spewed forth smoke and stench, making the 'air [...] greatly corrupted'. Their practice of boiling 'entrails and offals of beasts and divers diseased and unwholesome parts of Beasts' caused a public nuisance in 1821.[30]

Glue-making was slightly more complex. To make a glue bones

required, first, grinding, then boiling, cooling and reboiling, straining and evaporating until the syrupy matter could be set in tin moulds and left to dry.[31] Like size, glue utilised not just bones but also damaged pelts, furriers' shreds, horns, parchment clippings, salted ox feet, spetches or fleshings from fellmongers', leatherdressers' and tanners' yards, ears, tendons and clippings of skins. These were boiled to a jelly, clarified, cooled, sliced and dried 'into an untempting brown gelatine'.[32]

In his *Cyclopædia*, Rees stated that 'bone is one of the most curious and valuable of the animal substances, though considered vulgarly as little better than refuse, and scattered about without care'.[33] But everything in the bone trade was about to change. Since the end of the eighteenth century mills powered by water had ground, boiled and dried bones into a fine meal used as fertiliser. Crushed bones were 'regarded as one of the finest manures for light turnip soils'. William Beckett erected two bone mills in Hull, both powered by steam. By 1822 he had secured a regular supply of bones, which could be 'ground to any size required'.[34] Demand for ground bone increased during the second quarter of the nineteenth century, and bones and bone fertiliser were imported from Germany and Prussia. By 1846 it commanded such high prices 'that adulteration with slacked lime, sawdust, and the like, [was] not infrequently resorted to'.[35]

Wild reports of 'a million bushels of human and unhuman bones' being shipped to Hull to be ground into fertiliser should be read with proper incredulity. In the 1820s and 1830s people jittered about the fate of dead bodies; the horrible crimes of the body-snatchers Burke and Hare did not help, but their notoriety was partly fuelled by existing concerns. In 1822 the *Westmorland Gazette* reported on the arrival of bones (from hero and horse) from Leipzig, Austerlitz and Waterloo, 'and of all the places where during the late bloody war, the principal battles were fought'. The combatants' bones were said to then emerge from Yorkshire mills in 'a granulary state' before heading to Doncaster to sell on to farmers. 'It is now ascertained beyond doubt, by actual experiment, upon an extensive scale', the paper suggested, 'that a dead soldier is a most valuable article of commerce.'[36] Similar stories circulated seven years later.

The *Inverness Courier* included an article recycled by the *London Courier and Evening Gazette*, entitled 'Traffic in Human Bones', detailing a 'ship laden with bones from Hamburg' sailing into the Moray Firth. The cargo, 'collected from the plains and marshes of Leipsic', was said to be 'part of the remains of [...] brave men who fell in the sanguinary battles fought betwixt France and the Allies in October 1813'.[37]

These stories had become popular legends by 1832, a year in which the fears around cadaver misuse had reached a frenzy owing to cholera and to the Anatomy Act, which gave more freedom to engage in human dissection for medical purposes. That year the *Cheltenham Chronicle* cited vague 'Vienna papers' to copper-bottom a story about German human bones being imported in 'myriads of tons', with mills erected in 'the vicinity of the North Sea' to 'pulverise them'. British farmers were depicted casting the bones across their 'humid, cold, and poorest land', bringing land in Nottinghamshire, western Holderness '&c' 'into the highest state of cultivation'. Then comes the fresh-coined proverb: 'One ton of German bone dust saves the importation of German corn.'[38] What should we make of these accounts? That *Cheltenham Chronicle* article simply repeated the words of a humdrum article about animal bone-dust from the *Repertory of Inventions* (1832). The only modification was to the headline, jettisoning 'Bone-Dust for Cultivation of Grain' for the more sensational 'Importation of Human Bones'. That misled readers to infer that the 'myriads of tons' of bones were all human. They were not.[39] This all distracts from the more mundane truth: much animal bone was milled for use on poor soils from the late eighteenth century until about 1880.[40]

A Pomeranian straw man

Women performed the initial process of sifting the rubbish in the dust heaps, and women also sorted the rags as they entered the paper mills, shaking the loose dirt from them, cutting them up and stripping away any pins, hems, seams and buttons. Rags were put into different piles, ranging from superfine to coarse, depending on their qualities and how worn they were. Blue rags were set aside to make blue paper. Very coarse pieces and old paper made pasteboards. In cities, papermakers could be more select in the type of rag they pulped than was possible in country mills. Next the rags were washed and soaked or boiled, then soured or bleached, beaten to a pulp, and size was added to stop ink bleeding into the finished sheet (this was skipped for blotting paper); thence the pulp was taken to the vats, where men using moulds would scoop and shake the pulp into sheets, which were pressed between felts, dried, sized again, dried again, pressed and checked (all by men). If the sorting had been 'imperfectly performed', the pulped matter would have 'cloudy parts floating in it, owing to some masses being less reduced than others'. [41]

Mechanical developments changed the process of papermaking. From 1799 the vatman, the coucher and the dryer, who previously worked on a small scale, 'had to bow before that splendid aggregation of machines known as the paper machine'. Experienced papermakers working without machines had acquired a sense of how much size to add to the vat according to the particular constitution of the rag pulp. Imperfections crept in with a reliance on crude paper recipes for machine mixes, which wrongly assumed homogeneity in the ingredients. Resulting papers were badly dyed and unevenly sized, and so broke in the machine, making the engines froth. [42]

Pins or buttons might blight a batch of paper: 'as the rags were beaten, the button would be crushed to a fine powder, producing thousands of tiny particles that would disfigure every sheet.' [43] James Simmons, a Surrey papermaker, kept a personal diary on paper of his own making in which he detailed some work-related struggles. In March 1838 he wrote: 'I am often cast down by world troubles, sometimes I think [...] the Machines should so improve as to make better goods and at a cheaper rate than the hand made.' The diaries reveal details of Simmons's working life, his travels to secure rags and glue and to contract orders and his practical difficulties. In December 1838 the firm was

troubled in keeping the paper clean because the low Rags we are working have now a good deal of the mackintosh stuff in it, with the India Rubber between, which is difficult to sort out and when beat with the stuff [rag pulp] it goes through the Knotters [a machine to sift away lumps] & causes the paper to be full of black specks which will not pick out.

The gummy hitch recurred in August 1841, causing more financial distress.[44] Other papermakers were similarly affected: lumps of rubber were 'a source of much greater annoyance to the paper maker than is readily conceived'. The large black specks left in the paper 'enlarge[d] considerably' under the pressure of rollers.[45]

A surge in bankruptcies in 1788 had been chalked up to the failures of linen speculators, 'whose *money* and *goods* were so helpful to each other, the *rags* serving to make paper, and the paper imposed on the public for money'.[46] This ironic comment anticipates later critiques of shoddy goods and betrays a fear of flimsiness. The linen rags that made banknotes were soft replacements for hard metal coins: intrinsic and face values differed. Watermarked banknotes made from linen rags had been introduced in 1697, made by Rice Watkins in Berkshire.[47] By 1810 metallic coins had been supplanted by paper money as the key means of payment.[48]

The most assiduous rag savers in the late eighteenth century had been the Dutch. At that time the British would not acquire this ingrained habit of their neighbour, because they were 'able to live more comfortably than the people of other countries, and think the saving of rags not worth their notice, or think them so trifling a value, that a great part is burnt or destroyed'.[49] In 1790 the *Caledonian Mercury* asked readers with servants to ensure all 'fragments of old linen' were collected rather than burned.[50] Rag shortages meant that 'sufficient quantities cannot be obtained', and that rags became sufficiently valuable to be worth stealing. In 1802 John Bennett was transported for seven years after stealing rags from a paper mill in Hampshire.[51] On the orders of the Newcastle magistracy a notice was placed 'in all the public places of resort' in the city, directed to 'Genteel Women who amuse their idle hours in working'. These, it was supposed, 'frequently throw scraps of linen and cotton of various kinds into the fire'. The placard somewhat sardonically requested that 'every lady will preserve these trifles, and direct their maid servant to sell them'.[52]

With a shortage of rags, some paper mills stopped operating altogether; others switched to the manufacture of flour.[53] A London printer stated that the high price of paper in 1801 meant that 'he could not print the detached Numbers to complete sets of works he had advertised to publish, without being at a greater Expense in printing the same than what he sold them for'.[54] The Napoleonic Wars exacerbated the lamentable scarcity of rags and reduced imports of paper. Newspapers appealed, in the name of national production, for rags to be collected.[55] When prices increased, the public were asked both to reduce their consumption of paper and to 'husband both the rags and waste paper'. When channels of precious supplies from the Netherlands and Hamburg were blocked by Napoleon in 1808, it was argued that the 'use of ceremonious envelopes of letters, and of thick writing papers, ought to be generally discontinued [...] in public offices'.[56] The *Hereford Journal* looked forward optimistically to the arrival of Maltese and Italian rags and complained that newspapers 'are suffering cruelly from the dearness of paper and are subject to the intolerable and oppressive grievance of being confined to a fixed price'.[57]

Britain relied on rag imports because there were insufficient domestic supplies, but also because papermakers preferred continental cloths with a higher linen content than the cotton-loving British supplied.[58] According to the publisher John Murray, cotton per se was not the problem. Instead a toxic mix was to blame: low-quality cotton rags allowed to slip in by unscrupulous makers, plus too great a reliance on chlorine bleach and bulking matter such as 'gypsum or pipe clay'. The resulting 'wretched compound' cracked when folded. One piece of paper 'fell to pieces' in the post. Murray found a Bible from 1816 'crumbling literally into dust'. The brittle yellowed paper tasted astringent and gave off 'a volatile acid, evincing white vapours with ammonia' when heated.[59]

All such concerns might have been dispelled had one Pomeranian immigrant been a little better with his money. Matthias Koops, a former mapmaker, came close to solving the paper crisis in the initial years of the 1800s, only to be thwarted by a rocky credit history. Securing patents for paper made from de-inked and re-pulped paper, and for paper made from well-macerated straw,

Koops also worked on using wood pulp as a substitute for rags, which were already in short supply on account of Continental imports being curtailed. Koops's first patent for 1800 applied hot water to 'printed and written paper', which was then boiled in 'solutions of American potashes and lime water', to varying strengths depending on whether the paper had been 'printed with German or English ink, or ordinary writing ink'.[60] This was his most revolutionary process, but one that he himself suggests was partially inspired by Henry Clay's reuse of paper in papier mâché; Koops also acknowledged a similarity to the fabrication of paste boards.[61] It was followed by a recipe for paper made from 'straw, hay, thistles, waste & refuse of hemp & flax, & different kinds of wood & bark'.[62] Apparently, 'music notes' written on straw paper were particularly easy on eyes forced to rely on candles for light. From May 1800 Koops churned out de-inked and re-manufactured paper (the 'Regenerating Paper Manufacture') from the Neckinger Mill in Bermondsey, under the management of Elias Carpenter.[63] That summer the company put appeals in the newspapers for wastepaper which would be de-inked and refabricated. Prefiguring later entreaties by Victorian papier-mâché makers, one appeal for paper stated:

> Public Offices, Gentlemen of the Law, Merchants and others who have always quantities of Waste Paper and old Books unsold, and which are often burnt, not wishing to have them exposed to the Public, find now an opportunity to convert them into Money, being certain that they will be torn to pieces and re-manufactured.

The appeal also sought the backing of stationers, who were reminded that escalating paper prices were bad for business. Other trades would benefit: printers who were having to slow production on account of finding paper scarce, and booksellers and bookbinders who had been encouraged 'with patriotism' to slow their production.[64]

Koops was frantically busy the following year, trying to set up his Straw Paper Company. His massive manufactory was erected in

1801 at the Chelsea end of Millbank.[65] This required more protection and investment than he was able to secure. Koops had been naturalised on 1 April 1790, but shortly afterwards went bankrupt. Deals brokered as a result of his difficulties in 1790 returned to haunt him a decade later.[66] He called his old creditors to a meeting in June 1801, and they agreed he could pay his debts in instalments.[67] In an episode revealing bravado and rashness, Koops and his partners criticised the patenting system and sought 'an exclusive property' in his straw paper invention 'for ever', to cover enormous costs, all with 'no counsel, nor any witnesses except themselves'.[68] Elias Carpenter, giving evidence to the Parliamentary Committee considering the case, puffed the employment potential of the new manufactory, claiming that it was projected to employ one thousand people. He also explained how the price of paper would drop, passing around sheets of paper made from straw and wood pulp. Koops was granted permission to 'transfer or assign the said recited Letters Patent [...] or Shares thereof' to up to sixty people.[69]

Reneging on the 1801 deal, Koops failed to pay his final instalments. In October 1802 creditors pounced on his household possessions and business interests.[70] This would have surprised nobody acquainted with Koops's activities since arriving in England. In addition to his map work, Koops, an 'uncertificated bankrupt', had managed a short-lived insurance scheme.[71] Then he published his *Historical Account of the Substances Which Have Been Used To Describe Events and To Convey Ideas, from the Earliest Date to the Invention of Paper* (1801), a work of self-promotion and a critique of the patent system, padded out with a sprawling history of papermaking. The first chapter of the first edition was printed on straw paper; the second edition contained paper made from de-inked paper.[72] All this bespeaks a man of self-confidence. Koops was also bent. In court in 1790 he was described as a grifter who cheated tradesmen across Europe by manipulating bills of exchange. For a while he had dabbled with dying cottons using Turkey red (a process using waste bullock's blood[73]) but argued, unsuccessfully, that he was not liable to bankruptcy laws.[74] Koops cropped up again in 1797 at the centre of a bizarre case of tampering with a bill of exchange, concerning the prince of Monaco. This bill had been reduced 'to a blank' using 'Chaddick's Solvent for the Stone', an ink extractor.[75] De-inking was in Koops's blood.

A further reason for Koops's demise was the failure of the joint-stock company. Limited status did not protect his capital of £71,000 (as with Fritz Viëtor eighty years later).[76] The creditors to whom two instalments were due successfully disputed the fractional shares sold in the company.[77] By December 1802 the Straw Paper Manufactory was no longer a going concern (despite shares being advertised in late October). Koops's 'very obedient though distressed' workers penned a joint appeal, but the mill ceased to operate. It was recorded that 'a great number of wet packs of paper are spoiling in the vat-house and pulp is rotting in the vats'.[78] With rents unpaid, the partners were served notice to quit on 25 December 1803. Not very Christmassy. Koops was declared bankrupt for a second time, and his stock and machinery were auctioned off.[79] In his diary the painter Joseph Farington details gossip about Koops's business, heard from a bookseller: 'the partners had each lost several hundred pounds.'[80]

The proprietors of the building, the engineers Joseph Bramah & Son of Pimlico, auctioned 'all the utensils comprehended in that stupendous undertaking, late the Straw Paper Manufactory' in 1806. Bramah was not uninterested in papermaking. He had invented a hydraulic ('hydrostatic') press which had been used in the mill. In 1805 he applied for a patent for improvements in paper-making (mostly mechanisation) but did not take them further.[81] The site was used for other sales. In October 1805 it warehoused 'several thousand loads' of topsoil which had been 'for above a century under rich cultivation required for liquorice' before it was auctioned off. During the following year parts of the site were sold. After the remaining structure failed to attract any attention in 1807, the huge straw elephant was broken up.[82] Materials auctioned in 1808 included 200,000 'prime stock bricks', plus thousands of floorboards. The appropriately named Mr Biggs oversaw the sale, which also offered two massive coppers (each capable of containing 300 gallons), £100 of potash and two black horses.

Despite having twice been bankrupted, Koops managed to claw in some personal fortune. When his widow died in 1815, she left sizeable bonds for the East India Company, over £2,000 in annuities and 'diamonds Pearls and Trinkets'.[83] Koops may have been unreliable and foolhardy, but in the field of papermaking he was a true visionary. Papermakers did not again get busy with de-inking and recycling paper or making viable paper from straw – both of which reduced dependence on rag imports – until the middle of the twentieth century, despite experimentation during the later nineteenth century. A Parisian, Louis Lambert, claimed in 1826 to have improved Koops's recipe, mocking previous efforts to use straw to make a 'harsh, coloured paper which never came into general use', and sneering at the fact that Koops's straw paper works 'terminated unsuccessfully'.[84] In 1863 Benjamin Lambert, unrelated to Louis, boasted that he had invented the recycling of wastepaper (a decade after he dabbled with poudrettes of human waste), but it is simply unbelievable that Lambert did not know about Koops's work.[85] The Pomeranian's sticky reputation smeared his ideas.

Nobody wants readable cheese

The Brontës were interested in oddities, and the reuse of body parts gripped them with a ghoulish delight. They read avidly about dissection in the papers.[86] Perhaps they read in 1829 about a calling-card case and a pocket-book that had been covered with tanned human skin removed by William Burke, the executed murderer and infamous partner to William Hare.[87] Some of the books in the Haworth Parsonage in west Yorkshire were salted and stained, having been salvaged from the sea. The Brontë siblings learned a sort of thrift from this, their mother's library. They hardly ever wasted paper, reusing scraps, wrappings, sheet music and wallpaper to make their own tiny books. Branwell Brontë's 'Battell Book', made in 1827, is encased in part of a blue sugar bag. The cover of 'Albion and Marina', by his sister Charlotte, was crafted from a wrapper, the reverse of which reads: 'Purified Epsom Salts, SOLD BY WEST CHEMIST & DRUGGIST, Keighley'. An issue of Branwell's *Blackwood's Edinburgh Magazine* (**fig. 46**) is composed on fragments from two adverts sewn together (one for *the Life of John Wesley* and one for Thomas à Kempis's *Imitation of Christ*).[88]

Other people also looked to limit the uses to which new paper

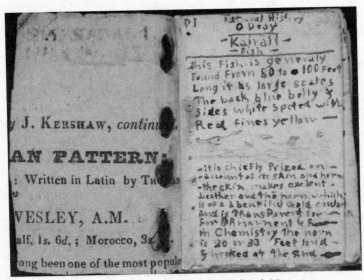

46. Branwell Brontë's *Blackwood's Edinburgh Magazine*.
Houghton Library, Harvard University, Cambridge MA.

might be put, including two friends, Joseph Hume and Henry Constantine Jennings, whom we have already met trying to set a ship alight in 1822. Jennings paid attention to various trades involving recycled materials: soap-making, tanning and papermaking, which incorporated grains and spent brewery mash.[89] In 1820 he had started to petition the government to exercise control over the stationery supplies used by printers. Among the abuses he highlighted were the fact that printers swapped government paper for inferior supplies, and that there were

> several immense stores of printed Parliamentary Papers, the magnitude and expense of which defies the power of computation; the greater part is at once a mass of useless matter; but which (with what has been yearly sold from Mr. Hansard's, and other places) must have cost this country many thousands of pounds.

The best remedy Jennings could foresee 'for this immense unnecessary accumulation will be, *not to print more than is really wanted*'.[90]

A Select Committee did indeed make inquiries about stationery, although its members claimed this had nothing to do with the intervention by Jennings. Evidence showed that high-quality paper and other items of stationery were being misappropriated – the Keeper of their Lordships' Boardroom was singled out for selling on perquisites of 'broken sheets and quires which were left upon the Board-room table' at the end of meetings. Such actions were henceforth prohibited. Printers were called in to explain their approach to paper supplies. In his evidence, Joseph Hume tried to distance himself from his friend and got into difficulty with the press in attempting to explain how Jennings was able to make his specific allegations. To prove that supplies of paper could be ill gotten, Jennings had needed to acquire samples, exposing himself to accusations of theft and exposing Hume to accusations of receiving stolen goods. The press saw this agitation as unpatriotic, describing Jennings snidely as Hume's 'tool' and 'his dear friend'; the pair were painted as, among other things, 'base agents of revolutionary France'. Newspapers described Jennings's claims as wild, vague and extraordinary – but it was decided that the stationery office was to be restructured.[91]

Later in his career, Jennings made his own contribution to the papermaking industry, with 'improvements in the manufacture of paper and artificial parchment, and of gelatine, applicable to the sizing of the same' in 1859.[92] Possibly these papers were for use as wrappings. Back in the early nineteenth century, pork butchers, grocers, tallow chandlers and cheesemongers made use of wastepaper supplies.[93] If a book did not sell, the publisher could cut losses by selling off the remainder 'as waste paper to the trunkmaker or the tobacconist'.[94] The fact that businesses were happy to buy up wastepaper meant theft was profitable. Some stolen books were fated to wrap cheese. A Bible and Thomas Gage's *Survey of the West-Indies* were spotted by an eagle-eyed vicar in a Westminster cheese shop in 1802.[95] In 1827 a vestry clerk heard rumours that 'part of our parish records were in circulation, with pieces of cheese', and indeed he recovered them from a local cheesemonger.[96] The printer of *Much to Blame* found himself 2,200 sheets short of the novel in 1829, and discovered the missing proofs at William Mills's cheese shop in Leather Lane.[97] Books from the legal publisher Butterworths were stolen for resale as waste in 1831, and some missing pages from Longman's publication of Rees's *Cyclopædia* were seen in a London cheese shop in 1812.[98] Sarah Pettit was found guilty of stealing paper consisting of pages from three different periodicals in 1802. A cheesemonger bought these. During Pettit's trial at the Old Bailey, the question of how wastepaper could be defined was argued. The cheesemonger and the bookbinder did not agree. The bookbinder asserted that 'waste paper is that which accumulates about the shop, rumpled and dirty', but the cheesemonger's definition was broader, and included jumbled up printed sheets.[99] John Churchward, who stole books in Exeter in 1827, and William Webb, who swiped a ream of paper in London in 1832, were given prison sentences and were held on the *Captivity*, a prison hulk formerly called HMS *Bellerophon*.[100]

There was self-interest in the definition of wastepaper. The condition and type of paper mattered to the cheesemongers: mucky or heavily printed paper was less desirable than clean paper. Nobody wants readable cheese. A black-lettered Bible, printed in 1597, was rejected in 1832 by a cheesemonger on account of its dirty 'thin dark paper'.[101] A similar discussion ensued during the trial of Sarah Corbyn for stealing official London Corporation papers, still tied with red tape. A worker in Samuel Harrison's cheesemongery on Fore Street said, '[w]e call it waste-paper, whether it is written on or not'.[102] Thomas Devereaux, a cheesemonger trading from Tothill Street in 1812, paid less for paper dirtied by 'writing and accompts'. Another prized pages from John Campbell's *The Lives of Admirals* because it was 'clean paper'. In 1810 Isaac Pettitt tried to palm off stolen sheets from Longman's first edition of Jane Porter's historical novel *The Scottish Chiefs* to a cheesemonger, who rejected it for being 'not the paper that exactly suited my trade[;] it would not hold the brine'. He preferred 'old ledgers or bill books'.[103] This was a very particular way of sizing up paper.[104]

This trunk (**fig. 47**), made after 1815, includes many features of reuse common to trunk-making. It includes pages from the tenth

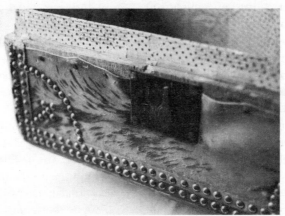

47 (a and b). Trunk made in the early nineteenth century, and detail.
Author's collection, photograph of detail by Taryn Everdeen.

volume of John Aikin's *General Biography*, published in 1815. People travelling by land or sea could purchase all manner of containers – chests, boxes, writing desks and cases for ladies' needlework. After the projected popularity of a translation of a German study in metallurgy proved unduly optimistic, the value of five hundred volumes, 'handsomely printed' in 1791, plummeted: 'by far the greater part of the impression […] became food for the Trunk-makers.' Likewise, some adaptations of Jacobean plays, due to be published in 1792, were instead 'cancelled and consigned to the trunk-makers'.[105] As their work was often commissioned and their manufactures unique, trunk-makers were in a good position to make effective use of waste materials. By contrast, mass production introduced in the latter half of the nineteenth century would lend itself less well to pragmatic recycling, since it placed greater emphasis on the reliability of components and was tasked to meet consumer expectations for consistency and similitude.

Salvaging the 'Royal George'

The Royal Navy – whose officers and sailors were familiar with trunks and ditty boxes of all kinds – has, historically, recycled names for ships. There have been eight vessels called HMS *Royal George* since 1714. J. M. W. Turner painted the fifth ship of that name, a first-rate launched in 1788, when she was broken up in 1822. This image is the only one of Turner's extant works to be applied to the reverse of a hardboard panel he had previously used as a sketching board. According to the curators at the Tate: 'traces of watercolour paint in many colours, blobs of animal glue, and knife and tear marks in the remaining paper, form the evidence for this alternative use.'[106] As Turner painted, the third ship to be called the *Royal George* languished underwater off Spithead, as it had done since a terrible accident in late August 1782.

In 1780, two years before the disaster, forty-six naval units had been copper-bottomed to protect their hulls against the ravages of the 'worm' (the boring mollusc *teredo navalis*) and make them more resistant to accumulations of weed.[107] 'Copper-bottoming' became figuratively adopted as a term indicating reliability. Previously, ship sheathing combined planking with animal hair and pine tar. As we have seen, the Navy experimented with coal tar as a defence against dry rot, but there was also considerable naval interest in using coal tar as a ward against the worm.[108] A desire to break the Baltic stranglehold over tar supplies, exacerbated when alternative sources dwindled after the American Revolutionary War, fuelled development of coal tar as an alternative to pine tar. Early coal tar promoters included several prominent Scotsmen, including the eccentric autodidact Archibald Cochrane, the ninth Earl Dundonald, an ex-naval officer and a keen chemist. His cousin John Loudon McAdam managed one of Cochrane's tar works.[109] Before the use of coal tar became widespread, its use on the bottom of boats was initially undermined by the more common use of copper as an anti-foulant. Copper sheathings had the advantage of being reclaimable and reusable.[110]

The *Royal George*, launched in 1756, was the flagship of the British fleet and a 100-gun first-rate ship of the line. For her construction some 3,840 trees were felled – according to some estimates 100,000 cubic feet of English oak, from 110 acres of forest. She sank off Spithead in 1782, while undergoing maintenance, despite being one of the ships copper-bottomed two years previously.[111] Copper-bottoming was supposed to reduce the regularity with which the *Royal George* had to be careened (turned almost on her side for her hull to be inspected and cleaned). Unfortunately, sheathing wrought new problems, in addition to a rather partial protection against the worm.[112] The copper had a corrosive effect on the iron bolts. Thomas Hancock later put forward a solution to the corrosion. In 1825 he proposed coating the copper with either cork or using a rubberised anti-fouling paint – an attempt to rubber-stamp the copper-bottom. Thomas's brother Walter continued this work with others, 'sheathing a great number of ships' until they 'got into litigation amongst themselves and the business ceased'.[113]

Before sheathing problems became apparent, a confident Lord Sandwich had boasted that 'Copper bottoms need fear nothing'.[114] Not so the *Royal George*: over 800 lives were lost when she capsized, including workmen and many women and some children who were visiting crew members while the ship was docked. A few days later 'bodies would come up, thirty or forty at a time', and were towed to land by watermen.[115] Attempts to raise the *Royal George* failed, and so, with each slack tide, bits and pieces of this giant were salvaged. Items made from leather and silk were dredged up, plus 'innumerable tallow candles'. The remaining wreckage was blown up by the Royal Engineers in a massive explosion in 1840. Salvage then continued apace. A single dive that year recovered timber, glass, cables, cask staves, a cutlass, part of a bedstead and a telescope.[116] Some items recovered between 1839 and 1840 were listed:

> Seven brass cannons, weighing 32,424 pounds, valued as old metal at upwards of £1300 – 16 iron ditto; 95 cwt of copper; 46 cwt of mixed metal; 75 cwt of lead; 235 cwt of iron: about 300 fathoms of timber [...] 90 fathoms of 24-inch rope cable; a fire hearth with copper boiler complete [...] ink-stands, one of ebony, and the other of lead [...] an ivory folding cutter, a large lump of red sealing-wax; a fragment of the handle of a penknife [...] real dragon china ware, blue-and-white [...] sundry wine glasses; several small punch glasses: also, two salt cellars, with horn egg spoon, and three cruet bottles. Several dozens of wine have been brought up, the contents of which proved any thing but palatable. [...] a woman's gipsy hat [...] the hood and collar of two silk cloaks [...] torn fragments [...] shoes, shoe buckles, skulls, human bones, a chequer board, broken crockery, an old fashioned wooden quadrant, made by Cole of London – besides a confused heap of heterogeneous articles. [...] Surgeon's implements [...] a pair of satin breeches and a large satin waistcoat with flaps.[117]

One notable item recovered was a typeface cast by W. Caslon of London in 1780, set into a stamp and marked: 'P C* Durham'. Lieutenant Philip Charles Durham had been the watch officer when the ship sank. He had survived to earn a dubious reputation for good fortune, and had his stamp returned to him. This is particularly poignant as recent research concluded that 'an alert officer of the watch would have prevented the tragedy'. The court-martial verdict at the time instead gave the blame to the lamentable state of the ship's timbers.[118] Still, billiards can be played in Burghley

House on a table made by J. Thurston & Co. from those same timbers; the zookeeper George Wombwell was laid to rest in a coffin in 1850 fashioned from them. Other materials were also reused. A lighthouse lantern was made from a gun.[119]

Using innovative diving techniques and underwater detonations of gunpowder, shipbreakers recovered many guns and hull timbers. In 1839 a capstan was recovered, only to be destroyed by fire a year later in an incident described in the last chapter. The ship's bell was reused in the cupola of the dockyard chapel in Portsmouth. An immense stack of timber measuring 120 feet by 35 feet by 7 feet was recovered between May and October 1841. Portsmouth surgeons the brothers Henry and Julian Slight wrote a book about the ship, at least nine editions of which were produced, with end boards 'ACTUALLY SAWN from the Leviathan TIMBERS' of the long lost ship.[120] Trinkets and snuffboxes made from the remainder, sold from Portsmouth dockyard, appear periodically at modern auctions.[121]

The Office of Ordnance advertised ad hoc sales of old artillery in the *London Gazette*. In 1803, 2,800 tons of old iron guns were put up for sale.[122] Some, their ends stoppered with old cannonballs, were reused as bollards on city pavements. By the time of the Battle of Trafalgar, in 1805, a Scottish company in Falkirk (Carron & Co.) made most of the light cannons carried on ships' upper decks, called 'carronades' after the company. After the Battle of Waterloo, in 1815, a deal was brokered with the government to ensure that captured French and Spanish cannons would not be repurposed by the British Navy (this would have had a disastrous impact on British armaments manufacture), but would instead be sold off to make bollards.[123] In all likelihood, it would have been difficult to retro-fit this artillery as they required different ammunition from that held in British arsenals. As street furniture they had propaganda value. Many surviving cannon bollards are close to ports, dockyards and military depots.

Old cannon also possessed value as scrap metal, especially artillery with bronze content. In 1801, eighty years before Fritz Viëtor set up business on Fann Street, the eventual site of his new warehouse was the location of a local recycling accident, when men at

48. French naval cannon reused as a bollard.
Photograph by Toby Sleigh-Johnson.

49. A long gun bollard, Bankside, London.
Photograph by Toby Sleigh-Johnson.

50. Stereoscopic photograph of Nelson's Column, Trafalgar Square, which incorporates copper from the *Royal George*. Author's collection.

Mr Eldridge's brass foundry who 'had been employed all night in melting down old pieces of ordnance' reached a cannon that still held a charge of powder and a ball. The cannonball smashed through a garret wall and landed in a room occupied by sleeping children.[124] Again, some cannonry was redeployed symbolically. Nelson surveys Trafalgar Square atop a plinth incorporating bronze from *Royal George* cannons (**fig. 50**).[125]

Somewhat better known than his painting of the fifth HMS *Royal George* is Turner's painting of *The Fighting Temeraire*. HMS *Temeraire* was broken up by the Beatsons, a firm of shipbreakers based in Rotherhithe.[126] The timbers, described as being 'as sound as they were the first day she was built', sixty years earlier, were reused in manifold ways. *The Times* reported that some oak provided a false limb for 'a jolly old tar who had lost his starboard leg at Trafalgar'.[127] Some were turned into church furniture, initially for St Paul's Chapel-of-Ease, Rotherhithe, from where they were moved to nearby St Mary's Church in the mid-twentieth century, when St Paul's was demolished.[128]

George Bellamy, a naval surgeon, administered medical aid to the crew of Nelson's flagship, HMS *Bellerophon*, during the Battle of the Nile in 1798.[129] As we have seen, in her retirement

Bellerophon became *Captivity*, a prison hulk, allowing the name *Bellerophon* to be recycled for another ship.[130] The original *Bellerophon* was broken up in 1836. The guns must have gone to Falmouth, as the town council there offered them up for salvage in the Second World War.[131] Bellamy used some timbers from his old ship to build a neo-Gothic cottage, Burrow Lodge, in Plymstock. By this time, the construction of buildings using shipbuilding materials already had a long history.

🗑 🗑 🗑

During the years between and around the Napoleonic Wars, Britain rode high on a tide of recycling arrogance. Many materials that could be recycled were not. European tips for repurposing were ignored. Potentially revolutionary designs for papermaking developed by an ingenious Pomeranian fell by the wayside. Eventually, brickmakers started to eschew the cinders and ashes with which they had previously added bulk to bricks, and dust collection was rendered unprofitable. The Navy, initially embracing waste products from the gas-lighting industry, shunned them in favour of other timber protection. Enthusiasm for reuse ebbed in an era characterised by missed opportunities for material reuse. Only during the period of war is there evidence of an acceleration in reuse endeavours, and these were mostly undertaken by women.

Then there was something of a reawakening of interest during the two decades up to the mid-1840s, at least in some quarters. When *Chambers Journal* took a long look at developments in reuse in 1846, it noted that soot, previously 'thrown to the winds', was now 'carefully bagged, and sold at so much per bushel', that gasification wastes, once considered only as pollutants, had become valued and that urine was tanked and sprinkled onto farmland. It continued: 'a large portion of our mercantile navy' had swapped 'pointless ballast' for guano conveyed back from the Pacific and the Atlantic, giving value to 'a substance which in our boyhood had no mercantile value whatever'.[132] The press got excited about other, more sinister, foreign wastes that were said to be reused. How well did the ground Hamburger bones and the desiccated French *merde* play into British pomp and glory days? Practices of reuse through-out the period participated in symbolic exchange: captured cannons became bollards; fragments of a wrecked ship went into Nelson's column; timbers into billiard tables. Not since the Reformation had reuse been charged with such national meaning. One newspaper account of the supposed traffic of human bones could not resist combining Shakespeare and Nelson in its sign-off: 'To what base uses we may return, Horatio!'[133]

10

Once More unto the Breeches (1720s–1780s)

In the summer of 1738, Thomas Westbeer, 'a Hackney Writer', was employed along with several others to sort through records in the Six Clerks' Office on London's Chancery Lane. He had the task of selecting those that 'were old, and proper' to archive in the Tower of London. Westbeer had other ideas, and smuggled records out under his clothes. According to witnesses at his trial, he hid 'great Parcels of engrossed Parchment' under the bed in his Charterhouse Lane lodgings before preparing them for sale. First, he carefully removed the stamps from 'Records made since the Commencement of the Stamp Act' and sold some of the parchments to 'Axtell', a dealer of books in Moorfields. Some of the parchments were crafted into tape measures for tailors, by cutting them 'long-ways, through the Middle', and sold to 'one that makes Cloaths to the Charter-house, for 8d'. The largest pieces went to a turner to make drum heads, and the smallest put aside for glue-making. Other records with names scratched from them could be re-used 'for [*pro forma*] Common Law Writs'.[1]

Westbeer had gone off the rails a few years previously, having fallen out with Gabriel, his father. In his will of 1742, Gabriel Westbeer described Thomas as his 'most disobedient and most undutiful son', leaving him five shillings so he 'may not say I forgot him'. Gabriel hoped that Thomas would quit his vile ways and lead a more sober life. Much of the legal wrangling, before the Old Bailey in 1739 and before King's Bench in 1740, centred not on

Westbeer's qualities but instead on the actual value of the parchments in question. Westbeer was released in 1740, after his defence
successfully argued that no felony had been committed. The manuscripts were 'of no use but to the owner', and 'not supposed to be
so much in danger of being stolen, and therefore need not be
provided for in so strict a manner as those things which are of a
known price, and every body's money'. The records themselves
were deemed to be public, and therefore the property of nobody,
not even the king.[2]

During the mid-Georgian period parchment and vellum
increasingly gave way to paper as a writing medium of choice.
Quality papers were made by James Whatman, who established a
paper mill in Maidstone in Kent in 1740. His 'wove' papers had a
uniform finish and weren't bumpy or watermarked. Skins were still
used, often to wrap up or cocoon items against the elements. The
start of the Industrial Revolution created industrial wastes for
future generations to deal with. William Bass established his Bass
Brewery in Burton upon Trent, which later provided yeast used to
make Marmite in the twentieth century.[3] And in 1762 many sperm
whales died after getting stranded along the English east coast. In
February a stinking whale carcass was found in Broadstairs, near
Margate, in Kent. The press commented that people were 'at a loss
what to do with it'.[4] This is such a different story from that told in
1918 from Bawdsey in Suffolk, when a beached whale was cut up
for glycerine. The patriotic songs 'Rule Britannia' and 'God Save
the King' herald from the mid-Georgian period. Britannia might
have ruled the waves, but she lost a colony with the American
Revolution (1765–83). In America frugality became a patriotic
statement in the 'Age of Homespun', and in letters Benjamin
Franklin's sister Jane Mecom shared a recipe for making 'crown
soap' using ashes, limestone, 'clean hard tallow' and bayberry wax.[5]

Protective containers

If old paper was eagerly snapped up by cheesemongers in the nineteenth century, in the eighteenth century some cast-off writing
material had been prized for an altogether less smelly purpose.
Among Thomas Westbeer's eager customers may well have been

51. Fishskin-covered spectacle case. Wellcome Library. CC.

goldbeaters. They often snaffled up old vellum manuscripts and used them to make gold leaf.[6] Called 'goldbeaters' skin', vellum and cow intestines cushioned the gold during beating. In the 1750s Philip Tuten, a goldbeater based in London, sold goldbeaters' skin and also bought old gold and silver items to melt.[7] Thomas Collett, of Long Lane in Southwark, made 'all sorts of Abortive Writing, and Binding Vellum; Drum-Heads, Writing Parchment and Forrill', and stocked the 'best skins for knap sacks'.[8] Skins and intestines were utilised for protective cases. Stray dogs were rounded up, to be skinned. Just before Christmas 1749 a man was found drowned in London's Fleet ditch. He was 'one who us'd to traverse the Common-Sewers, in Search of dead Dogs for the Benefit of their Skins'. Dog-skin toothpick cases were advertised in 1780.[9] Fish skin was easily mouldable into curved shapes and so used to clad small etuis, cases for mathematical and laboratory equipment, knives and pencils. Pocket-sized globes and small pocket cases could be compassed by fish skin.[10]

Vellum and parchment eventually gave way to paper as the prime writing medium. During the eighteenth century Britain became dramatically less reliant on paper imports than before, with many paper mills established in the countryside. By 1720 only a third of the paper supplies came from imports.[11] The papermaking techniques we have seen in previous chapters were developed at this time. Local merchants began collecting rags to sell to paper makers. The diarist Thomas Turner (1729–1793) wrote about adding this trade to his shopkeeping profession in East Hoathly in Sussex, while in 1714 Ellis Edward of Bicester also kept a shop filled with various scraps, including a huge pile of 'Linin Raggs & wolen rags' valued at more than £13, equivalent to nearly £2,000 today.[12] A careers manual from 1761 described the ragmen:

> Mean as this business may seem, there are some Rag-men who employ some thousand pounds in it. They buy linen Rags of the Rag-gatherers, &c, and even export them from abroad. These they deposit in warehouses; and, after they are sorted, sell them to the Paper-makers.[13]

The apparent wealth to be made in this trade belied the difficulties. The scarcity of home-worn rags that dogged later periods was felt from the eighteenth century.[14] Imports of rags rose steadily between 1725 and 1775: in April 1772, 24 tons of fine rags were sold by the ton at the Bristol Exchange Coffee House.[15] Imports then dipped again, before increasing between 1786 and 1790, and then dropping back by 1800. The Netherlands and France were the chief early suppliers. After 1750 the majority came from Germany, while other rags arrived from America, Scandinavia, Eastern Europe and Russia.[16]

Papermakers had long experimented with their recipes. In 1787 Samuel Hopper made paper that was 'free from specks & iron moulds' by combining the 'best rags' with alabaster, talc and plaster of Paris, plus the 'best white sugarcandy pounded'.[17] A few years later he reused 'leather cuttings, shavings or parings of every kind of leather whatsoever' to make an artificial leather. Tweaks permitted additions of 'junk or hemp' (junk was rope that was no longer usable as cordage) and fine clay. This substance could be used

for covering the fronts, backs, sides, and tops of coaches,

chariots, postchaises, sedan chairs, and trunks, and for making band, hatt, and other boxes, waiters, tea-trays, inkstands, ink potts, snuff and tobacco boxes, and other things, mouldings, cornices, ceiling and other ornaments for rooms, and for binding of books.[18]

William Cunningham applied a marine acid 'gas' in his process of making 'tea crown & Bible crown paper', boiling his rags in a 'potash ley' in a copper pot. He specified that coloured rags should be 'soured' first to destroy particles of iron that remained in the printed cloth and spoilt the colours.[19] These were the final years of souring, which usually involved soaking in soured milk; bleach was developed as a superior and faster alternative by a Scottish chemist at the end of the eighteenth century. Bleaches meant less attention could be paid to the colour of the rags at the sorting stage.[20]

Clement and George Taylor, from a papermaking family in Kent, also worked on bleaching rags using substances such as pearlash and 'dephlogisticated marine acid' (derived from 'manganese, sea salt, and oil of vitriol').[21] They attracted criticism for taking out a patent in 1792 which they believed gave them exclusive rights to bleach paper. Other papermakers (especially Scottish ones) thought differently because they had already adopted similar processes.[22] McHackles were raised. A group of Edinburgh papermakers pressed out their ire into a publication accusing the Taylors of pretension by 'falsely arrogating themselves the merit of being inventors' of a long-known method of bleaching, and defiantly printed this on bleached paper.[23]

Most of the imported rags went to make paper, but some were turned into buckram. While previously just a type of linen or cotton fabric, by the mid-eighteenth century buckram was a coarse gummed fabric made from strips of old sheet, pieces of sail or other linens, dipped in size (made from bones and other animal wastes). Paste stiffened the fabric, making it useful to upholsterers, staymakers, saddlers and tailors. Fashioned also into dust covers for book jackets, buckrams were sold as pieces, or in broader lengths, depending on the size of the reused fabric. Buckrams lacked homogeneity.[24] Buckram stiffeners often prepared the paste or 'glutinous matter' themselves, and could use 'all sorts of Linnen-cloth, as to render it of several Degrees of Stiffness'.[25] Joseph Daker stiffened buckram on London's White Cross Street, historically the location of the storage grounds of night soil men.[26] The Ledger family in South-wark made a small fortune sewing the business up. Robert Ledger, a buckram stiffener and dyer, established the Horsleydown printing works in the 1760s and had a workshop along Maze Pond. Edward and Henry Ledger advertised widely. In 1788 the *Norfolk Chronicle* included a request for any non-tanned or oil dressed 'Coloured or White Leather and Collar-Makers Shred, Parchments, Scrowls, &c', which items could be worked into the buckram mix if sent by post. The same article boasted that the Ledgers were 'the only Persons in London who manufacture and sell' various 'Buckrams, Glazed Linens, Pocketings, Dyed Linens, Brown Hollands &c'.[27] The three-volume set of Dickens's *Master Humphrey's Clock* from 1840 has end boards made from pasteboard, covered with buckram. Prior to the nineteenth century books were purchased as a wrapped bundle of printed sheets, which were then taken to be bound. Like the trunk-makers, bookbinders made tailored items, individually crafted to order. Because books and trunks were made for different customers they did not need to look identical, so there was much scope to reuse paper, vellum, parchment and leather.

Hardwares

Old metal vessels beyond the help of tinkers had a use in the trunk-makers' workshop. Cut into strips, they made good edge protectors.[28] A trunk made in the first half of the eighteenth century by Henry Nickles, who operated from premises next to the Cheap-side Conduit in London, is now in the Larvik Museum in Norway. It has metal edging strips and a paper-covered interior, and is a perfectly preserved example of what a typical trunk looked like for most of the eighteenth century and well into the nineteenth.[29] Samuel Pratt, who made trunks in the early eighteenth century, had a trade card that announced, 'Old Trunks taken in Exchange'. Presumably Pratt harvested these trunks for parts.[30] The trunks he received had already gone through a process of reuse inherent in their creation.

Small bits of metal could also be broken down to make lead solder, or simply added back into the crucibles and smithy pots. Metal resale values have ensured that they were (and still are) commonly collected or purloined. As a boy in London in the 1780s, Francis Place (who became a tailor) grubbed around in the gutter behind his house. He found enough iron to be able to 'purchase the materials for a paper kite'. He later financed the acquisition of a pair of skates by salvaging iron and lead from the fire-ravaged ruins of houses off the Strand.[31] The Sheffield Smelting Company was established by John Read, a refiner who moved to the city in 1760 and recognised the potential in the abundant waste metal found there. By 1775 he had appointed agents in London and Birmingham, and lugged waste to his smelting works by the canal. Much of the metal he processed came from sweepings, some from the sandy residue left by refining. Royal Mint wastes were also processed by his firm.[32]

John Read wasn't picky about the types of waste metal he smelted, but some traders sought out particular metals. Stubs were nails taken from old horseshoes, 'procured from country farriers, and from poor people' who picked them up 'in the great roads leading to the metropolis'. Twisted gun barrels made use of these stubs, thought to be the toughest and softest iron available, 'farther purified by the numerous heatings and hammerings' they had undergone. The repeat workings removed impurities. Tumbled in a drum, the stubs emerged free of rust. Twenty-eight pounds of stubs made an average single barrel, which saw scraps combined with steel. Women sorted through the stubs 'to see that no malleable cast-iron nails, or other impurities, [were] mixed with them'. The added steel was often itself recycled, pieces of old coach springs being favoured.[33] The scientist and architect Robert Hooke was informed that the use of 'old Shoos hoofs' for gun barrels went as far back as 1674.[34] Eventually not enough stubs could be procured to satisfy demand. Instead, 'pieces of the best iron [...] about the same size of stub nails' were used.[35]

During the eighteenth century some silver was cast into drinking sets for tea, but as tea-drinking became ever more popular during the eighteenth century, silver tea sets became rarer, comparing unfavourably to imported porcelain chinaware which conducted heat less well. As chinaware became dominant, silver tea sets were melted down and the silver was redeployed.[36] English clay was not of sufficiently high quality to mimic the imports of Asian porcelain, which became more desirable. New chinawares were prey to clumsy fingers. The riveting processes adopted by John Nike and his ilk in the Victorian period were developed in the eighteenth century and changed little until they were abandoned in the early twentieth century following the invention of adequate ceramic adhesives.[37] Meanwhile, there were recipes for simple (and probably ineffectual) cements made from animal glues, garlic juice, dairy products and even snail slime.[38] An early eighteenth-century method for cementing glass or chinaware 'a good way' called for egg whites to be combined with quick lime, the powder of burned flint, resin and lime juice, applied with a feather to broken edges. Apparently, when held together by a warm fire, the edges would bond well.[39] For a brief period, some menders burned china back together – coating the broken edges, clamping them together and re-firing the object.[40]

Edmund Morris, a china repairer in mid-eighteenth-century London, riveted china and also perfected 'a Peculiar Art' of mending chinaware, 'which was never before found out in this Kingdom'.[41] Riveting china required great skill: for each rivet, two holes were drilled into the china at a slant, a task that necessarily called for enormous care so as not to damage the vessel further. Morris was so confident in his own facility with such delicate work that he offered an impressive guarantee: 'if any of my work should come to pieces within 20 or 30 years I will repair it without any further expense.' One of Morris's competitors, Richard Wright, plying his trade from Middle Moorfields, claimed to make broken china vessels watertight, fixing 'Handles, and Spouts to Tea-Pots; Rivets and Cramps [to] all sorts of China'. In 1745 Wright drew attention to his 'new and peculiar method of sewing and rimming china'. He also sold broken shards of china to gardeners to bulk out soil.[42]

China-riveting, like other mending trades, was not lucrative. Peter Johnan, a London china-riveter, died in debt in 1772 with few saleable goods. At his own instruction the funeral was frugal.[43] Like the later menders of John Nike's era, eighteenth-century riveters were associated with itinerancy and immorality.[44] Some people set their minds to make china more robust, and so less in need of repairs. Calcinated bones were added to the mix. Thomas Frye, a former mezzotint engraver from Dublin, had a factory in Bow, close to the Essex slaughterhouses from where he sourced bones. Ash from incinerated bones made up almost half of the material in a tough, thin china for which Frye secured a patent in 1749. Frye resigned due to ill health in 1759. It is possible he had been poisoned by the industrial processes involved in making his chinawares.[45]

Mending the breeches

Probate inventories indicate something of a 'consumer revolution' in the eighteenth century: things such as tea sets became more widely owned. Most people did not benefit immediately from the expansion in the circulation of stuff, but they were able to buy up the leftovers of the rich.[46] Clothing was altered or adapted for new bodies or changes in fashion.[47] Poorer citizens rarely bought new clothes and made do with cast-off garments. In London the second-hand market was centred on Tower Hill and the Barbican (which would later be home to Fritz Viëtor's papermaking and pianoforte-vending concern). Bridget Green was one tradeswoman who, in 1744, washed and repaired fine lace, supplemented laces, 'foots Silk, Worsted, Thread or Cotton Stockings as neat as in the Loom', and also turned stockings into gloves.[48] Most material transformations went unrecorded. Repair work, however skilful and inventive, was quite often frugal opportunism rather than a distinct occupation or trade, combining with many other attempts to make ends meet.

Clothes could only be refashioned a few times before becoming threadbare and tattered. Provided enough good fabric remained, smaller items of clothing could be redeemed: a garment for a child, or a cloth cap. Piece brokers bought up 'shred and remnants of cloth' from tailors and sold them on. A careers manual from 1761 described these brokers as being 'generally decayed taylors, or some cunning men, who have crept into the secrets of this profitable, and often very dishonest trade'.[49] Tailors who were ageing or who had lost acuity or dexterity could refashion themselves into botchers, who repaired and modified clothes. Some tailors were known as 'cabbage' by association with the remnants of fabric, 'cabbage'.[50] Old shoes were rejuvenated or modified by cobblers, or 'translators'. 'When shoemakers are decayed' versified 'The Three Marry Cobblers' in an old ballad, 'Then doe they fall to our trade'.[51] Footwear beyond repair could be boiled down to make glue or size.

Hard-wearing breeches made from skin and hide would last a long time but were difficult to keep clean. Many were passed down begrimed and greasy. A newspaper advertisement from 1750 claimed that 'Foul leather Breeches are clean'd and narrow Breeches are stretched in a new invented Machine'.[52] The radical tailor Francis Place was a leather breeches maker in the 1790s, a time when the trade suffered a decline on account of changes in fashion. Working men started to prefer corduroys and velveteen breeches, while 'Gentlemen rode in Corduroy and Cassimere Breeches, and leather was no longer commonly worn by any class of persons'. Place was forced to diversify into breeches made from 'stuff' (a thick woven cloth). He also made 'Rag-fair breeches', inferior articles made from 'skins, which were damaged either by sea water or worms'. Ingenuity was needed to work around the flaws when cutting out. Place also cleaned the leather breeches of specific clients.[53] Thomas Parker, 'Rat Destroyer', was a veritable jack-of-all-trades. He offered wardrobe revival services, and in 1791 advertised that he 'cleans, turns and dresses' hats and 'alters their crowns in the nearest manner to the present fashion'. Parker also killed vermin, surveyed land, drew maps, helped with astronomy and navigation, dyed hats and mended china.[54]

All the extra tea-drinking saw sugar consumption rise. In the mid-eighteenth century, wealthy citizens with rotting teeth could try a tooth transplant or a spot of tooth whitening. In the 1770s local papers across the country carried adverts for supposedly famous German dentists with various names, who could whiten teeth. In 1777 the *Manchester Mercury* carried an advert for Mrs Simon from Hanover, who rendered 'teeth as white as Alabaster, though they were black as coal', without the use of instruments or brushes, which, it was stated, ruin the teeth.[55] In Maidstone and Southampton, nearly identical adverts to those drawn up for Frau Simon called attention to 'the famous' Mrs Bernard from Berlin.[56] In 1780 Mancunian mouths could seek whitening from Mr Davise from Mecklenburg-Strelitz, whose advert repeated many claims made by the female dentists.[57] The evocatively named Mr Suddon pliered his dental trade in London and Bath, keeping his hands busy throughout the year. In 1778 Suddon boasted that he could insert anything from 'one Tooth' to 'an entire set'.[58] Another female dentist, Mrs De St Raymond of Soho, declared herself to be skilled at setting grafts and human teeth.[59]

The 1780s were apparently busy years for the dentists of London. Mr Wooffendale relocated to Piccadilly from Liverpool in 1787; his advert concluded with a simple request: 'Wanted several hundred front teeth.'[60] A competitor, J. Spence, hand-picked replacement teeth to suit their mouth-fellows, using teeth 'to suit in colour and use the other real Teeth they are connected with'.[61] Mr Moor, an Oxford-based dentist, offered a Guinea for two front teeth in 1777, to be replaced with two teeth made from 'Sea-horse bone' (actually hippopotamus tusk). A decade later Moor worked in Manchester, where he was busy 'ingrafting human Teeth on the old Stumps' and appealing for 'human teeth of the upper jaw'.[62]

While France was setting out on political revolution, in England there was a revolution in artificial teeth. One such revolutionary was Mr Patence – presumably he went about things differently from Mr Suddon. Patence claimed to have provided dental care to Prince Oginski, the king of Poland, and to the 'Empress of Russia', preferring artificial teeth over recycled gnashers so as to save his patients from the very real 'hazard of contracting the venereal disease, which when once attained by human teeth, no human art can cure'.[63] In 1804, *De Chemant's Dissertation on Artificial Teeth* drew attention to 'the dangerous consequences arising from the use of Teeth extracted from dead bodies, or made of any animal sub-stance'. Instead, readers could satisfy themselves with De Chemant's 'incorruptible mineral paste for Artificial Teeth'.[64] Experiments with real tooth transplants continued in the face of such fears. A man in Congleton literally gave his eye tooth to help his brother.

Both patients 'waited upon a medical gentleman residing not 100 miles from the post office, who accordingly extracted the carious tooth, and transferred the sound [supernumerary] tooth of his brother to its place, which latter operation was performed with comparatively little pain'. It seems the transplanted tooth took, gums growing up to embrace it with presumably fraternal affection.[65]

The name 'seahorse' could be applied to many creatures at this time, but it generally referred to the hippopotamus. Clearly, some dentists used tusks to make teeth, but other hard minerals grown on animals also had applications. Uses were found for the fragments of horn, bone and ivory, by-products from other trades, and those substances could be materially transformed again. Harts' horns were rasped into thin shavings used by apothecaries, who also made and sold 'Harts-Horn Drops' – used among other things as smelling salts.[66] Neatsfoot oil, made from the shinbones of cattle, was also used to salve damaged ankles, and as a leather softener. In his *Pharmacopoia Extemporanea* (1701), Thomas Fuller outlined the recipe for a spinal liniment made up of the urine of a healthy person, red wine and neatsfoot oil (this was also thought to work on rickets).[67]

A milieu of experimentation absorbed and provoked fears about the dismemberment and re-memberment of humans. Johann Conrad Dippel, one supposed inspiration for Mary Shelley's *Frankenstein*, was born at Castle Frankenstein on 10 August 1673. The monster in *Frankenstein* was crafted from recycled body parts gathered from charnel houses and dissecting rooms, plus slaughterhouse waste. Dippel, a theologian and alchemist, acquired fame for his oil – later called Dippel's Oil. He once tried to exchange the recipe of the elixir for ownership of the castle in which he had been born. Dippel's Oil could be made from those parts of animals containing gelatinous substances, and oils from bone-burning could also be made into it. The best source was horn of stag. Essentially, Dippel's Oil was distilled hartshorn oil. Used as a 'powerful sudorific' to induce sweat, it was also thought to moderate epilepsy, hysteria and convulsions, and to counteract rabies. It had applications outside of quackery too. A pigment maker, Johan Jacob Diesbach, combined used potash, which according to one source, was left over from Dippel's oil-making, to synthesise a pigment later known as Prussian blue – of blueprint fame. In 1733 Dippel wrote a pamphlet claiming to have discovered another elixir, which would keep him alive until he was 135 years old. He died the following year, aged sixty.[68] When a Prussian blue factory was listed for sale in August 1787, the particulars gave an insight into the business. There was room to manufacture starch on the side, and paraphernalia available for 'rectifying HARTSHORN'. Interested parties might approach 'A. B.' at Mr Story's Soap & Candle Warehouse, 57 Queen Street, Cheapside.[69] Horns and hooves (plus leather-clippings, remnants of flock and dust from wool mills) were also burned to create prussiate of potash, later described by Charles Babbage as 'that beautiful yellow crystallized salt'. This substance gave colour to calicos.[70] Clara Matéaux, whose father was a dyer, later wrote about the use of distilled oil of deer horns in dye works, 'ammonia having a peculiar effect upon vegetable colours'.[71]

Paper reusers

Chrysal; or, The Adventures of a Guinea (Dublin, 1760) is a story fictitiously purported to have been discovered serendipitously among a pile of waste papers in a chandler's shop, set aside for the wrapping of butter, mustard, brick-dust, small coal, tea, sugar and snuff.[72] We've already seen that butchers, tallow chandlers, cheesemongers and trunk-makers used wastepaper for packaging. One man, John Ratcliffe, a ship's chandler based in Bermondsey, became 'infatuated' with the black letter books sold to him as wastepaper for wrapping, and 'imbibed his love of reading and collecting, from the accidental possession of scraps and leaves of books'. Ratcliffe, a 'very corpulent' man in 'a well-powdered-wig', had 'remarkably thick' legs: 'He was a very stately man, and, when he walked, literally went a snail's pace.'[73] As Ratcliffe gradually shuffled from wastepaper reuse to rare book collection, he amassed hundreds of valuable books, including works written by Chaucer, published by Wynkyn de Worde, as well as modern works by Pope and Walpole, plus dozens printed by William Caxton.[74]

Ratcliffe's library was the result of his intimate acquaintance with treasured books at the point when they seemed about to become worthless. Snippets fuelled curiosity; accidental readings educated. An advert by Christie's for an auction of his books in 1776 explained that Ratcliffe had spent three decades amassing his library, which comprehended 'the largest and most choice Collection of the rare old English Black Letter in fine preservation and in elegant Binding'.[75] This was despite a fire a few years previously, during which – 'incapable of assisting himself' – Ratcliffe stood outside his house, 'lamenting and deploring the loss of his Caxtons'. Reassuring him they were safe, a neighbour produced the 'fine curled periwigs' he had taken to be Ratcliffe's treasures.[76] An edition of *Le Doctrinal de Sapience*, translated by Caxton in the late fifteenth century, sold for £144,500 in July 1998. But this was not the first occasion the book had passed through Christie's auction house. In April 1776 it fetched nearly £9, being snapped up by a bookseller on London's New Bond Street, after Ratcliffe's death.[77] In his curious will Ratcliffe requested that his body be disembowelled and embalmed. The body itself would be placed in the vault of the church of St Andrew by the Wardrobe, along with any remains of his two dead children. His wife would possess his property off Fleet Street – for as long as she remained unmarried.[78] Ratcliffe liked to keep things together.

Other paper reuse was on a much smaller scale than Ratcliffe's collection. Although it was humdrum and quotidian, it would have amounted to a significant redeployment of material nationally. This was paper reuse for correspondence, the kind that my Nan would have approved of. In 1779 Benjamin Adams, an East India Company merchant based on London's Lime Street, sent a letter to Samuel Ayres, the captain of *The Terrible*, 'a fine new frigate, built in the river for a privateer, mounting 28 guns and to carry 200 men'.[79] *The Terrible* was commissioned by the Royal Navy to cruise the seas, capturing prize vessels and disrupting enemy trade. Adams was an agent for Ayres.[80] To wrap his letter, Adams reused an old legal document from 1732 which recorded the transfer of property in Alnwick, in Northumberland, from Thomas Athy senior to his son. Adams's brother Thomas was an attorney in Alnwick, and was possibly the source of the document used as the wrapper.[81] Ayres docked in Cowes in early March 1779, read the contents of the

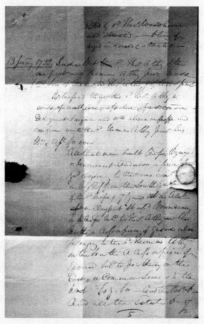

52 (a and b). Letter wrapper from 1779, reusing a legal manuscript from 1732, and reverse of the same letter, showing the legal manuscript. Author's collection, photographed by Taryn Everdeen.

letter and then (in his own handwriting) returned the wrapper (with the same, or changed contents) to Adams.[82]

Many of the crew of *The Terrible* made pre-cruise wills, bequeathing any prize money to named individuals. Countersigned by Captain Ayres, these remained on land for safe-keeping.[83] Later in 1779, *The Terrible* took prizes, including various American ships filled with timber and oil. William Purdy, a London broker, sold the 'entire cargo' of the *St John*, a prize taken by *The Terrible* in October 1779, including: casks of wine and brandy, bay salt, drugs, Irish linen, silk, wool and eighteen pairs of silk stockings.[84] In early 1780 she added to her haul an American brig laden with salt and was reported to be in 'pursuit of a large French frigate-built vessel'.[85] Shortly afterwards, *The Terrible* foundered near Bermuda, and 183 men died.[86] One of these was Captain Ayres, who had left to his brother Edward his 'watch, buckle and buttons' in a will written while he was 'considering the dangers and perils of the seas'.[87] At the end of 1780 'officers and seamen, or their legal representatives', were instructed to apply to Adams 'to enable them to receive their shares of the Prize Money'.[88]

Adams's letter also encourages us to consider another reuse of materials: the capturing and salvaging of cargoes from ships. Privateers had long known the benefits to be gained from taking ships and their contents, and teams worked to rescue objects and materials from shipwrecked vessels. Some people profited from deliberately wrecking ships to take their cargo; the prize was worthwhile. There were rewards for salvaging goods but, at this time, no financial rewards for rescuing people.

1780: The State Tinkers

The best-known protest orchestrated by the American revolutionary patriots against the British government was the Boston Tea Party. In 1773 three trade ships in the harbour at Boston were boarded by protesters angered by taxation without representation. Chests of tea were tipped overboard. The saltwater rendered the tea unsaleable, unsalvageable. During the American Revolution (1765–83) many Americans (especially women) undertook a quieter sort of revolution, by making do with materials at hand so as to reduce

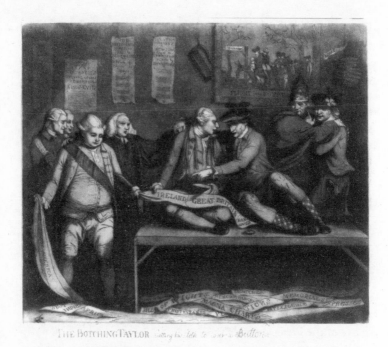

THE BOTCHING TAYLOR *cutting his cloth to cover a Button*

53. *The Botching Taylor Cutting His Cloth to Cover a Button*, 1779. British Cartoon Prints Collection (Library of Congress, Washington, DC), LC-USZ62–45441.

the young state's dependence on imports.[89] Meanwhile in England a disastrous foreign policy, a fractious government and economic distress led to civil unrest. Anti-war propaganda focused on George III and his key advisers, all of whom were seen to be recklessly endangering the nation. Two satirical prints from 1779–80 use the imagery of destructive repair to critique the efforts of politicians to keep Britain intact and the Empire safe. *The Botching Taylor* (1779) has George III sitting on a tailor's table in 'Taylor's Hell', slashing the cloth of Great Britain.[90]

It was also around this time that the Royal Navy embarked on its copper-bottoming scheme. The experiments were administered by John Montagu, fourth earl of Sandwich and the First Lord of the Admiralty. Notwithstanding his later fame as a portable bread

THE STATE TINKERS.

The National Kettle, which once was a good one,	The Master he thinks, they are wonderful Clever,
For boiling of Mutton, of Beef, & of Pudding,	And cries out in raptures 'tis done! now or never!
By the fault of the Cook, was quite out of repair.	Yet sneering the Tinkers their old Trade pursue,
When the Tinkers were sent for, —Behold them & Stare.	In stopping of one Hole—they're sure to make Two.

Published Nov.r 10.th 1780 by W.Humphrey N.º 227 Strand.

54. James Gillray, *The State Tinkers*, 1780. British Cartoon Prints Collection (Library of Congress, Washington, DC), LC-DIG-ppmsca-10750.

snack inventor, Lord Sandwich was corrupt and ineffective. In *The Botching Taylor* Sandwich holds a document headed 'A Scheme for ruining the Navy'. In another cartoon lampooning attempts to reform the Navy in 1780, James Gillray depicted *The State Tinkers*, three men armed with tinkers' tools breaking up a large damaged and patched bowl. 'Ld Sandwich' wields a mallet. George III watches. Some verse beneath reads:

The National Kettle, which once was a good one,
For boiling of Mutton, of Beef, & of Pudding,
By the fault of the Cook, was quite out of repair,
When the Tinkers were sent for, – Behold them & Stare.
The Master he thinks, they are wonderful Clever,
And cries out in raptures, 'tis done! now or never!
Yet sneering the Tinkers their old Trade pursue,
In stopping of one Hole – they're sure to make Two.[91]

Tinkers mended buckets and other metal vessels and often carried an old tin pot which contained fire.[92] Often regarded as beggars, they were the poorest of all the itinerant hawkers. They were synonymous with patchy attention, thought to 'make more noise than work'.[93] The final lines from the verse accompanying Gillray's *State Tinkers* were recycled from *A Pleasant New Song about a Joviall Tinker* (1616): 'He would stop one hole and make two,/ was not this a joviall tinker'.[94]

John Dunton gave a typical description of the ineffective tinker in *The Informer's Doom* (1683): 'a drowsie, bawdy, drunken companion, that walks up and down with a trugg after him, and in stopping one hole, makes three'.[95] Ballads from the 1680s run with another notion too: that tinkers sorted lonely housewives. Prefiguring a much later pornographic trend for housewife/repairmen scenarios, in *Room for a Jovial Tinker, Old Brass To Mend*, from the early 1680s, sexual relations between 'your Ladyship' and the tinker start with a broken cauldron. The winking lady plies the tinker with drink, and the pair lie on the bed:

The Tinker did his work full well,
The Lady was not offended,
But before that she rose from the bed,
Her Cauldron was well mended.[96]

A less explicit ballad of 1680, *The Jovial Tinker, or, The Willing Couple*, plays on a similar theme: maids with holes to mend, the tinker hammering away. The ballad weaves a metaphor for pregnancy into a classic attack on the deficiency of tinkered repairs: 'her

Kettle fell in two:/ he knockt her till she Gig'd again, as boys they us'd to do.'[97] Gillray's image insinuates that the First Lord of the Admiralty was shafting the nation.

🗑 🗑 🗑

By the late eighteenth century Britain had lost a major colony. The old country was patched and tinkered with, and feeling sorry for itself. But if recycling trades were largely disreputable and therefore an increasingly important satirical resource, they could also offer prospects for social mobility in some of the ways trumpeted by Dickens in later times. John Ratcliffe made fame and fortune from a collection of desirable books, after clocking the potential value of tomes destined for wrappage. The British have long shown a fascination with people who appeared to be poor but were actually rich. They marvelled in 1768 at the tale of Elizabeth Woodbridge, an old woman from Saffron Hill, 'who had for many years dealt in old rags and wastepaper, by which she acquired upwards of three thousand pounds'.[98] These enterprising figures, often on the edge of things, were harbingers of that industrious lot who would transform the national economy in the nineteenth century.

Wastepaper was used to line bandboxes (containing ruffs and bands), pie dishes, charter boxes and trunks.[99] In his illustration of *Beer Street* William Hogarth included, front right, the image of a pile of books destined for Mr Pasten, 'the trunk-maker in St Paul's church-yard'. Texts bound for oblivion included: *Modern Tragedies*, vol. 12; George Turnbull's *Treatise on Ancient Painting*; and *Politicks*, vol. 9999. The frontispiece for Pope's *Dunciad Variorum* depicts a donkey laden with volumes to be recycled.[100] In 1740, upbraiding the Prime Minister Robert Walpole (not for the first time), the novelist Henry Fielding referred to 'eleemosynary' copies of the *Daily Gazetteer* that had been distributed for free and paid for by Walpole's administration. They were turned into box linings by 'some prudent Housewife'. Thinking forward, Fielding imagined Walpole's 'Glories', uncelebrated at the time, re-emerging 'in an ancient Trunk or rotten Bandbox'.[101] Today's news, tomorrow's trunk-lining.

11

Uncivil Speculations (1630s–1720s)

A careless bit of recycling helped to trigger a riot in Edinburgh in July 1637, which in turn formed part of a chain of events leading to the English Civil War (1642–51), via the Bishops' Wars of 1639 and 1640. In the mid-1630s, keen to introduce religious reforms to Scotland, William Laud, the archbishop of Canterbury, worked towards the imposition of a new prayer book based on the English one. Years of dithering and amendment caused significant confusion and delay. Earlier draft versions printed in Edinburgh between 1633 and 1635 were withdrawn before publication, and the printers' proof sheets were sold to businesses in Edinburgh, including grocers. Robert Baillie, a Scottish clergyman, wrote of the delays that blighted the printing of various drafts, noting that pages from an early version were 'given out athort the shops of Edinburgh, to cover spice and tobacco'. People unwrapping their pepper saw hints about what their new prayer book would be like.[1] Other proof sheets were retained by the printers and used to stiffen the covers of the published version of the Scottish Prayer Book of 1637.[2]

At the time the Church of Scotland still had bishops, but combined these with a reformed theology which brought in presbyteries and synods. The presbyteries were more democratic, and the two systems – the episcopal and the Presbyterian – ran awkwardly together, creating the potential for administrative conflict. Until the 1630s most Scottish churches used a liturgy from the mid-sixteenth century to guide the governance of the church: John Knox's *Book of Common Prayer* rather than the Anglican *Book of Common Prayer* followed in England. Laud was determined to bring the Scottish church service more into line with services in England and to suppress the more extreme reform ideology in Scotland.

The new prayer book, 'Laud's Liturgy', was part of this plan. Single printed sheets of corrected proof from earlier versions circulating as wrappings in Edinburgh shops may not have been illuminating separately, but shoppers could compare their wrappings. Discarded snippets, perused by observant shoppers, fuelled speculation about the work and deepened a groundswell of Scottish fears about the imposition of a 'foreign' prayer book. On 23 July 1637, Jenny Geddes, an Edinburgh market trader, threw a stool at the minister in St Giles's Cathedral, objecting to the use of the new prayer book, starting a riot there. Afterwards, the Scots signed a National Covenant in 1638 to protect their church. According to this covenant, all prayer books needed to be signed off by the Parliament of Scotland and the Church of Scotland. The new prayer book was banned from Scottish churches, and Scottish Covenanters accused Laud of 'popishness', of bringing back Catholicism through the back door.[3]

Early modern binders regularly used broken-up medieval manuscripts and printed papers to line spines, create pastedowns (the inside linings covering the end-boards) or strengthen other parts of books. Usually, such manuscripts had been superseded by versions in print or were not to the taste of the religious reformers. Once supplies of manuscript waste ran dry, binders used printed waste instead, and they also reutilised the wastepaper from their own businesses (accounts, inventories and correspondence): the prayer book proofs were fair game. In September 1891 the Irish-born bishop of Edinburgh, John Dowden, picked at a copy of the 1637 prayer book, rummaging through the leather binding and pulling out older versions of the text. He encouraged others to do likewise, 'to sacrifice [the covers], though it be with a pang, to the advancement of liturgical inquiry'.[4]

In the turbulent years spanning the 1630s through to the 1720s, roughly the lifespan of Isaac Newton (1642–1727), whom we encounter soon, many later recycling and reuse practices were developed. Materials were stretched in the British civil wars, also known as the Wars of the Three Kingdoms (1639–51), during which Charles I was executed in 1649. The English Civil War saw the Parliamentarians under Oliver Cromwell (the 'Roundheads') pitted against the Royalists (the 'Cavaliers'), who were loyal to the monarchy. These years saw plenty of imaginative and ingenious recycling. In 1660, at the end of the Cromwellian Protectorate period, the monarchy was restored. The 1670s were marked by anti-Catholic hysteria, which saw the enactment of Test Acts – laws that limited public office to professionals who conformed to the established religion and so barred Catholics and nonconformists – and also the fictitious 'Popish Plot' (1678–81), for which Titus Oates, a clergyman, was found guilty of perjury after fabricating a plot to assassinate Charles II. The Bank of England was established in 1694, and there was a 'Great Recoinage' in 1696, which aimed to replace the hammered silver coins in circulation. A joint-stock company called the South Sea Company was founded as a public–private partnership in 1711 with the aim of reducing the national debt. The company was granted a monopoly to trade in the South Seas, but foreign conflict meant the company never gained any significant profit. Stock in the company rose as it focused more on the government debt, with a peak inflating the share price in 1720: this was 'The South Sea Bubble', followed by a collapse which burst the bubble. Other companies also sought to raise money from investors around this time, and some projects and companies made extravagant claims about their ventures. These were also nicknamed 'bubbles'. A few of these outlined schemes involved the reuse or recycling of materials, as we will see.

Kite-Flying Puritans

Cromwellian troops ripped books apart during the English Civil War. A raid on Winchester Cathedral in October 1646 led to books being sold to a London grocer, and various 'large parchments' from the muniments room getting turned into kites 'withal to fly in the air'. Troops had raided four years earlier, when they used the bones of Saxon kings and bishops to break stained-glass windows, burned some records and scattered others in the gutters. The chapter clerk gathered up remnants as best he could, recording the vandalism in his memoranda.[5] Destruction was not limited to the Parliamentarians. In 1642 Royalist troops under the command of Prince Rupert plundered the Buckinghamshire home of the Parliamentarian lawyer Bulstrode Whitelocke. They wantonly squandered his supplies, littering their horses with 'sheaves of good wheat'. That waste was at least more recoverable than the books and estate records in his study, which were torn up. Some were used as splints to light pipes of tobacco.[6] Parliamentarians set up residence in St Paul's Cathedral in 1643, using parts of the building to stable horses. An Inigo Jones portico was 'shamefully hew'd' to make lean-to shops for artisans and seamstresses.[7]

Equipping the armies during the civil wars consumed vast resources. In 1645 it was said of the Parliamentary commander that 'no man can endeavour to be more frugal than [Sir William] Brereton, but it is incredible how much is spent'. Uncertain access to supplies of gunpowder caused headaches for both sides.[8] Gunpowder-making had been limited by a monopoly before the wars. Manufacture of gunpowder required supplies of saltpetre, or nitre, the collection and sale of which had not been so constrained.[9] English saltpetre production had started in 1561, the business of a German engineer, backed by William Cecil, the Secretary of State. Saltpetre came from decaying matter sourced in various ways: including from unearthed cottage floors, dovecotes, privy cesspits and stables. This matter was boiled, then refined using urine and wood ash. The liquor that eventually leached out was boiled again. On cooling, potassium nitrate crystals could be skimmed off.[10]

Many buildings were denuded of their flashing and gutters before battles and sieges, so that the lead could be used by musketeers. Destructive rain dripped in. Oxford's corn market collapsed in 1644 after its lead had been stripped to make shot.[11] During the third siege of Lichfield, in 1646, cannonballs were made using lead from the cathedral roof, enhanced with scrap iron.[12] In Wales the Roundheads used metal pipes from a Wrexham church organ to craft musket balls, and lead was stripped from St David's Cathedral for bullets.[13] In 1643 the parishioners of St Peter and St Paul in Deddington, a village in Oxfordshire, were reported to have sold broken church bells to be melted down to make artillery for the Royalist army.[14] Shot cast in an ad hoc fashion using things close at hand in a hurry could be jagged, and wounds inflicted by them were apt to turn septic. Rumours developed, suggesting these 'poisoned bullets' had been tampered with deliberately, chewed, boiled in vitriol or rubbed with sand.[15] The truth was that the shot was just badly made.

Despite the wars, people still continued with their humdrum, domestic recycling. Indeed, with supplies limited by bad roads, people would have been more creative with their material reuse. Like many small and seemingly insignificant items (buttons, tokens and gaming pieces, such as chessmen, for example), toys were often the products of reutilisation and home crafting. Most toys from the early modern period were made this way, using whatever was to hand. A hoard of seventeenth-century toys found in a disused stairwell in Market Harborough church are mainly of wooden construction, supplemented by clay, bone, leather and fabric. The construction of balls, their exteriors made from small patches of leather sewn together, lent itself well to the reuse of scraps, both for the exterior cover and for the interior filling. Spinning tops could be made from leftover bits of wood, bone and horn – and sometimes had small metal tips too. Ankle bones from sheep and other hoofed animals were deployed in the game of 'knucklebones', or jacks.[16] Toys found in a Lincoln well, the result of seventeenth-century backfill, reveal much recycling in their creation. One item had multiple lives, from being a stave in a bucket, then reused as a sneck (a swivelling door catch), before a final reuse as a child's bat. A leather ball was stuffed with shredded scrap leather, perhaps the shavings left by a currier after paring hides, or gathered from the floor of a cordwainer or cobbler. Crudely fashioned ceramic discs made from cut-down broken crockery shards represent another form of recycling in toymaking.[17]

There are several examples of toy whirligigs made from reused tokens recorded on the Portable Antiquities Scheme database. Tokens were made from base metals, such as lead, and were used as a form of alternative currency in the early modern period, especially when the number of minted coins in circulation dipped. Two holes were made in the tokens and the edges crudely serrated. When twisted, a cord threaded through the holes would make the disc spin, creating a whirring or buzzing sound. These toys were based on medieval 'buzz bones', which used discs made from pig bones.[18] One surviving whirligig was made from a halfpenny, and, more intriguingly, another was fashioned from a lead alloy toy frying pan.[19]

The mental image of Cromwellian troops flying kites made from documents is difficult to conjure up and runs counter to our sense of Puritan behaviour. It turns archives into playthings: kites were for schoolboys. Such a symbolic use reduced official and formal materials into fripperies. The well in Lincoln that contained the toys was probably filled in deliberately during iconoclastic activity during the Civil War. Other materials were adapted and redeployed to pay for fighting. Recycling played a role in playing and in fighting, as it has in more recent times too: scrapping with scrap.

Liquidating

Full-size utensils made from precious metals were liquidated to fuel the Civil War conflict. Parliamentary and Royalist forces both absorbed collections of plate in 1642.[20] Establishing his mint in Oxford in 1643, Charles I asked the colleges to loan him their plate, to be melted down. Magdalen volunteered all the college plate, with the exception of the Founder's Cup, just shy of 300 lb. of precious metal.[21] The composition of these precious metals was analysed – assayed – using cupels: little porous cups made from finely ground bone ashes moulded into small vessels. Deer antlers were thought to make the best cupels, but fish spines could be used. These vessels were filled with tiny portions of weighed metal to be tested and were then wrapped in lead. When melted in earthen ovens, the lead removed base metals, leaving the precious metals refined.[22] Bone ash absorbed impurities without affecting the precious metal.[23]

Examples of church bell liquidation during the conflict may

have planted an idea in the mind of one man bent on managing welfare relief in 1649. In his treatise *The Poor Mans Friend*, Richard ('Rice') Bush, suggested that London's church bells could be liquidated, which he estimated would raise £9,660 for distribution to the poor. He hoped that such plans might 'remove us from many evils, disorders, clamours and tumults, and procure on us the blessings promised to them that consider the poor'. Bush further proposed that all citizens could donate old clothes, shoes, boots and hats to door-to-door collectors, for distribution among the needy, and might, for one year, 'forbear altering their apparel into other fantastic fashions'. These suggestions appear particularly altruistic when Bush's employment, as a draper, is factored in: his own business would be undercut by a reduction in cloth sales.[24] Bush died by the end of the year, and the bells survived mostly intact.

A portion of the plate held by families and institutions survived the civil wars unscathed only to be liquidated during the Great Recoinage of 1696, at a time when Isaac Newton was warden of the mint. Recoinage was complete by the time Newton became the master of the mint at the end of 1699. Citizens were induced to sell their plate and bullion to the state. Recoinage was needed because people had lost faith in their coins, which were often clipped or damaged. The bad coins drove the good ones from circulation and into hoards. At staggered intervals, various denominations of clipped coins were stripped of legal tender, called in and liquidated. The scheme was not a success because people simply hoarded the new coins, and silver declined as the fundamental measure.[25] The cost of buying up old coins and plate to provide the silver for new coins was defrayed by a tax on window glass.[26]

In 1695 Parliament started to excise the manufacture of glass. It was an unpopular move. Glassmakers stressed about threats to their livelihoods. *The English Manufacture Discouraged* (1697) showed how the manufacture of glass added value to English raw materials such as kelp, ashes and coal, and considered the recycling of the manufactured glass: 'many Hundreds of poor Families keep themselves from the Parish, by picking up broken Glass of all sorts to sell to the Maker.' Additionally, the import of potash used in glass making added to customs revenues, and the end-product, the English glass, reduced imports of finished glass from Venice and elsewhere.[27] Following debates and petitions, the excise was rescinded before the century was out. During the period when there was an excise on glass, it is likely that small workshops were set up 'hidden in back yards and cellars, where broken glass was re-melted and converted into small goods' without paying duty.[28] A dealer in broken glass was a 'lower kind of tradesman', as a London weekly newspaper had it in the 1750s. Broken glass had an appraisable value.[29]

Greasy bubbles and speculative recycling

The soap-boiler was one of five tradesmen identified as 'greasy companions' in *The Figure of Five*, a strange book published in 1645.[30] Soap-boilers used wood ashes and lime – for hard soap, tallow usually added – for soft soap, oil – but they could make use of pretty much any fat that came their way, however rancid.[31] In London the civic authorities insisted that tallow was to be used for candle-making when supplies became scarce, to the detriment of soap-makers. Wrangles over tallow sales continued for years and had previously turned nasty when the butchers who supplied much of the rough fat were accused of mixing stocks with water, guts, butter and 'other corupte thynges'.[32] Alkaline lye, used in soap-making, was made from ashes treated with water and lime. Waste from the lye making was sold on as soaper's ash. In Bristol this was used by the city glassmakers.[33]

The other 'greasy companions' identified in 1645 were the butcher, the cook, the tallow chandler and the kitchen stuff wench (the only female greasy companion).[34] 'Kitchen stuff' dealers sold pots of fat to soap-makers and tallow chandlers. Kitchen stuff was waste fat collected by cooks and housewives and sold on to women who called door-to-door, tub in hand, who paid by weight.[35] Kitchen stuff was much greasier than broken victuals (leftovers often collected from kitchens), being essentially the drippings of fat that issued from cooking. It could be made into very poor-quality candles. In 1683 John Dunton described 'knavish' chandlers who made candles by alternating dippings into good tallow and 'filthy dross' (kitchen stuff), producing candles apt to drop and 'waste away', which greatly hindered 'the poor workman' during the night.[36] In a supposedly witty jest published in 1674, a 'young wagg' calls over a woman who bought up kitchen stuff, and presents her with a pot half full of faeces. The woman 'according to custom put her arm in the pot'. She secured her revenge by smearing his face with the contents. The 'joke' hinged on the meaning of what stuff it was that 'dropt from flesh'.[37]

Other people were besmearing the projectors, who were seen to profit unfairly. The practice of projecting to secure patents and monopolies gathered pace in the seventeenth century, and some schemes involved reusing or recycling. After 1624 only novel inventions were supposed to secure patents, but abuses saw monopolies of essential substances and products, many of which were not novel inventions. Several treatises from the 1640s outlined various dire problems caused by monopolies and patents on various materials, including soap, starch and rags. *A Pack of Patentees. Opened. Shuffled. Cut. Dealt. And Played* was written in 1641, a year when Parliament denounced monopolies as corrupt. This anonymous tract shuffled together various patentees in a corrupt pack, their devilish trades described as dishonest. The ware of the soap patentee 'doth stink like him', and the products of the starch-maker are adulterated with bran. Leather patents pushed up the price of leather, but it was the tanners and the cordwainers who were blamed for higher prices. Boots and shoes being dear, they were worn to holes, but cobblers had lost custom years before (when leather prices were lower), and so had pawned their tools and lasts, rendering repairs more difficult.[38]

These patents were not in the interests of the consumer. Another anonymous work, *Hogs Caracter of a Projector* (1642), described projectors as vermin. Corner-cutting processes to make money dishonestly are outlined: the projector is 'a rare extractor of Quintessences' who draws up plans to make the wastes from beer, 'ale, wine, tobacco, brick, tiles, soap, starch, alum, cards, dice, and lobsters' into 'the pure spirit of gold' on exchange for fines and fees for patents. Being the type to hear the call of battle but too afraid to fight, the projector remains in the city, rather than join an army, and buys up 'all the Parsnips and Carrots that comes to London to make Dildoes' for the wives, old women and 'poor whores' left behind.[39]

The years leading up to 1720 saw rampant speculation on all manner of schemes, some of which would have involved recycling – had they come to anything. Daniel Defoe invested in a patent salvage scheme to recover treasure from ships wrecked off the Cornish coast.[40] The dramatist Aaron Hill virtually bubbled over with ideas. In 1716 he put his mind to a plan to fashion fuel for paupers from 'the very sifting of the Wharfingers Coal-Heaps'. Small coals crushed to a fine powder could be added to 'black Owsy' river mud to make a cement moulded into 'Culm-Balls'. The 'size of large Cannon Bullet', the balls would be large enough not to fall through grates: a 'service to [the] poor' who scavenged large lumps of burned coal.[41] Hill had other irons in the fire. He suggested that the nation's potters buy up 'old broken China', to be ground down and refashioned into new porcelain clay – a process that permitted chinaware to make 'some kind of Amends for its Brittleness by giving a Value, even after 'tis broke, to that which is now but unprofitable Rubbish'.[42] The South Sea Bubble burst in 1720, spewing forth a ferment of criticism about speculation.

One of the schemes highlighted as worthy of mockery in the wake of the South Sea Bubble bursting in 1720 was a scheme to fatten pigs.[43] In 1642 the fattening of beasts without grass, hay or grain had been presented as one of the horrifying schemes to project the nation into a dire future.[44] By the eighteenth century, for 'the sake of cheapness', urban pigs ate 'greaves', a waste product from tallow chandling. Greaves were described as a 'black contaminated sulphurous substance' and could also be fed to dogs. Dissolved in 'whey, or the washings of ale-barrels', greaves fed 'prodigious Numbers of Hogs', with some claiming that 2,000 hogs were kept at a distiller's yard in Deptford. Critics argued that hogs fattened on greaves 'are blown up with such nasty Filth which runs from most Brewerys and Distillerys', and even that the pigs were 'always kept Drunk'.[45] In 1733 Lewis Smart, a distiller who kept several hundred pigs near Tottenham Court Road, argued that his pigs were a 'public convenience', as they consumed the waste products of his trade.[46] In common with later economic crises, the bursting of the South Sea Bubble had an impact on how material values were perceived and affected enthusiasm for material reuse.

Patching and piecing

Five hundred Millions, Notes and Bonds,
Our stocks are worth in Value,
But neither lie in Goods or Lands,
Or Money let me tell ye.
Yet tho' our Foreign Trade is lost,
Of mighty Wealth we vapour,
When all the Riches that we boast
Consist in Scraps of Paper.[47]

So concludes a poem written by the satirist and publican Ned Ward. While Hill, Defoe and others were preoccupied with their projected fancies (which turned out to be just 'scraps of paper'), an unknown woman was busy with another recycling project which made actual use of scraps of paper: the earliest datable English patchwork coverlet was completed in 1718.[48] Patchwork quilts used up patches of cloth plus scraps of paper: the paper was used as templates for the individual shapes, and the pieces of material were then tacked over them, creating patches that could be sewn to other patches. The templates were usually removed before the quilt was completed, but a silk quilt held at the Victoria & Albert Museum in London still contains them. They include part of a newspaper report from 1705 detailing the life of the infamous per-jurer Titus Oates, in the year of his death.[49] Templates were also retained in the 1718 coverlet, which was made primarily from old silk dresses.[50] It is possible the paper was kept in for a practical purpose, to make the quilt warmer.[51] Alternatively, their retention may have been sentimental. Some of the templates in the coverlet are fragments of manuscripts – personal letters in various hands and household accounting materials, a few of them dating to the 1680s and 1690s. Others carry snippets of printed material, one from a parliamentary speech from 1707.[52] Caution is required when using the templates to date the quilts themselves, as the paper could

55 (a and b). Mocked-up early eighteenth-century patchwork.
Mudlarked pin courtesy of Lara Maiklem.

have been in circulation for decades by the time it was used this way. Quilts sandwiched wadding between a patchworked top and backing fabric: the wadding might also consist of recycled materials.[53] Since patchworking was a domestic activity usually performed by women who made no written commentary about their endeavours, it is hard to gauge its popularity. Research does show that thefts of patchwork recorded in the Proceedings of the Old Bailey increased rapidly after 1780.[54] By that time, some wealthy dilettantes were buying new fabrics for the purposes of patchwork, or to supplement recycled stock.[55]

Not all sewing was intended to craft new items: some simply repaired broken ones, before they fell into rags. Moths left holes needing to be mended, and garments were holed through wear and tear.[56] Darning techniques practised in the seventeenth and eighteenth centuries were little different from those taught by Eleanor Griffith in the twentieth century. In a complex satire of materialist philosophy, Alexander Pope (writing as 'Martinus Scriblerus') used the famous avarice of the seventeenth-century merchant Sir John Cutler to adduce a modern reimagining of Heraclitus' well-known dictum that 'no man ever steps in the same river twice'. Instead of the river, Scriblerus uses the ludicrous example of Cutler's black woollen stockings, which the merchant bade his maid darn so frequently with silk that they 'became at last a pair of silk stockings' – a play on Theseus' paradox, which asks if the gradual replacement of all pieces of a given object turns it into a new thing.

Had Pope himself been more familiar with the darning needle, he would have pointed out that this story was less than watertight. Stockings need darning in specific places, especially the heels and the toes – places where the body applies its most active pressures. Most of the stocking would have remained woollen, although often it was darned. Men like Pope had maids for darning, so the detail probably escaped his attention.[57] Ned Ward made a similarly ignorant comment about darned stockings decades previously: a pair of coarse yarn stockings donned by a member of a fictional parsimonious society called the 'Split-Farthing Club' were worked over so much that they were left with too little of the 'first knitting' to 'shew [...] original Contexture'.[58] Pope and Ward might have thought that a dull point – but it illustrates that many of the most durable descriptions of historical repair had only a tenuous connection to everyday practices, which are themselves very poorly recorded.

A waste of paper

Quilting templates were just one use for papery scraps.[59] As with other eras, paper was twisted into pipe lighters, used for kindling, wiping or similar expediencies. At some uncertain point a late seventeenth-century map by the Dutch engraver Gerard Valck was scrunched up and used as a draft excluder in an Aberdeen chimney.[60] Draft manuscript pages from Randle Holme's *Academy of Armory* (1688) were used to fill gaps behind a mantelpiece in a house owned by his family.[61] Paper was reformed into worlds. Halved into hemispheres, wastepaper was pasted into caps, enclosing the wooden frame that bisected artificial globes. In 1717 the geographer and mapmaker Bradock Mead, who later worked under the alias John Green to shake off a chequered past, outlined plans for making such objects at home. Making the globes involved creating a ball of wood and some wire to make a sphere, around which brown and white wastepaper would be pasted 'till you judge it to be the thickness of Paste-board'.[62]

A project to finance a pasteboard manufactory featured in satirical mementoes authored in the wake of the 1720 crisis. In a pack
of playing cards (printed on pasteboard) designed to commemorate
the folly of the Bubble, a 'Past-Board Manufactory' adorns the two
of clubs. A vignette concerning the scheme appropriates lines from
a Thomas Sheridan poem of 1718, about the pasteboard silhouette
of a man's head. 'As empty sayings flow from Windy Fools', starts
the passage, 'So Pastboard Bubbles rise from Paper Skulls.'[63]

In the early seventeenth century pasteboard had been composed
of sheets of paper pasted together. The name stuck despite changes
in production. By 1720 it might more accurately have been called
pulpboard, as it was constructed from low-quality paper – often
the shavings from book-edges – pulped and pressed together, like
later strawboards and millboards.[64] A legal wrangle concerning an
unsettled bill in 1771 furnishes more details about the process of
pasteboard-making. Robert Fleming pressed pasteboards in-house
for his bookbinding business in Edinburgh. Iron rollers commissioned from a Kentish millwright arrived too late. Consequently,
no board could be made after September; presumably it wouldn't
dry adequately.[65] By the end of the eighteenth century a cluster of
pasteboard mills along the Thames in Buckinghamshire marketed
their products in nearby London.[66]

We have seen already how wastepaper could be used as wrapping in various shops (cheesemongers, grocers, butchers,
fishmongers and chandlers). In his *Dunciad* (1728), Pope refers
mockingly to the fate of dull books, 'greas'd by grocer's hands'.[67]
Shopkeepers and servants would routinely assess scraps of paper for
their many potential uses, mentally weighing them up for size (the
gluey content). Wastepaper was also used domestically: 'maids
would know which sheets could seal foods, and which might be
best put aside for wiping.'[68] More mundane printed matter went
the same way; many a broadside sheet became a 'privy token'. Sir
William Cornwallis kept 'pamphlets and lying-stories and two-
penny poets' in his privy.[69] A ditty called *Bumm-Foder, or,
Waste-Paper Proper to Wipe the Nation's Rump with* (1660) draws
scatological connections between the Rump Parliament and human
waste. The poem describes itself as being fit only to wipe with.[70] An

56. William Hogarth, detail from *Masquerades and Operas (or Bad Taste) of the Town*, 1724. The Metropolitan Museum of Art, New York, Harris Brisbane Dick Fund, 1932, Accession Number: 32.35(80).

anti-Catholic poem decried legal exemptions made during the reign of the Catholic James II. Test Acts were tattered 'into thrums' by the exemptions, becoming worth less than the paper they were written on. (Thrums were the fringe of warp ends left after weaving.) The laws became metaphorical waste, to be 'us'd with Pasties and Plums': '*Magna Charta, Magna Farta*, made Fodder for Bums.'[71]

Have you ever started a new notebook, with pages put speculatively aside, often alphabetically, for future jottings? I have. Finding that I had left scant space for one category, and too much for others (nothing at the end, under X, Y and Z), I was afflicted with pangs of guilt for my wastefulness. Paper was sufficiently affordable in 1612 for Isaac Newton's stepfather, Revd Barnabas Smith, to have done just this, to keep track of his theological musings. Many of the hundreds of pages were left blank, and Smith did not find much to say about 'Frugalitas', beyond one line from Proverbs 6: 6. Momentarily abandoning alphabetic order, Smith skipped to 'Prodigalitus' for his next category, which remains blank – one leaf of wasted paper. Newton must have snatched up this notebook in 1664, when he left Cambridge to escape the plague. He repurposed it for his own notes on calculus and continued to use the book until the 1680s or later. In space Smith had set aside for 'Antichristas', Newton put down some telescopy notes; the timid label 'Cateche-sis' inscribed by Smith is immediately followed by a more bombastic 'Geometria' in Newton's hand. Even once Newton had finished with the book, many blank pages remained.[72] Others also willingly reused the notebooks of others.[73] People were careful with paper in the seventeenth century; they reused when they could.

The period from the 1630s until 1720 was a whirligig of activity, with speculations about religious motives, liquidations and kite-flying puritans. In *A Pack of Patentees* (1641) the rag-man raked kennels and dunghills for rags, hoping for 'the linnen which the Hangman gets', and pulling in all rags except shrouds.[74] Demand for rags to make paper eventually outstripped domestic supply, necessitating the imports detailed in earlier chapters. In turn this prompted trade controls such as the Burying in Woollen Acts of 1666–80, which prohibited the use of linen for shrouds, for 'lessen-ing the Importation of Linnen from beyond the Seas and the Encouragement of the Wollen and Paper Manufactures'. This not only protected linen supplies but also rendered a portion of the nation's wool forever unrecyclable.[75] The early eighteenth century saw rampant speculation, with designs for reuses that never mat-erialised. When the bubble burst, some would have reached for the smelling salts, made from hartshorn. Others were busy with more mundane reutilisations, making pasteboard and soaps, patchwork-ing and feeding pigs. It was apparent that not everyone who dipped into the country's waste pulled out a handful of crap or a fistful of bubbles.

12

Leading the Reforms (1530s–1630s)

In 1565 Richard Longlandes made a bridge so that his sheep could go to pastureland. Their bridge was constructed from broken-up bits of rood loft from St Andrew's in Boothby Pagnell in Lincolnshire. (A rood loft is a gallery in a church, above a screen separating the chancel from the nave.) Two years previously Longlandes had burned bright images of Mary and St John, which had previously adorned the rood screen, in a spectacular pyre, when he had served as a churchwarden. That year he also oversaw the destruction of one of the altar stones, which was reused to pave the church floor. Francis Paynell took away the other, using it as a hearth stone in renovations on his Norman manor house. Longlandes sold a cross, a pair of cruets for water and wine, a handbell and a sacring bell to a brazier in Grantham, a town five miles to the north-west.[1]

But Longlandes was not doing anything wrong. Church paraphernalia such as these became bric-à-brac after being made redundant by liturgical changes around the Reformation. Parishes had made or bought up replacement items during the short reign of the Catholic Mary I in the mid-1550s, but the fudgy Elizabethan religious settlement of 1559 rendered them superfluous once more. Churchwardens from nearly two hundred Lincolnshire parishes returned an 'Inventory of Monuments of Superstition' in 1566, recording the fate of 'popish ornaments' brought back under the Marian regime. These indicate how *verboten* church movables had been defaced, destroyed, reused or sold during religious rummage sales, at which village elite and churchwardens appear to have had first dibs. In Folkingham, eight miles to the east of Boothby Pagnell, the rood loft, with images of Mary and John, had burned earlier, in 1560. The churchwardens there were quicker to obey the new rules.

John Townesend, a tinker who lived in nearby Haconby, bought up various metal objects from the Folkingham churchwardens, including crosses, a pax (an object kissed during Mass), a pyx (a small box used to protect the Eucharistic Host), a pair of censors (which held incense), a 'ship of brass to put frankincense in' and brass candlesticks. It was noted that Townesend put these to profane uses. Townesend also obtained religious objects from Rippingale, a village six miles south of Folkingham. Two years later, in 1562, Townesend became the churchwarden for St Andrew's in Haconby, where he oversaw various sales in his own parish. The vicar took away a stone holy water vat to use as a trough for his pigs. A villager, Thomas Carter, bedecked his horse with a reutilised sacring bell.[2]

These extensive church bazaars took place across England in the 1560s. Altar stones – perhaps the least portable of such items – repaved many church floors. In Barkston, Croxby and Bitchfield in Lincolnshire they formed parts of village bridges. I spent a weekend hunting for these and other religious bridges early in 2019: a few can still be discovered, usually downhill from parish churches at nearby streams. Most are forgotten, disused or buried; many are completely lost.[3] Elsewhere in the county, altar stones became stepping stones, stairs, sinks and cisterns.[4] A symbolic degradation was evident in much reutilisation of church property: stones from the graves of priors of Durham Cathedral were built into a 'washinge howse [...] for women la[u]nderers to washe in'.[5] Back in Lincolnshire, in Horbling, part of the rood loft became a weaving loom. In addition to the rood screens and lofts, which were often given to the poor of the parish, presumably as fuel, other large wooden (or stone) structures were also destroyed or taken away. Easter sepulchres were elaborate wooden closets from which hangings were suspended during Passion week and Easter-tide. The sepulchre from the church at Burton was used as firewood in melting lead for church repair. In Croxton, the sepulchre was reused as a 'shelf for to set dishes on', and North Witham's sepulchre was transformed into Francis Flower's clothes press.[6]

Previously sacred items were put to mundane uses in Lincolnshire, as elsewhere: handbells became mortars; the pyx from Markby church became a set of scales; one from Bonby church served as a salt cellar; others became toys. The Alford churchwardens described these items simply as 'trash'. In Gayton, Somersby, Winterton and Hareby such things were melted down into solder for church repairs. An especially imaginative cruet reuse was evident in Wroot, where they were re-employed in spinning, as spindle whorls. Much of the plain cloth fabric removed from Lincolnshire parish churches was torn and given to the poor. More elaborate fabrics were fashioned into stomachers and a purse. The cloth that had covered Branston's pyx was being used by John Storr's wife to wipe her eyes in 1566. In Denton doublets were made from vestments. Elsewhere vestments became cushions, clothes for children or actors and bed hangings. Beds in Haconby were furnished with purple velvet cushions and canopies made from erstwhile vestments. An alb and banner cloths plus the vestments from Withern were all sold to a tailor in Belvoir.[7]

All the senses had been employed in the pre-Reformation Catholic liturgy, but the sense of hearing became paramount for Protestants. Catholic eyes had contemplated the lights, tapers, torches, images and icons. After the Reformation, those in the pews could still enjoy stained glass in some churches, where the expense of plain replacements justified their survival, but much else went. Churches had formerly been suffused with frankincense, perceptible to the nose. Rosary beads, relics and pax had been handled or kissed. Sacring bells had rung when the Host was elevated during Mass. Injunctions of 1547 and subsequent legislation and practice deleted many sensory experiences, reducing the involvement of senses other than hearing in a religious context, and making redundant much of the paraphernalia of worship.[8] Those items were secularised through reuse.

A waste of plunder

The Dissolution of the Monasteries permitted a massive redeployment of stuff. As Margaret Aston put it, 'England acquired a whole suite of ruins' from which materials were removed at the behest of the crown during a 'huge process of ecclesiastical *bricolage*'.[9] Patchwork construction, adaption, extension and repair of buildings led to the incorporation of parts of other buildings, or of other things. Building materials themselves are often made with a degree of reutilisation: we have seen how bricks were made with ashes, cinders and other bits of rubbish. Occasionally bones and clay pipes can be spotted in them. Wall plaster could contain animal hairs or small coal dust.[10] Parts of buildings were reused in the construction or extension of other buildings, and some formerly Catholic institutions were recycled wholesale, reused for the new Protestant faith or secularised. Four hundred years before 'superfluous railings' were identified as materials to be liquidated for the good of the nation, 'superfluous buildings' left after the Dissolution were eyed up for demolition. In 1539 the King's Commissioners identified buildings forming parts of the religious houses that were now redundant, such as churches, chapter houses and hospitals.[11] Ruined buildings became supply centres, from which anyone with the means, need or inclination might filch useful remnants.

Owing to the idiosyncratic nature of the English Reformation, there was an immediate and wholesale reuse of the old parish churches. Elsewhere, on the Continent, brand new Protestant churches were built. However, in England the tendency was to transform existing churches. Parishes were instructed to make internal changes: for instance, they were to turn their chalices into communion cups. The Victoria & Albert Museum holds a covered communion cup made from a chalice in the early 1570s by John Jones, an Exeter goldsmith.[12] Much ecclesiastical copper and brass ornamentation was flogged.[13] The churchwardens of Long Melford in Suffolk sold three hundredweight of metal in 1548. In 1551 it was agreed that the churchwardens of Great Yarmouth 'shall deliver unto Mr Bailiffs the brasses that were upon the gravestone in the church, to be carried to London to melt for weights and measures […] for the use of the town'.[14]

Monumental brasses were levered off church floors and re-engraved; the majority of brasses engraved between 1545 and 1575 have both medieval and early modern work on them.[15] Known as brass palimpsests, for want of a better term, these were usually flipped around, and so had medieval engravings on their obverse sides. A few, including one in Gunby, Lincolnshire, that memorialised Thomas Massyngberde and his wife, Joan, reused the side that had originally been inscribed.[16] John Fitzherbert's younger brother, Sir Anthony, a lawyer, is commemorated in Norbury church in Derbyshire on the brass for a medieval Croxden prior, reversed and re-inscribed in 1538.[17] Symbolising the Dissolution in one object, the inscription for William Hyde and his wife, of Denchworth in Berkshire, was made on the reverse of a French memorial to the laying of the foundation stone of Bisham Priory in the 1330s. They had died in 1557 and 1562 respectively. In Margate, in Kent, the brass on which a memorial to Thomas Fliitt and his wife, Elizabeth Twaytts, was inscribed in 1582 appears to have been imported from Flanders, reutilising an elaborate medieval Flemish design bedecked with vine leaves and grapes and a boy catching

57 (a and b). Brass palimpsest of Elizabeth Blount (obverse and reverse). Courtesy of the Trustees of the British Museum, London.

58. Despenser retable, Norwich Cathedral. Photograph by Maud Webster.

butterflies.[18] Flemish tomb-makers based near the Strand in London bought up brasses from dissolved religious houses, intent on using the obverse sides for new engravings. This depiction (see **fig. 57**) of Henry VIII's mistress Elizabeth 'Bessie' Blount was engraved in 1540, on the reverse of a crowned Virgin Mary from the early fifteenth century.[19] Bessie Blount probably had less massive hands in reality.

The survival of medieval religious items and structures was haphazard: some were actually saved by being put to more mundane and limited reuse. A medieval painting of the Kiss of Judas in the church of St Mary the Virgin, Grafton Regis, Northamptonshire, was turned around during the Reformation and reused as a panel for notices. An altarpiece in St Luke's Chapel, Norwich Cathedral, survived the Reformation as well as later despoliation during the civil wars because it had been reused as a plumber's table.[20]

Bemoaning ecclesiastical reuse in the 1630s, just before further despoliation during the civil wars, John Weever, a Laudian-leaning author with his own axe to grind, likened plundered religious buildings to shipwrecks, 'dasht all a peeces'. To Weever, the recycling of marble memorials and brass, purloined through 'greedinesse', was 'the foulest and most inhumane' of actions. He describes marble tombs being 'digged up, and put to other uses', inscriptions ripped off for the 'greedinesse of the brasse', 'dead carcases, for gaine of their stone or leaden coffins, cast out of their graves'. He portrays the royal commissioners, chief agents of the Dissolution, as grave-digging, gold-finding, sacrilegious robbers.[21] In an ecclesiastical history the clergyman Thomas Fuller lamented the altar cloths adorning private halls, daily bibations from old chalices, horses quaffing from 'rich coffins of marble'. The best stuff had been 'transported beyond the seas'.[22]

Monastic libraries were ransacked during the Reformation. Whole collections were burned, and some items became commercial wrappings.[23] Service books from Roche Abbey in Yorkshire were used to patch wagon covers.[24] Richard Layton, a principal agent of reformation, wrote about the ignominious banishment from Oxford of the works of the Catholic philosopher Duns Scotus, made 'a comon servant to evere man, faste nailede up upon postes in all comon howses of easement': toilet paper.[25]

Only one in every ten thousand broadside ballads has survived since the sixteenth century. The majority were put to other uses: lining pie dishes, absorbing dirt or binding books.[26] The 'Water Poet' John Taylor quipped that the *Commentaries* of Julius Caesar were 'good for nothing but stoppe mustard pots', echoing similar suggestions in Thomas Nashe's *Unfortunate Traveller* (1594), where a page could light a pipe, become a napkin or wrap velvet slippers or mace.[27] Wastepaper would also wrap fat or soap sold by chandlers or on a 'flaxwife's stall'.[28] The sale of wastepaper to shopkeepers became a factor in bringing Thomas Cromwell's Great Bible to publication in 1539. Richard Grafton and Miles Coverdale, the printer and a translator, went to Paris to work with a French printer, but part of their printed stock was confiscated as heretical by the Catholic Inquisition and sold to a haberdasher for use as wrapping, 'being about four dry-Fats full'. Cromwell's agents bought some of this back and sent it to London, to join sheets smuggled out via the English ambassador.[29]

Clearly there was much material reuse at this time, but if we take a broader view of Reformation reuse it is clear that much recyclable material was not reused, especially that formerly owned by religious houses. This is evident from the sheer number of ruined institutions across the country, with a high concentration in

Yorkshire. The costs of transporting materials from these ruined sites rendered comprehensive reuse unviable. Some stones from Fountains Abbey were reused to make a nearby mansion house, but much was left to waste and ruin.[30] In August 1539 the royal commissioner John Freman explained that to demolish the remaining Lincolnshire religious houses would be costly, as there was no ready market for the released supplies. He suggested instead the removal of the bells and lead, which he was about to effect, and then leaving the buildings open to the elements, assuming that, if any person nearby wanted stones, they might come to fetch them.[31]

Vacant monasteries were stripped of their roofs, and the lead was recast into blocks ('pigs') and set aside, as property of the Crown.[32] Lead from Jervaulx Abbey, removed in the sixteenth century, was found lying by the ruined west wall in 1923; windows in York Minster were re-leaded using this fortuitous supply.[33] Roger Burt has noted that a surge in the 'velocity of circulation of [...] British monastic roofs' increased lead stock across Europe. This had a 'devastating effect' on lead prices, triggering a crisis in Continental lead-mining industries. Only with growing use of lead shot in the late sixteenth century did the stock of lead shrink and its value surge. All previous re-applications had left the lead recyclable, but henceforth, despite evidence of shot being collected from battlegrounds, a full recovery would have required a rummage through corpses. Eventually, there was a struggle to sustain the stock.[34]

Thrifty with thrift

The venerable notion that the early modern period was characterised by thrift and frugality lacks supporting evidence. In a groundbreaking article on recycling in pre-industrial England, Donald Woodward concluded that 'little could be allowed to go to waste'.[35] Yet citizens in the 1560s had neither the ingenuity nor the need to reutilise materials in as comprehensive a manner as people did in the 1860s. People had fewer things to waste, but they nevertheless did a good job of wasting stuff. Dogs ate bones; cattle did not eat cattle; nobody wore dog-hair jumpers. Early modern people were wasters, but who can blame the historians for imagining the opposite, when so many contemporary authors stressed thrift?

Fuelled by biblical moralising about wilful waste making woeful want, early modern espousers of 'waste not, want not' often just recycled medieval advice in pursuit of their readers' money. Prodigal waste was part and parcel of elite conspicuous consumption in the sixteenth century; refuse and surfeit were signs of power. White and claret wine ran 'at certain conduits plentifully' during the afternoon of the coronation of Henry VIII's new queen, Anne Boleyn, in June 1533.[36] Waste was built into quotidian semiotics of display: although maids were passed down clothing from their mistresses, they were not ordinarily able to wear these for fear of being criticised or punished for dressing above their station.[37]

Most of the texts through which we gain insight into this period were written by the elite, and it was in their interests to persuade people beneath them to marshal their meagre resources effectively, for otherwise they would be unable to pay rents, leases or fees, manufacture profitably, fight in armies or sail the seas. These texts were written by men, but men were rarely interested in what domestic thrift – woman's work – might really involve. Recycling a wide range of late medieval sources, John Fitzherbert cobbled together a *Boke of Husbandry*. He devoted a section to guiding couples through a thrifty year, seemingly unaware that much of their energy would have been spent lighting fires and collecting water every day. Fitzherbert had no need to attend closely to such necessities: other people lit his fires, other people fetched him water.[38]

Gervase Markham, another unconvincing hawker of recycled thrift advice, made a living breeding horses and churning out so many books that in 1617 the Stationers' Company took the unusual step of making him promise never to write any more books on the diseases of livestock. Markham plagiarised himself and others, insisting that 'English thrift' legitimised such practises. In his *Countrey Farm* readers are advised to prolong the life of agricultural and domestic matter. From manuring to candle longevity, the book identifies various frugal ways to maximise the lives of things. It was first published in Latin, not by Markham – indeed it first appeared before he was alive – and was then packaged as a French text in 1564 and translated into English in 1600. Markham filched parts for

English Housewife (1615), and then published his own version of *The Countrey Farm* the following year, representing its strictures as an English form of thrift. 'Partly because of this recycling', notes Wendy Wall, 'Markham's reputation has not fared well.'[39]

Ultimately the sources are lacking for a comprehensive review of early modern recycling. Half a dozen times Woodward laments the dearth of contemporary records that would bolster his argument. We can assume that there would have been considerable eking of stuff in rural lives, and people manufacturing goods and processing materials would have followed (often medieval) methods through which the by-products of other trades were reutilised. It is clear that there was a symbiotic reuse of waste materials in various trades, and that there was also competition for some reusable materials. Uses for by-products are in evidence, but there were missed opportunities to reuse and recycle. Not everything was valued in the way we might expect, or in the ways that some contemporaries would have us believe.

In *Christ's Tears over Jerusalem* (1593) Thomas Nashe described a Jerusalem in which 'witherd dead-bodies serve to mend High-waies with'.[40] Nashe was jolting his readers to think about acceptable limits of reuse, not outlining early modern norms. In earlier chapters we encountered similar shock stories: fertiliser made from bones of soldiers killed during the Napoleonic Wars; reports of German corpses being used to make glycerine. Each story forced people to reflect on the contest between economic rationality and other value systems. In all of these cases it was 'others' who were presented as people capable of a corporeal recycling unimaginable to the English. These stories often circulated in economically difficult times. From the mid-1550s onwards people had been compelled to think more carefully about prices: the currency was debased; foreign exports became more expensive; and economic crises widened and deepened poverty. The poor received homilies about thriftiness.

Thomas Tusser, a rhyming farmer, published *A Hundreth Good Pointes of Husbandrie* in 1557. His advice was versified for ease of recollection. In this, and in later editions, Tusser emphasised the importance of thrift. Giving advice on scouring, he came up with the terse lines 'No scouring for pride' and 'Kepe kettle whole side'. Scouring was necessary, but too much was harmful; it was 'pride without profite', which robs 'thine hutch'. Speed and convenience also played second and third fiddle to thrift: 'set tubs out of Sunne,/ for mending is costly, and crackt is soone done.'[41] The message was that thriving came with thriftiness: 'The thrifty [...] teacheth the thriving to thrive.' The modern concept of 'thrift' implies the need to economise owing to a lack of resources, but thriving suggests superabundance and prosperity. Pre-modern notions reconciled the two words: thrift meant thriving, with the latter eventually assimilating notions of prosperity and fortune. The assumption that all early modern folk were naturally thrifty is partly undermined by a wealth of contemporary suggestions that they be thriftier.[42]

Officials like Sir Thomas Smith, a politician and diplomat, buoyed by humanistic commonweal zeal, spent their energies advancing middlemen and projectors. Smith hoped to harness avarice to the interests of the nation, increasing manufactures and decreasing imports. Patents, monopolies and state-sponsored projects permitted, in Lorna Hutson's words, 'the nationwide mastery of national resources; specifically, with the patriotic motivation of self-interest in the undertaking of economic projects'.[43] Many claimed industrial advantages and boasted of employment potential. Smith recorded his ideas in *A Discourse of the Commonweal of this Realm of England*, written while in exile in 1549 and first printed in 1581. There he argued that 'trifles' imported from abroad could be made in England. This would reduce imports of a huge range of items, including glass objects, cards, puppets, toothpicks, knives (from Turkey), aglets (metal tips which prevented the fraying of lace ends), buttons, pins (from the Netherlands) and paper. Irritated most of all by the circumstance that many imports had processed cheap raw materials sourced in England, even English waste products, to which value was added only through labour, Smith complained that 'of our fells they make Spanish skins, gloves and girdles; of our tin, salts, spoons and dishes; of our broken linen cloth and rags, paper both white and brown'. England ought to get herself into a position to add value to its own resources, both raw materials and reutilisable matter.[44] From early in Elizabeth's reign, patents were drawn up for various chemical and mechanical processes, some of which involved recycled or reused materials. These included soap-making, pin-making and alum manufacture.

Starch, another product protected by a patent, was manufactured from wheat, or the bran of wheat, thereby creating an alternative use for a foodstuff. Much starch had previously been imported from the Netherlands, but domestic production increased. By the seventeenth century some starch-makers combined their trade with pig-fattening: the swine were fed on bran meal left behind after refining.[45] Starch was used for the satisfaction of vanity: stiffening ruffs and conditioning hair and wigs. The fashion for wearing starched ruffs accelerated during the 1560s. Ruffs, which proclaimed that the wearer had someone to tend to them, supported various employments: the making of ruffs, the making of starch and its application to the wealthy be-ruffed. Although he himself had once been depicted atop a mule, wearing a starched ruff, William Cecil argued in Parliament that starch sated only the rich man's vanity, and that it caused hunger, taking wheat from the poor man's bread. It was probable, he thought, that some people would resort to bread made from acorns. Ruffs had reached 'such unconscionable length' as to become not only ludicrous but also physically obstructive. To avoid taxing ruffs, starch-making was instead limited; a monopoly was granted in 1588.[46] By the end of the sixteenth century it became clear that certain men were profiting handsomely (and parasitically) from patents or monopolies. A notoriety and controversy attached to them. There were mutterings about monopolies in Parliament in 1571, and a decade later abuses were glaring. The projecting spirit had led to monstrous and corrupt schemes: monopolies had been (mis)granted for the sale of necessary substances, such as salt, or for non-innovative developments.[47] The barrister Edward Coke, a key author of an anti-monopoly bill in 1621, punctured any remaining sense that there was truth to the commonweal lie about the idealistic nature of those projects: 'sometimes when the public good is pretended, a private benefit is intended.'[48]

A German jeweller benefited from the monopoly culture. John

Spilman repaired the coronation jewels of James I, but he made much of his money making crisp white paper: a real alchemy, turning paper into gold.[49] He was granted a patent to make paper in 1589 by a government aware of the need to import foreign methods.[50] John Tate had made coarser paper in the late fifteenth century, by repurposing an old corn mill. In 1588 Spilman likewise took over a wheat mill and a malt mill. Later, in Kent, fulling mills which had been used in the woollen industry were converted into paper mills.[51] Thomas Churchyard wrote about Spilman's enterprise:

> Through many handes, this Paper passeth there,
> before full forme, and perfect shape it takes,
> Yet in short time, this Paper yucke will beare,
> whereon in haste, the workeman profit makes.
> A wonder sure, to see such ragges and shreads,
> passe dayly through, so many hands and heads.[52]

Spilman, who hoped that 'Lynne Ragges, or shredes of Lynnen, or Clypinges or shavings of papers or Lether' would be banned from export (this wish went unfulfilled), used rags 'fitt for making all sorts of white paper', which would 'set Thowsandes on worke'.[53]

Greedinesse

It was in the aftermath of the commonweal humanistic reforms of the mid-sixteenth century, which dwelt on marshalling resources, that the pamphleteer and playwright Thomas Nashe wrote *Pierce Penniless* in 1592, as a nuanced and sensitive reaction to the English crisis of resources. Nashe's writings ridiculed possessive and greedy members of the elite, with their impersonal approach to governance.[54] Georgia Brown has detailed some of the ways in which Nashe revels in prodigality, how he 'brazenly proclaims that he has experienced poverty as a consequence of his unthriftiness and this guarantees the value of his text'.[55] In 1571 Henry Bedel preached a sermon in London, exhorting his congregation to pity the poor. People knew they ought to be charitable, he observed, but they generally were not; the 'purse is able, but the heart is not frank'.

One reason for this was niggardliness, which 'causeth many to profess such a needless necessity that it is kept from the poor'. Instead of charity, 'profuse prodigality wilfully doth waste'.[56]

With such sentiments as a backdrop strode Greedinesse in Nashe's *Pierce Penniless*, a Dickensian character before his time, wearing shoes made from crab shells (presumably still with grabbing claws attached), 'toothd at the toes with two sharp six pennie nailes, that digd up every dunghil they came by for gould'.[57] The allegory is highly specific. Nashe was satirically recalling a licence granted to William Treasorer in 1581 to export old shoes to the continent. Treasorer, an organ-maker who had also worked under Edward VI and Mary, had first been granted this right in 1555. Treasorer could export 'one Hundred thousand lasts of ashes' and 'foure Hundred thowsand dozens of old worne Showes' in consideration of his having made a 'newe instrument Musicall giving the sound of Flutes and recorders'. He had also promised to repair the organ at Greenwich. Treasorer, a German by birth, had come to England in about 1540.[58] Donald Woodward unpacks some of the uses to which the cast-off shoes were put, noting that they were shipped to France from Yarmouth. He cites Henry Belasyse's mid-seventeenth-century praise of English leather, held in 'great esteem abroad in so much that whole shipfuls of old boots brought out of England, and set up again after the French fashion, afford great gain to the merchant that bringeth them over'.[59] The footwear was repurposed. In 1598 one 'Ede Schetts' was granted the same rights that Treasorer had enjoyed, Treasorer having died four years previously. This was probably Edmund Schetz, a musically inclined page in the queen's household.[60] Rights to export ash and shoes appear to have become a perk for court musicians.

Greedinesse's cupidity is bottomless. Carrying a crowbar for a staff, he wears clothes that are all recycled: his hood is made from inscribed parchment, is 'buttond downe before with Labels of wax' and lined with fleeces sold by fellmongers; his cape is lined with cat skin and tasselled with fish hooks rather than aglets (one of Smith's 'trifles'); he wears breeches made from the 'lists' of woollen cloth. These were the selvages, or waste borders, which gave protection to buyers of cloth, their presence indicating that none of the length had been removed. Greedinesse is a projector. He has been granted cloth selvages 'by letters patent [...] to the utter overthrow' of artisans who used to make use of them in the production of bow cases and cushions (presumably as padding).

Nashe may have had in his mind the export licence recently granted to Simon Furner, a merchant, and John Craford, a scrivener, 'of carrying out of list and shreds of cloth'.[61] This duo had created havoc in 1592. In being granted the right to export refuse horns they undermined the trade of the horners, 'to their undoing in general'.[62] The arrangement appears to have been conveniently beneficial to Elizabeth. On gaining a twenty-one-year licence to 'transport listes, purrels [the selvages on kersey, a coarse woollen cloth], and shreddes of woollen cloth and all manner of hornes', Furner and Craford were required to pay most of their fee for this to the queen's cannoneers at Ostend, for services rendered between 1585 and 1588, when Elizabeth had supported the Dutch in their revolt against Spain (by 1592 the gunners had still not been fully paid).[63]

With breeches bulging 'like beerbarrels' with bonds of credit and forfeitures, Greedinesse has gained much from the misery of others. He is interested only in materials and situations that permit him to 'wrest an odd fine out'. Lorna Hutson has remarked on the 'legal security' emblazoned on his clothes, suggestive both of 'the proverbial hypocrisy of the hooded capuchin and of the snug comfort of legal privilege written on *parchment* (sheepskin) securely soldered up with the royal wax against the chill of political or economic uncertainties'.[64] Greedinesse is wed to Niggardize. In 1570 Richard Porder, a London clergyman, had sermonised that greediness and niggardliness were two sides of the same coin. The niggard, he suggested, was not 'so costly a spender', but he still desired to fill his own coffers at the expense of others: 'the greedinesse of their desyre is such: that they will stille fede the same desyre with gaine.'[65] Niggardize dresses in recycled wares too: 'a sedge rug kirtle', a loose pinafore-type garment 'that had been a mat time out of minde'; a shawl 'borrowed of one end of a hop-bag', both recalling projects centred on fen drainage to maximise hemp yields.[66] Niggardize accessorises too, with an 'apron made of Almanackes out of date'

and an old pudding pan 'thrumd' with nail pairings for her hat.
Everything is saved – even spit. Snot is barrelled up to use as grease
on woollens, a reference to the projects for finishing woollen goods
in England.

Together, Greedinesse and Niggardize formed the childless
house of Avarice. Nashe's brilliant critique presents hyperactive
reuse as an unholy alliance of the public and the private: what
Greedinesse brings in, Niggardize marshals. The house is 'vaste,
large, strong buildt, and well furnished' except for a small galley
kitchen and a cellar light on beer. Spiders – representing the nation's
weavers in this parody – cannot thrive here, where every resource
is reserved for thrifty Avarice. Rats and mice – small traders – carry
off Greedinesse's 'wel dunged' greasy codpiece.[67] The fecundity of
the household – the nation – was at stake. Hyperbole aside, in both
man and wife Nashe exposes excessive and protected reuse as indic-
ative of material immorality. Both halves of Avarice squander their
time engaged in waste extraction; Greedinesse sieves dunghills and
'shop-dust' by the cartload, just to 'gaine a bowd Pinne'. The projects that triggered Nashe's ire continued long after 1592, and though
some may actually have fostered industrious employment and
redistributed wealth, they remained deeply controversial. Monopolies were debated during the tempestuous Parliament of 1601, when
they were blamed for the inflation of the 1590s. One MP complained that key commodities 'are ingrossed in the Hands of those
Blood-Suckers of the Common-Wealth'.[68] Surely this was not what
Thomas Smith had intended half a century earlier? Indeed, many
of the schemes had not conformed to early hopes that innovation
would be protected and rewarded. Some were simply convenient
rewards or promises. Nashe leaves us in no doubt that an obsessive
ring-fencing of resources is damaging and wasteful. His target is
economic privilege granted by the state.[69] Abundance, not want,
caused this avarice.

🗑🗑🗑

What would Tusser, Markham and Fitzherbert have made of this?
The English community was taught from pulpits to be frugal and

thrifty, to save for the future. These were the ideals disseminated also by Tusser and his ilk. But by the late sixteenth century the rhetoric of austerity too obviously clashed with the protection afforded to projectors.[70] For a long time after Henry VIII fell out with Rome the English sought to shed a reliance on alum from the Papal States and Catholic Spain. This eventually led to the manufacture of English alum by processing burned shale with acid and human urine or potash made from seaweed. Sir Thomas Chaloner, whose father had purchased Gisborough Priory in north Yorkshire after the Dissolution, rather conveniently believed that shale found locally was especially suited to making alum. In 1607 he obtained a monopoly for thirty-one years, arguing that the business would create jobs, not only in manufacture but also in shipping coals, urine and alum. Continental manufacturers were brought over to kick-start the business and train employees.[71] Chaloner himself was eased out of control in 1609, when James I proclaimed alum a royal monopoly. Tons of barrelled urine were delivered to the works to be fermented to make lant (aged urine), to be used in industrial processes. Some urine came from the countryside, some from London. Coal was conveyed by sea from Sunderland to the works. The ships then took the alum to London, whence they returned with urine. The urine of poor labourers was thought best. Both urine and alum were also used in dying, as fixatives.[72] Ultimately, alum manufacture under royal monopoly was a failure: clothworkers were forced to accept an inferior product at a higher price; dyers complained that this alum 'wanted some time to lie and dry'; and much alum was smuggled in from abroad.[73]

It is thought that the shipping of urine for the alum monopoly originated the phrase 'taking the piss' – which is pretty much what Thomas Nashe thought of the whole sorry business.[74]

Conclusion

To Singe a Goose

The most poignant memento of my Nan's saving and recycling is a coupon for ½ lb. of tea – on the reverse of which she had scribbled a sum. This sum is overwritten with notes about plants. Needing more space, she had then used the face of the token, writing cross-ways over the coupon details (**fig. 59**). This wasn't the only example of her obsessive reuse of even the most seemingly trivial scrap: most of her notes were inscribed on chocolate bar wrappers which were

59 (a and b). Both sides of
a tea token reused multiple
times by Evelyn Bradbury.

kept flat, ready for use, under the cushion of her armchair (fig. 60). The number of times this one coupon was used sums up my Nan – to reuse this coupon several ways shows the lengths to which she went to eke out the life of everything, no matter how small or insignificant it was. Going through her things after she died in 1992, we found several packets of unopened writing paper.

Ardent reusing can seem out of place in a world full of material goods, and the degree to which Nan took it bespeaks a strange relationship with things. Endlessly kept for best, some things are never used. Are those then wasted? Although she was fond of plants and wildlife, Nan reused things not for environmental reasons but because she could not abide waste. She was sixteen when the Second World War began, and was a married young mother at its end. In these formative years of household management she took to national salvage with aplomb. She saved and mended for the rest of her life. Many sheets in her bungalow had seams down the middle, a throwback to the 'make do and mend' period, when worn sheets were torn in half and hemmed edges were joined together to make use of unworn parts. Nan was from Birmingham, an exemplary salvaging city during the First World War. Her mother had imbibed some of the recycling vibe alive in that city, and would later pass this on. Nan was also influenced by her father – Cyril, an engineer – who made many of his own tools. I inherited quite a few of them; using them lends more satisfaction to my own repairs.

Nan would have despised the recent trend for 'upcycled art', partly because her reuse was intensely pragmatic, partly because she would have found these decorations generally ugly. One can, of course, find examples of meaningful upcycling. The font of Norwich Cathedral (fig. 61) is formed from a vessel used for making toffee in the city's Nestlé Rowntree chocolate factory, which closed in 1994. This rather nicely flips various civic reuses to which defunct Catholic church paraphernalia was put in the sixteenth century. An eBay search turns up vendors who have upcycled or 'reclaimed' metal windows. One such item was described as 'Reclaimed semi-circular mirror upcycled from a Crittall window'.[1] Allotment holders sometimes reuse Crittall windows to make Heath-Robinson-style greenhouses on their plots. But pianos upcycled into

60 (a and b). Both sides of a Yorkie Bar wrapper used as scrap paper by Evelyn Bradbury. Author's collection.

drinks cabinets would have made Nan's lip curl. What she would have made of the long history of reuse and recycling is impossible to know, but I do know that she would have been cross with me for not writing it entirely on the backs of previously used papers.

Recycling presents no linear history of improvement. There is

61. Font, Norwich Cathedral, made from a toffee vessel.
Author's collection, photograph by Maud Webster.

no neat story of progress, but instead one that pings and pongs stickily like an over-napthaed rubber ball. Economic, military, industrial, urban, environmental and demographic pressure on material resources did not always elicit consistent responses. Patterns of consumption, the availability of raw materials and commercial products, and the costs of transport and labour all affected the quality and quantity of material reutilisation. Neither was there a straightforward decline from easily detectable halcyon periods. Donald Woodward concluded his article about early modern recycling with the assumption that the period had seen unusually high levels of recycling, triggered by 'population growth, rising prices, and a growing economic polarization of society', speculating that the eighteenth and nineteenth centuries saw 'lower levels' of recycling.[2] This is contradicted by evidence that reuse in mid-nineteenth-century London was greater in scale and variety. Recycling has taken different forms since the sixteenth century. Environmental concerns now have a greater impact on modern recycling; it is organised by local authorities, and geared towards specific cultural values. Historic recycling was generally ad hoc. It made more use of private networks of material redistribution and was less focused on moral or global judgements than on economic and resourcing concerns.[3] In making sweeping statements about the course of salvaging history, do we consider the percentage of all materials that were recycled, or the volume of reused stuff, or the time spent reusing things?

Victorians, who marvelled at how science permitted reuses that their forebears had never so much as imagined, would certainly have taken issue with Woodward. A detailed article from 1846 – probably written by Lyon Playfair – pointed to 'astonishing progress' made in reutilising wastes such as bones crushed in the bone mills, formerly 'thrown aside as offal, fit only to be buried out of sight'. 'How our forefathers would have laughed at the prediction of bone-mills, and British soil fertilised with ship-borne bones from Germany', remarked the article's author, before considering wastes from gasification, soap-making and sugar-refining: night soil, soot, urine. These reuses would, apparently, have astonished their ancestors, who had simply looked on these as 'nuisances to be rid of'.[4] Chemistry is cast as 'the grand economiser' and 'Nature's housewife, making the best of everything'.[5] 'Pressure of population' is identified in a key article from 1868 as the main force driving the reuse of things 'that not very long ago were utterly unutilized'.[6] James Platt, the man whose company made the buttons turned into a whistle, wrote a book about the economy in 1882. He quoted heavily from an article written eighteen years earlier, lifting passages on 'the utilization of hitherto useless things', and observing that

> what our forefathers neglected or despised we have learnt to appreciate; what they threw away we carefully gather up. Nothing is too small or too mean to be disregarded by our scientific economy. The seeming rubbish and fag-ends of creation, which our ancestors would gladly have thrown over the garden wall of the world into the limbo of chaos or of space, are now converted to profitable purposes, conductive to the greater comfort and prosperity of life.[7]

At the end of the century, a report about arsenic recovered from copper-mine waste took up the refrain: 'What our fathers and grandfathers threw away, […] we find profitable to work for something it contains which was unknown or disregarded by them, and has since acquired a new value'.[8] Thriving through thrift was represented as new, but we cannot be sure if it really was. What we can say is that there *was* an increase in the attention paid to reuse by the male literati in the nineteenth century.

In gauging the meaning of their own accomplishments, scientists and businessmen of that era overlooked the fact that increasing industrialisation produced increasing quantities of waste. There was simply more matter to reuse – so much that it caused nuisances such as heaps of alkali and other hazards.[9] Some inventions were spurred by finding uses for pollutants.[10] But this was often interpreted as a moral revolution, rather than a pragmatic response to changed circumstances. A similar tendency can be identified in the aftermath of the First World War, when Henry Spooner lamented 'past folly and extravagance'.[11] Frederick Talbot was so focused on industrial and commercial reuse that he ignored the humdrum reuse of materials by 'the man-in-the-street and the woman at home', remaining resolutely fixated on the accumulation of all the nation's waste with the aim of boosting its commerce. He complained that poor people often just burned their waste.[12] It is a classic example of the recurring lesson of history: the possibility of reusing wastes in novel ways saw them earmarked only for those purposes, while criticism was attached to other uses. Whereas citizens were told to avoid burning any paper during the Second World War, fifty years earlier newspaper readers were advised to squeeze soaked pulpy paper into balls, which, when dried, could be burned on the fire.[13] Thrifty reuse in one context is selfish waste in another.

A blame game

Most British people have only recycled when they have had to, and when they have done so, they have often preferred to embrace it on a small scale. In August 1940 a young widow pondered the durability of the nation's enthusiasm for salvage: 'I wish I could think that it were likely to survive the war', she wrote. The diarist wondered why metal, paper and bones were not collected in peacetime: 'It is really no trouble to sort out tins and papers and bottle-tops, in fact it has become almost automatic by now. Why does it take a war to make us sort these things, and to make the government arrange to use them?'[14] Maybe the answer to this is that only in wartime are materials in such short supply, and the need to be careful with them can be painted as patriotic; at other times, when people are free to waste as they can afford, they do so. In fact, consumers are often urged to buy new to support economic growth.

National and cultural identities have often shaped attitudes to reuse, and such differences are sometimes used to determine the moral value of those attitudes. In general it seems the Scottish have been better recyclers than most Brits (except, maybe, Yorkshire folk or Brummies). Londoners appear to have been the worst household recyclers of modern times, despite their access to a thriving market for reused matter. When, in 1886, rates at which domestic refuse was salvaged and reused across the country were compared, London fared badly in comparison with Glasgow.[15] Recycling entrepreneurialism or promotion appeared to be common among émigrés or their offspring; immigrants arrived with continental methods to apply to discarded matter. The Germans were the real alchemists, turning waste into gold, and British thrift was no match for German frugality.[16] This apparent truth spawned a durable mythology. 'Now the German, when he encounters a waste', noted Talbot with annoyance and admiration, 'does not throw it away or allow it to remain an incubus.' 'Saturated with the principle that the residue from one process merely represents so much raw material from another line of endeavour', Talbot went on, 'he at once sets to work to attempt to discover some use for the refuse.' Talbot blamed British culture for wastefulness – and for missed opportunities, such as in the dye industry, where Perkin's ingenuity had been transformed into German profits.[17] Captured German battleships were recycled by Thomas W. Ward of Sheffield, and other breakers, after the First World War. Soon the surplus was so unmanageable that the British government resold some redundant ships to German companies, who towed them away for demolition: HMS

Hercules was scrapped in Kiel in 1922.[18] In both world wars other countries assumed that German supplies would run short long before they did. Marshalling of resources that placed reuse at the centre helped the Germans to fight for longer.

If geography played a large part in people's attitudes to recycling, so too did class. Among those in the upper echelons of society, reuse was most prized when items were kept within a family or possessed a secure provenance. The heritage of such objects is bound up with the prestige of ownership; from this, value attaches to the modes of acquisition and redistribution: thus auction houses and antique dealers are deemed facilitators of acceptable reuse, and someone homeless sifting through bins is deemed merely pitiful.[19] Fashion dictates that value changes more or less arbitrarily: mauve crinolines; xylonite shirt collars; plush Wombles. The actual utility of such objects often remains unchanged; they are just less sought after and possess less cultural cachet. At that point they are cast aside for less fashionable (poorer) people to use. Charles Babbage presented this as a democratisation of the things themselves.[20] The new plastic items that appeared at the end of the nineteenth century also inaugurated a form of this: Bonzoline billiard balls which looked like ivory; celluloid combs in imitation of tortoiseshell; pianos bedecked with *bois durci*. Henceforth poorer people could afford to have more new things because new products were not made from scarce and expensive resources. *Needing* any old thing has become a mark of social failure, the ability to replace it with a new item a form of class distinction.

It was always 'other' people who were blamed for wasting and not reusing: the rich blamed the poor (who processed much of the waste); adults blamed children (though children have long been important agents of recycling); and men blamed women (who gathered much of the waste).[21] With all the talk of forefathers, and what they might think, it is easy to forget that it was women who performed most of the industrial recycling legwork. What would our foremothers have thought? Female labour was cheap, and women undertook the majority of domestic recycling and reuse as well as much commercial and industrial sorting and separation. Female involvement was largely overlooked when it occurred, and it has been underplayed since, despite Susan Strasser's valiant efforts to highlight feminine reuse in *Waste and Want* (1999).[22] Talbot pointed to thousands of tons of string, which, 'thrown into the waste-bin, become a tangled mass'. Untangling the string 'was a task calling for weeks of labour and infinite patience', so he suggested that prisoners might profitably be put to work on this, freeing string for reuse 'in the overhaul and repair of [mail] bags'. He reasoned that prisoners' time and labour 'does not count'.[23] Likewise women's time. When officialdom did address women, it was to hector them to do better, not to congratulate their efforts, and yet women kept many recycling trades going. Consider the dust heaps; the papier mâché factories; the rag-sorting centres; the shoddy mills; women sieving the arsenical pyrites; and all manner of collecting, sorting, sifting and processing.

Filthy obsolescence

The value of repair work was, and still is, limited by the price of new items, and also by an aversion to paying for things to be returned to an original (but no longer perfect) state. When goods became cheaper, repairers struggled.[24] It was not financially viable to charge twopence for a bucket repair if a new one could be bought for the same price. The concept of built-in obsolescence was actively promoted in 1932 by Bernard London, a Russian-American estate agent who reasoned that, as the Depression was man-made, so it could be solved by man. He advocated policies to remove incentives to reuse items and urged manufacturers to create goods with a limited lifespan, or what is sometimes called 'contrived durability'. This, he believed, would animate consumption, necessitate production and stimulate the economy.[25] From the 1980s onwards, more businesses pursued policies of contrived durability. Such items include technical gadgets with tamper-resistant screws and otherwise well-built domestic goods with flimsy plastic handles or unsecured circuit boards.[26]

Consumers have been slow to embrace recycled products. Gordon Hall Caine, in charge of a dwindling paper supply during the First World War, lamented that some British papermakers shunned wastepaper. Printers also demanded spotless sheets, with 'neither dirt nor speckle', placing further limits on reuse.[27] Historically, idiosyncratic national tastes have added new processing demands – for white toilet paper, clear plastic bottles and glass bottles in fixed colours.[28] It is difficult to argue against disgust. Some reuse induced disgust through sensory nuisance: pig-fattening; soap-making; bone-boiling. Fritz Viëtor never drew attention to the slaughterhouse blood in plaques he sold to make pianos fancier. Meat and bone were concealed in animal feed. This idea of the hidden and the fake, of the disgusting cloaked by a thin but desirable veneer, has informed public scares that have affected recycling – for better and for worse. There have been times in British history when people were less willing to share or recycle things because they feared communicable diseases.[29] This had implications for shoddy and paper production when fears were raised about pestilential traces.[30] At the close of the eighteenth century woollen rags from London were prized by farmers as a soil improver, but many were wary on account of 'the danger of catching the small pox in chopping and sowing them'.[31] In 1882 a doctor reported that 'certain effluvium-producing businesses' dealt with 'refuse matters', some of which were 'liable to be infected with the specific contagia of infectious diseases'.[32]

Even the present era can harbour a wide range of sensibilities that inform attitudes to salvage, saving and reuse: I know an academic who refuses to borrow library books. He sees my collection of wartime recycled dust jackets as a symptom of insanity, unsanitariness or both. It is not a new fear. A label appears inside a second-hand library book, inserted in the 1930s, when illnesses like scarlet fever, influenza and diphtheria circulated: 'Readers should report to the Local Librarian any case of infectious disease occurring in the house while a Library Book is in their possession.' In 1942 the Bristol Education Committee insisted that James Ross's book salvage team handle donated books carefully, to avoid the risk of infection.[33] Victorian book-fumigating machines were designed to remove contamination.[34] But dirt in one era is prized in another, and fear of contagion thrown to the wind (or never countenanced). Robert Hooke, the seventeenth-century scientist, architect and inventor, was obsessed with his own health and lapped up news that powdered human excrement, 'if blown into the eye[,] was an excellent powder to take of films and cloudes from the eye'.[35] While some of us are happy to collect items from the past, and fill shelves and cupboards with interesting objects rummaged from extinction, there are some things that everyone can agree are better left to the ages.

Pulped fiction

Contexts determine values. A glut of reusable waste causes prices to dip, and new inventions can increase demand, pushing prices up.[36] Before ammoniacal salts (a gasification waste) became commodified in 1840, bones were used to make ammonia. The value of bones had steadily increased; they were also used in manure-making. Rising prices saw bone neglected by glue- and size-makers, in favour of cheaper materials.[37] Some manufactures became unviable as waste was demanded by new industries. The market for kamptulicon flooring had already collapsed by 1903, when it was reported that cork wastes were more valuable than ever on account of the dwindling of Italian raw cork source (cork-yielding trees were increasingly being felled to make charcoal and potash).[38] Specific shortages of one material can have a knock-on effect. In 1815, when British meat consumption declined as a result of poverty, problems were experienced in industries relying on by-products of meat production. Stocks of tallow used to make soap and candles dwindled, and imports from Russia were much increased.[39] By then the term 'Russia' was used to describe all sorts of bad tallow: the 1770 inventory of Bristol soapmakers Farrell, Vaughan & Co. listed supplies of 'bad mix'd tallow, mostly Russia, being the produce of Bad Candles melted down'.[40]

In the eighteenth century wastepaper was valuable to cooks, shopkeepers and trunk-makers, but not to printers, for 'after it is printed [it] is of no intrinsick value at all'.[41] For makers of bespoke items, a variety of materials could be used to good effect. Industries that ground or pulped waste materials could weather some differences in quality, but those requiring homogeneity found recycled substances disappointing. Even cesspool waste varied, not just because of the length of time it had awaited collection, but also according to the diet of the donor: 'the sewage of Belgravia would be richer than that of Bethnal Green.'[42] Materials of varying qualities could not be easily incorporated into industrial processes so finely gauged as to require precision. Bits of old mackintosh in the rag mix spoiled paper. Uncertainties about the volume of supply triggered industrial wariness, especially in businesses requiring new machinery to process reused material. Investment was viable only if supply could be guaranteed.

The value of a piece of paper, or parchment, can be determined partially by what is inscribed on it: banknotes, deeds, autographs. However, sometimes the materials bearing inscriptions have proven to be more valuable than the words or images. We have seen this with brass 'palimpsests'. Traditional palimpsests were surfaces for writing or drawing which could be reused, erased or cleaned and written over. Just as sailors in the later Royal Navy would holystone the decks of ships like the *Royal George* to make them look new, so medieval monks abraded old vellum manuscripts with pumice stones and knives so that new scripts could be written atop. Ultra-violet light can now disclose former labours, which appear as ghostly under-layers.[43] In such items it is plain that the parchment was afforded a greater value than the texts written on it. The first folio of a collection of devotional texts attributed to the hermit Richard Rolle, copied out in the late fourteenth century or early fifteenth, has a blank rubbed patch left ready for reuse. This volume, produced on reused parchment, then passed through various hands.[44]

Secrets can be exposed when items are sold or donated for recycling. With this in mind, modern businesses shred paper before dispatching it. This is a trick that could have prevented a riot: as we have seen, problems with the Scottish prayer book in 1637 were fuelled by advance notification of the contents, because pages had been used as food wrappings. The Admiralty stripped gunnery gear from vessels before sending them for salvage in order to avoid artillery secrets and advances getting into enemy hands.[45] The quest to maintain secrecy got out of hand in one instance in 1881, when a firm of solicitors had the 'shrewdness' to go and witness the destruction of their waste at the papermaker (a service also provided by papier mâché makers). Just after leaving, smug with their prudence, one partner returned 'in a panic' because he had realised 'that amongst the pulped paper was a policy of insurance upon a ship which had been unfortunately lost at sea!'[46] The avoidance of regret has always been important to how we think about reuse.

Donors can also incur other more worldly costs, chiefly of transportation. The costs of driving glass to bottle banks in the late twentieth century were absorbed by householders and do not appear in assessments of energy savings through recycling. Offers by railway companies to transport salvaged material on favourable terms made the movement of materials more cost-effective during the Second World War.[47] A century earlier, London street waste was profitable only if carted to a destination within three miles, or if farmers bringing produce to market were willing to return with carts laden with it.[48] Furnace slag, the lumpy residue left after smelting metal ore, became a cumbersome waste for the Victorians. Land was purchased to dispose of it.[49] 'Hundreds of labourers', commented Simmonds, 'are engaged in conveying to remote and undisturbed spots the enormous piles of black, friable, clinker-looking stuff.' Some of this 'metallic encumbrance of the smelting works' was deposited in gullies, some dumped into ravines, some shot into the ocean.[50] That's both human and environmental costs.

Bashley Britten, a civil engineer, suggested in 1876 that iron-works and glassworks might be combined to utilise slag in the manufacture of the glass and remove the costs involved in shifting the slag.[51] By the time Britten lectured, and Simmonds published the second edition of *Waste Products and Undeveloped Substances*, blast-furnace slag had been put to several more experimental uses: as imitation stone or marble, as paving and road building material, as bricks, cement and fertiliser. Slag could be moulded and cast, and even blown into a mineral wool used for lagging.[52] 'Most of these schemes', noted Britten, 'have failed and been abandoned', chiefly on account of the cost of transportation. Iron manufacturers found slag to be 'scarcely remunerative'. They could not get rid of enough to avoid having to heap it into landscape-sullying piles of 'ugly refuse'.[53] The value of a reusable material is related to the value of alternative raw materials, and affected by processing, handling and transport costs. This is why an apparently profligate attitude to any one material is not a good indication of a society's wastefulness; glass bottles were abandoned in Victorian dumps because glass was generally cheaper to manufacture than to recycle.

National stewardship

The resource needs of particular eras are weighed up against the ambition to steward the past for those to whom its pieces are handed in the future. National crises led to temporary shifts in the conception of waste and reuse pressures. Patriotism elevated the notion that there were limited quantities of things, and those things needed to be used in ways that appeared fair from the perspective of the commonality. This led to a greater intervention in the waste habits of others, embodied in busybody uniformed salvage wardens. During recession, domestic belt-tightening and sudden introspection concerning household consumption becomes apparent. Flick-flacking patterns of diligence and *laissez-faire* recurred until environmental issues complicated the picture, most fundamentally after the late 1960s. A see-sawing between 'stewardship of my stuff' and 'stewardship of the nation's stuff' was henceforth complicated by a new obligation: 'stewardship of the Earth'. Recycling slowed in the scrabble for opportunities in the boom years of the 1980s. Picking up again in the twenty-first century, some recyclers were attracted to a new moral stance and a global outlook, albeit sometimes to counter guilt induced by ever larger carbon footprints.

Reuse also requires a consideration of the national heritage. Judgements as to the value of documents can later prove regrettable. It would be difficult to assess how many records of potential use to the historian were taken from councils and solicitors' offices to be dumped in salvage during the wars (let alone those reused to patch wagons or flown as kites by Puritans during the early modern period). Archivists drew attention to the 'indiscriminate destruction of historical material' that had taken place between 1916 and 1918.[54] The Mass Observation team observed that documents handed in by 'libraries and private people' during the First World War would have been 'of immense interest to anyone trying to reconstruct the atmosphere of the war years'. The only records wholly preserved related to 'big, diffuse, impersonal things'. Mass Observation's mission was to 'salvage history', literally and metaphorically.[55] These concerns were old. William Downing Bruce, a barrister and antiquarian, drew attention to the neglect of parish registers in 1850. Bruce had found evidence that some of the registers of Otterford, Somerset, where the churchwarden was a shopkeeper, had been used to 'enfold his goods'. In South Otterington in Yorkshire, register pages had been used 'to singe a goose'.[56] Registers had been repurposed as lace-making templates, or kettleholders. The parish clerk of Plungar, in Leicestershire, who was a grocer, reused the registers 'as waste paper for wrapping up his grocery commodities'. In Northamptonshire an 'old parchment register' had been 'sewed together for the tester of a bedstead'. Parts of other registers were used to 'direct [...] pheasants' or to 'write bills on'. Another parish register was hacked to pieces, the clerk being a tailor who, 'so often as he wanted slips of parchment for measures, made no scruple to cut out the old written register sixteen leaves', leaving only 'one whole leaf and two bits, containing entries from the year 1633 to 1636, and some odd dates from that time to 1645'.[57] A magnificently illuminated letter written by Elizabeth I to the Emperor Wan Li of China in 1602, which George Waymouth failed to deliver, wound up in a bran bin on a Lancashire farm, before being deposited in the Lancashire Archives.[58]

The MP Megan Lloyd George warned in 1943 that care should be taken to avoid destroying old records or books of value when rummaging around for paper salvage.[59] Despite attempts to forge links between archivists and wastepaper merchants, 'owing to a genuine misunderstanding' there had been 'a veritable holocaust of rate books, poor law records and other valuable documents in the custody of local authorities'.[60] Many Rutland Quarter Sessions records went for pulping.[61] Coroners' reports for Hastings made between 1868 and 1928 were sent to salvage in 1942.[62] A Home Office circular advised a reduction in the embargo after which coroners' reports could be destroyed, from fifteen years to ten.[63] Some archives benefited from salvage drives: the Wiltshire and Swindon History Centre holds a selection of poor law records 'from papers put out for salvage in 1945'.[64] Northamptonshire

Archives gained many items, including: a Crimean diary by a surgeon, William Alexander Barr, which makes reference to Florence Nightingale; ledgers kept by an Oundle cordwainer in 1872; and a list of shareholders for the Northamptonshire Iron Company, founded in 1853, a year after smelting resumed in the county.[65]

The Bristol Paper Drive of 1942 – which saw the set of *Master Humphrey's Clock* handed in – also vacuumed up many other treasures, including first-edition volumes of Edward Gibbon's *History of the Decline and Fall of the Roman Empire*. More amusingly, a 'lurid-covered thriller' entitled *Paper Salvage Crime* (by H. H. Clifford Gibbons, written under the pseudonym Gilbert Chester) was also donated.[66] A letter to *The Times* by 'Quickswood' urged caution: 'No readable book should ever be destroyed – least of all those which silly people call "trash": that is, the undistinguished fiction of a bygone age. These novels are valuable sources for the history of social custom.' I have been unable to find a copy of *Paper Salvage Crime*, no. 30 in the Sexton Blake Library series. Quickswood identified 'old directories' (of great value in my own researches) as jettisonable.[67] Over two million copies of unsold Mills & Boon novels were used in the surface of the M6 toll road at the start of this century. Richard Beal, project manager for Tarmac Central, remarked that 'the books are renowned for their slushiness but when pulped they help make the road solid'.[68]

Early plastics were made of recycled matter. Pulpware, *bois durci*, Parkesine, celluloid and the rest used cotton and paper waste. Gradually, plastics developed into soft and hygienic materials – not themselves easy to recycle but seen as perfect wrapping and packaging – replacing paper and becoming ubiquitous in modern life. At first a form of recycling themselves, plastics caused serious reuse and recycling problems. As well as the plastic littering our oceans, we can look back to earlier instances of difficulty. The addition of nylon and terylene fabrics into rag bags of the second half of the twentieth century undermined mungo and shoddy manufacture. In a guide to DIY repairs available early in the last century, householders were warned that xylonite-handled knives were particularly tricky to repair, on account of their flammability.[69]

The resale value of plastics was low. A dozen Bonzoline billiard

62 (a and b). Crystalate billiard balls, box photograph
by Toby Sleigh-Johnson, close-up by author.

balls, stolen 'in a drunken freak', were the subject of a curious debate
in the Old Bailey in 1907. Asked to value them, the secretary of the
Bonzoline Manufacturing Company argued that he 'did not recog-
nise the existence of second-hand billiard balls' since Bonzoline balls
'do not wear out'. One general dealer suggested the balls were worth
only about eight shillings; another declared them to be 'unsaleable'.[70]
Marketers of billiard balls made of Bonzoline (also 'crystalate')
stressed that they had 'all the merits of ivory', but 'without any of the
drawbacks': they didn't crack, they were a consistent colour, 'more
durable, more reliable and not subject to climatic influences'.
'Beware of imitations', barked a box of Bonzoline snooker balls, and
with extreme cockiness Bonzoline billiard balls described themselves
(via the tusk of an elephant) as 'better than ivory'. And yet the cracks
are showing on the crystalate balls (fig. 62).

'From here we grow'

From this *Rummage* some key developments and ideas have emerged, some of which may be instructive. First, and perhaps most importantly, *reusing* is more sustainable than *recycling*. Back in 1973 the charity Friends of the Earth campaigned to retain returnable bottles with deposits. Outside Scotland, the deposit bottle has all but disappeared, although there are encouraging signs that the political tide is turning back towards returnable deposit bottles.[71] The only significant bottle reuse is in the dairy industry, but only a tiny percentage of all milk is delivered in glass bottles by milkmen; most is purchased from supermarkets in plastic bottles. The Dagenham Closed Loop plastic milk bottle recycling facility opened in 2008 and recycled huge quantities of high-density poly-ethylene (HDPE) bottles. Most of the bottles containing milk sold in the UK had been processed through this company until it encountered difficulties in 2015, when a dip in oil prices triggered a fall in the price of virgin plastic bottles. The demand for recycled bottles declined. The company was taken over by Euro Closed Loop Recycling, but that company went into administration in 2016.[72]

Second, people in the past did not regard domestic recycling as something to be co-ordinated with local or national government. More reuse and recycling were carried out in the home, and more could be still. How much people recycle or reuse is bound up with how much they consume. The term 'recycling' is twisted by modern tabloids when detailing clothing worn by celebrities such as the duchess of Cambridge. These people only have to wear an outfit from their wardrobe twice to be congratulated for thrift.[73] While the prodigality of the rich might be a mark of honour, the allegedly wasteful poor have often been unfairly cast as the scum whose ways would sink the economy. In lambasting his 'wantonly extravagant' fellow citizens, Frederick Talbot had the temerity to complain that the poor wasted more because they consumed less. Assigning their wastes to rag-and-bone men, scrap merchants or wardrobe dealers was worthless, he claimed, simply being 'so trifling as to be deemed quite unworthy of consideration'.[74] No *Blue Peter* badges for them. But reuse habits were never simply about individual consumption patterns. There were multiple forces at work, especially when the whole nation experienced material shortages. Everyone was supposed to pull together during the war, to save resources. The patriotic dimension of recycling is deeply suspect; richer households could have provided more during both wars. They did proportionately less.[75]

Third, we ought to guard against the inconsistencies that creep into decisions about material reuse, both individually and as a nation. The British have shown themselves apt to interpret national need idiosyncratically. While homeowners and councils up and down the land were clinging on to their railings during the Second World War, a heap of used razor blades was amassed. This, despite the Controller of Salvage at the Ministry of Supply having stated that such collections 'involve an amount of effort that is incommensurate with the result'. It took a million blades to make just a ton of metal.[76] Undaunted, mainline railways had collected nearly four million blades by August 1944.[77] In Cirencester, where councillors doubted the need to collect municipal railings, a civic razor-saving scheme was deemed an almighty success.[78] In paying attention to such small details, they missed the bigger picture. A similar thing is at work in the modern world, where much emphasis is placed on recycling excessive packaging, distracting from bigger issues – domestic paraphernalia, clothing and other items that could be recycled or repaired much more than they are. Of course, the reason why the small wins are stressed is that, in order to engage with the larger problems, people need time, space and skills. Those resources are lacking in a society structured by the expectation that one will work all week to spend money on new delights at the weekend. Most would consider it eccentric to spend their weekends fixing machines and repairing clothes, and this is a problem.

Finally, material processing ought not to be seen as one simple activity – one substance or item turned into another. Holistic material reuse involves deeper considerations about what can be recycled or reused. *Rummage* has uncovered processes developed by people who went one step beyond a simple reuse. We've seen how the dust left by coke-making was processed into artificial fuel; how bone-black was revived to filter the colour from sugar (the bone-black being the first reuse, the revived bone-black a subsequent one); how Bismarck brown, a dye made from fuchsine wastes, was itself made through recycling. We've also touched on the wastes of the shoddy processes used for fertilisers, papermaking and dye production; suds from soaps (made with waste products) reclaimed and used as lubricants; mungo waste turned into flock wallpaper; and the combination of soap and another product of reuse, coal tar, to make medicated soap. If more modern industries focused on re-recycling processes, more waste could be recovered. A true recycling culture is multi-directional, interlinked and culturally engrained: it does not revolve around single issues or a single type of material, and cannot be wholly engineered from above.

I have emphasised cultural values through an examination of sentimental attachment, democratic possession, aesthetic judgement and different levels of functionality, avoiding calculations of financial worth and intrinsic value. Being worked on adds value to some reuse items that themselves might be acquired cheaply. This has long been a focus of industry, commerce and governments. It partly motivated attempts to wrestle recycling ventures back from the Continent in the sixteenth and seventeenth centuries. An object or substance can be rendered worthless in the context of excess or surfeit, or when reuse is made problematic owing to health hazards.[79] The reuse of discarded objects reveals contemporary idiosyncrasies and tendencies.

The word 'rummage' originated in the arrangement and rearrangement of ships' cargoes.[80] Many of the recyclers introduced in the course of our *Rummage* spent time aboard ships in their youth, where they may have noticed the ways sailors managed limited resources in limited space at sea. Some maritime developments seem close reflections of the state of recycling culture at the time. Take, for example, ships' bottoms, the location of many recycling initiatives using copper, paper wadding and coal tar to combat dry rot, the 'worm', or barnacles and algae. In the 1860s an admiral 'enquired whether Parkesine resisted projectiles, and whether it would be a useful insertion in the armour-plating of warships'.[81] A little later, in 1881, it was reported that artificial woods, such as *bois durci*, which often incorporated reused matter, were likely to be used to cover the bottoms of ships made from iron or steel.[82] By 1907 celluloid – another early plastic – had been used experimentally to protect ships' bottoms against parasites, but it was not cost-effective.[83]

One unexceptional year, hundreds of acorns dropped from an oak tree in an English grove. Some were eaten by red squirrels or birds; some were taken as 'mast' and consumed by pigs. Some grew into oak trees; and some of those trees were felled in the middle of the eighteenth century, hewn into the hull of the Royal Navy's magnificent flagship, the *Royal George*. Two pieces of timber from that ship now sit on my desk, having been salvaged from the sea in the early nineteenth century. They enclose a book whose pages were made from recycled linen rags in 1840, possibly incorporating rags from Germany, Russia or Italy.[84] The leather spine was tanned in a process using oak bark. The characters of the title, *Loss of the Royal George 1782*, were made from gold recovered from sweepings (**fig. 63**). An acorn grew into a mighty oak, which grew into a mighty

63. Henry Slight and Julian Slight, *A Narrative of the Loss of the Royal George at Spithead*, 4th edn (1841). Author's collection, photographed by Toby Sleigh-Johnson.

ship, which sank into a mighty ocean to be salvaged for other uses. Now the mighty oceans are clogged with plastic debris. Who will salvage that?

Notes

Bibliographical Abbreviations

BHO   British History Online

BM   British Museum

HOC   House of Commons

MOA   Mass Observation Archive, Mass Observation Online, Adam Matthew, Marlborough, MA, Observation Online, http://www.massobservation.amdigital.co.uk

NSC   National Salvage Council

OB   Tim Hitchcock, Robert Shoemaker, Clive Emsley, Sharon Howard, Jamie McLaughlin et al., *The Old Bailey Proceedings Online, 1674–1913* (www.oldbaileyonline.org, version 7.0, 24 March 2012).

ODNB   *Oxford Dictionary of National Biography* (online edn), www.oxforddnb.com.

TNA   The National Archives

V&A   Victoria & Albert Museum, London

1. Time Up for *Master Humphrey's Clock*

1. Bristol Archives, James Ross, 'National Book Recovery Drive', appended to Salvage Drive minute book, 1941–44, M/BCC/SAN/14/1; 'Good Books and Rubbish', *Times*, 9 January 1943, p. 2.

2. Walter Benjamin also threw focus on the remnants of daily lives ('to dwell means to leave traces') in *The Arcades Project*, trans. Howard Eiland and Kevin McLaughlin (Cambridge, MA, 1999), pp. ix, 9.

3. 'Reusing' and 'recycling' are imprecise terms. Brian Clapp has defined the latter as 'finding a new use for goods that have already passed through the productive cycle once'; many other definitions

are similar. Productive cycles have changed. They were – and are – often in flux or impenetrably complicated. For such reasons it is difficult to maintain a strict definition of 'recycling' across broad stretches of time. See B. W. Clapp, *An Environmental History of Britain* (London, 1994), p. 190.

4. Cited in 'Utilisation of Waste', *Review of Reviews*, October 1901, p. 402.

5. Jenny Jones, 'Is Your Recycling Being Incinerated?', *Guardian*, 8 January 2018; https://www.theguardian.com/commentisfree/2018/jan/08/recycling-incinerated-waste-china-ban-25p-coffee-cups

6. Frank Trentmann casts recycling as forging 'a virtuous link' between business and people – but not one in a closed loop 'or a self-sufficient local metabolism in which material energy keeps circulating through the same veins': *The Empire of Things* (London, 2016), p. 628.

7. Gervase Markham, *Cheape and Good Husbandry* (s.l., 1614), p. 97; Anon., *A Treatise of Oxen, Hogs, Sheep and Dogs* (London, 1683), p. 37.

8. Jon Vogler, *Muck and Brass* (London, 1978).

9. See Timothy Cooper, 'Rags, Bones and Recycling Bins', *History Today*, 56:2 (2006), pp. 17–18.

10. Arjun Appadurai, *The Social Life of Things* (Cambridge, 1986), 'Introduction', p. 5.

11. Eleanor Mayne, 'Recycling Rage', *Daily Mail*, 21 August 2006, p. 14 [FB04]; 'Join the Great Dustbin Revolt', *Daily Mail*, 24 April 2007, p. 1; Ian Gallagher, 'The First Recycling Martyr', *Mail on Sunday*, 22 October 2006, p. 1; Fiona MacRae, 'Puzzled by Recycling?', *Daily Mail*, 14 August 2008, p. 27.

12. 'Waste Paper', *Leisure Hour*, 30, July 1881, p. 420.

13. P. L. Simmonds, *Waste Products and Undeveloped Substances*, 2nd edn (London, 1873), pp. 31–2.

14. James Strachan, *Recovery and Re-Manufacture of Waste Paper* (Aberdeen, 1918), pp. 3–8, 17, 40–41.

15. F. H. Norris, *Paper and Paper-Making* (Oxford, 1952), p. 87.

16. Clare Taylor, 'PPA'S Deinking and Recycling Guidelines', www.ppa.co.uk, October 2015, p. 1.

17. See Beverly Lemire, 'Consumerism in Preindustrial and Early Industrial England', *Journal of British Studies*, 27:1 (1988), pp. 1–24; Ariane Fennetaux, Amélie Junqua and Sophie Vasset (eds), *The*

Afterlife of Used Things (London, 2015); Mark Blackwell (ed.), *The Secret Life of Things* (Lewisburg, PA, 2007); Madeleine Ginsburg, 'Rags to Riches', *Costume*, 14 (1980), pp. 121–35.

18. Elvis & Kresse, https://www.elvisandkresse.com/pages/our-story; see also Turtle Doves Cashmere, https://www.turtle-doves.co.uk/pages/turtle-doves-story; Anthea Gerrie, 'A Good Yarn', *Independent*, 11 January 2007, ['Extra'], p. 10; Annie Sherburne, http://www.anniesherburne.co.uk/about-1

19. David Adam, 'Green Idealists Fail To Make Grade', *Guardian*, 24 September 2008, p. 6.

20. George Monbiot, 'Black Friday Is Right', *Guardian*, 22 November 2017, https://www.theguardian.com/commentisfree/2017/nov/22/black-friday-consumption-killing-planet-growth

21. Author's own experience.

22. Robert Lilienfeld and William Rathje, *Use Less Stuff* (New York, 1998), p. 25.

23. 'Make Do Policy', *Birmingham Daily Post*, 30 November 1973, p. 20.

24. Susan Strasser, *Waste and Want* (New York, 1999), p. 10.

25. Hope Newell, *The Little Old Woman Who Used Her Head* (New York, [1935] 1973), pp. 100–02.

2. Stuffed Animals (1950s–1990s)

1. Census returns for 1881, 1891, 1901; 'Long Ashton', *Bristol Mercury*, 16 March 1898, p. 6; 'Long Ashton', *Western Daily Press*, 19 April 1901, p. 9; Elisabeth Beresford obituary, *Telegraph*, 27 December 2010, p. 35.

2. James Adair, 'My Family and Other Wombles', *Times*, 11 August 2007, [S3], p. 8.

3. BBC, *The Wombles*, 1:2, 6 February 1973; Adair, 'My Family and Other Wombles'.

4. HOC, *War on Waste – A Policy for Reclamation*, Green Paper, Cmnd 5727, September 1974, p. 2.

5. Margaret Wray and Michele Nation, *The Economics of Waste Paper Reclamation in England* (Hatfield, 1977), pp. 107–8, 266; 'A Valuable Load of Old Rubbish', *Aberdeen Evening Express*, 22 January 1975, p. 14; 'That's Not the Only Waste, Either', *Coventry Evening Telegraph*, 16 April 1974, p. 51.

6. 'Charity with Silver Lining', *Aberdeen Evening Express*, 29 December 1970, p. 5.

7. 'Charity Goods in Blaze Saved', *Coventry Evening Telegraph*, 5 January 1972, p. 21.

8. Vogler, *Muck and Brass*, p. 36.

9. Raymond G. Stokes, Roman Köster and Stephen C. Sambrook, *The Business of Waste* (Cambridge, 2013), p. 216.

10. A. W. Neal, *Refuse Recycling and Incineration* (Stonehouse, 1979), p. 40; Christine Thomas, *Material Gains* (London, 1979), pp. 86–8.

11. J. J. P. Staudinger, *Disposal of Plastics Waste and Litter* (London, 1970), pp. 14, 75–6.

12. 'Walking on Discarded Coffee Cups', *New Scientist*, 21 February 1974, p. 471.

13. Vogler, *Muck and Brass*, pp. 35–6; 'Knob Orders Boom', *Birmingham Daily Post*, 28 May 1973, p. 15; Ken Burgess, 'Point Was Not Wasted', *Coventry Evening Telegraph*, 25 May 1973, p. 3.

14. Department for Environment, Food & Rural Affairs 'Deposit Return Scheme in Fight against Plastic', https://www.gov.uk/government/news/deposit-return-scheme-in-fight-against-plastic, 28 March 2018.

15. John Windsor, 'Rubbish Reaches for the Moon', *Guardian*, 25 April 1973, p. 6; 'Bottle Lack Might Bring Back Milk Rationing', *Times*, 23 October 1973, p. 3.

16. David Jones, 'Packaging', *Times*, 23 June 1972, p. 14.

17. '330 Million Milk Bottles Lost in a Year', *Times*, 6 September 1961, p. 9; 'Enforcing the Law on Litter', *Times*, 31 July 1959, p. 4.

18. 'Bottle Lack Might Bring Back Milk Rationing', *Times*, p. 3.

19. Martin Sherwood, 'In Conference', *New Scientist*, 12 October 1972, p. 107.

20. Jeremy Bugler, 'Schhh! They're Returning the Non-Returnable', *Observer*, 4 April 1971, p. 8.

21. Martin R. Taylor, 'Mrs Donoghue's Journey', in Peter Burns and Susan Lyons (eds), *Donoghue v Stevenson and the Modern Law of Negligence* (Vancouver, 1991), pp. 4–20.

22. 'Monster Bottle-Washers', *Lancashire Evening Post*, 3 February 1926, p. 8.

23. Bugler, 'Schhh! They're Returning the Non-Returnable'; 'Friends of the Earth is Ten Years Old', *New Scientist*, 30 April 1981, p. 294; Jeremy Bugler, 'Schwepped into Action', *Observer*, 5 December 1971, p. 14.

24. 'Sch-h-h ... We're Not To Blame', *Birmingham Daily Post*, 25 October 1971, p. 16.

25. Bugler, 'Schwepped into Action'.

26. Stokes, Köster and Sambrook, *Business of Waste*, p. 164.

27. Andrew Porteous, 'A Case for a British Bottle Bill', *Solid Wastes*, 68:5, May 1978, pp. 207–14.

28. 'Plea To Return Empty Bottles', *Birmingham Daily Post*, 7 June 1974, p. 18; 'Drink Firm Say Money Thrown Away', *Coventry Evening Telegraph*, 20 May 1974, p. 32; 'Cheers! Now Where Are the Bottles?', *Coventry Evening Telegraph*, 4 June 1974, p. 41; 'Empty Attitude', *Birmingham Daily Post*, 6 June 1977, p. 5.

29. Windsor, 'Rubbish Reaches for the Moon', p. 6; advertisement, *Coventry Evening Telegraph*, 27 June 1973, p. 8.

30. Bugler, 'Schhh! They're Returning the Non-Returnable'.

31. 'Adding Up', *Birmingham Daily Post*, 19 August 1974, p. 1; Patrick Greenfield, 'Scotland Planning Deposit Return Scheme', *Guardian*, 5 September 2017, https://www.theguardian.com/environment/2017/sep/05/scotland-planning-deposit-return-scheme-for-bottles-and-cans; A. G. Barr, https://www.agbarr.co.uk/responsibility/our-environment

32. Jones, 'Packaging'; Margaret Allen, *Bottle Banks* (Manchester, 1980), section VII.

33. Ian Shepherd, 'Recycling Aid to Glass-Fibre Industry', *Birmingham Daily Post*, 9 July 1974, p. 19; 'Research Started on Recycling Glass', *Times*, 26 January 1972, p. 18; Thomas, *Material Gains*, p. 2.

34. 'Re-Cycling Smash Hit', *Coventry Evening Telegraph*, 15 November 1977, p. 7; 'Aberdeen May Get Bottle Bank', *Aberdeen Evening Express*, 26 August 1978, p. 18; Allen, *Bottle Banks*, section VII.

35. GMF, *Glass Recycling Directory* (London, 1974).

36. 'Our new savings bank for energy and raw materials', advertisement for the Glass Manufacturers' Federation, *Guardian*, 26 November 1979, p. 3; Department of the Environment (DoE), *Bottle Banks* (London, 1982), p. 7; 'Real Service', National Dairy Council advert, *The People*, 7 December 1980, p. 16.

37. David Jones, 'The Bottles of Britain', *Observer*, 14 April 1974, p. 17.

38. Edward Townsend, 'Recycling Glass: A Not So Simple Task', *Times*, 4 September 1974, p.19; DoE, *Bottle Banks*, p. 16.

39. Andrew Porteous, 'Recycling Technologies and Strategies', inaugural lecture, Open University, 12 October 1987, p. 11.

40. Allen, *Bottle Banks*, sections V–VII.

41. Jon Vogler, 'The Bottle Bank Con Trick', *New Internationalist*, 157 (March 1986), pp. 27–8; Vogler, *Muck and Brass*, p. 34.

42. Vogler, 'The Bottle Bank Con Trick', pp. 27–8.

43. B. W. Clapp, *An Environmental History of Britain* (London, 1994), p. 208.

44. www.findmypast.co.uk; 1891; 1901; 1911 census returns; 1939 register; 'Change of Address', *Hendon and Finchley Times*, 22 April 1927, p. 5.

45. Simone Doctors, 'Jon Vogler Obituary', *Guardian*, 23 April 2017, https://www.theguardian.com/environment/2017/apr/23/jon-vogler-obituary

46. Vogler, 'The Bottle Bank Con Trick', pp. 27–8.

47. Vogler, *Muck and Brass*, p. 5.

48. 'Diary', *Times*, 16 June 1975, p. 14; Simone Doctors, Jon Vogler obituary, *Guardian*; Vogler, *Muck and* Brass, pp. 19, 44; Thomas, *Material Gains*, pp. 118–19.

49. Neal, *Refuse Recycling and Incineration*, p. 12.

50. Porteous, 'Recycling Technologies and Strategies', p. 10.

51. Linda Millington, 'Salvage Operation May Be Called For', *Birmingham Daily Post*, 3 January 1974, p. 10.

52. 'Traders' Bid To Beat New Waste Charges', *Coventry Evening Telegraph*, 29 July 1975, p. 17; Muriel Tibbs, 'Solving the Great Paper Chase', *Coventry Evening Telegraph*, 14 November 1973, p. 10; 'Diary', *Times*, 16 June 1975, p. 14.

53. '£3m down the Festive Drain', *Coventry Evening Telegraph*, 19 December 1977, p. 2; John Carvel, 'At £25 a Ton There's Money in Waste Paper', *Guardian*, 30 March 1977, p. 16.

54. 'Derek's Charity Effort Wins Another Award', *Aberdeen Press & Journal*, 10 December 1979, p. 2.

55. Carvel, 'At £25 a Ton There's Money in Waste Paper'.

56. Advertisement, *Birmingham Daily Post*, 8 March 1973, p. 3.

57. Alan Murray, 'Waste Not, Want Not', *Aberdeen Press & Journal*, 15 November 1978, p. 14; 'Derek's Charity Effort Wins Another Award', *Aberdeen Press & Journal*, 10 December 1979, p. 2.

58. Wray and Nation, *The Economics of Waste Paper Reclamation in England*, pp. 89, 92, 108; Clapp, *Environmental History of Britain*, p. 199.

59. Emily Cockayne, 'A Case of Fiberite', www.rummage.work/blog/fiberite, 24 November 2018.

60. Ian Aitken, for Halton Borough Council, *How to Profit from Waste* (s.i., 1980), Appendix III.

61. See, for example, Tibbs, 'Solving the Great Paper Chase'.

62. Vogler, *Muck and Brass*.

63. Burgess, 'Point Was Not Wasted', p. 3; Thomas, *Material Gains*, p. 37; 'Waste Not, Want Not', *Coventry Evening Telegraph*, 12 September 1977, p. 29.

64. 'Council Clash on Waste Disposal', *Coventry Evening Telegraph*, 7 January 1976, p. 10.

65. Peter Rodgers, 'Waste Not, Want Not', *Guardian*, 27 July 1973, p. 10.

66. Jones, 'The Bottles of Britain'.

67. HOC, *War on Waste*, esp. pp. 1, 5, 8, 10.

68. 'Now "Mrs Steptoe" Joins In', *Birmingham Daily Post*, 20 November 1974, p. 19; 'Brass Tacks', *Spare Rib*, 48, July 1976, p. 26.

69. Lydia Kallipoliti, 'From Shit to Food: Graham Caine's Eco-House in South London, 1972–1975', *Buildings & Landscapes: Journal of the Vernacular Architecture Forum*, 19:1 (2012), pp. 87–106; 'Waste Matters', *Guardian*, 8 August 1974, p. 11.

70. R. H. Gustafson and J. S. Kiser, 'Nonmedicinal Uses of Tetracyclines', in Joseph J. Hlavka and J.H. Boothe (eds), *The Tetracyclines* (Berlin, 1985), p. 405; E. I. R. Stokstad et al., 'The Multiple Nature of the Animal Protein Factor', *Journal of Biological Chemistry*, 180 (1949), pp. 647–54; T. H. Jukes et al., 'Growth-Promoting Effect of Aureomycin on Pigs', *Archives of Biochemistry*, 26 (1950), pp. 324–5.

71. World Health Organisation, 'Antimicrobial Resistance: Global Report on surveillance 2014', http://www.who.int/drugresistance/documents/surveillancereport/en

72. S. T. Omaye, 'Introduction to Food Toxicology', in David H. Watson (ed.), *Pesticide, Veterinary and Other Residues in Food* (Cambridge, 2004), p. 15.

73. Board of Trade (Merchandise Marks Act, 1926) – Report of the Standing Committee, *Bone Meal and Flour: Hoof Meal, Horn Meal, Meat Meal, Meat and Bone Meal and Carcase Meal: Dried Blood*, November 1930, Cmd 3729 (London, 1930), p. 3.

74. Advertisement, *Swindon Advertiser & North Wiltshire Chronicle*, 6 June 1883, p. 2; advertisement, *Lincolnshire Chronicle*, 23 February 1929, p. 11; advertisement, *Market Harborough Advertiser & Midland*

Mail, 24 October 1930, p. 4; advertisements, *Lincolnshire Standard & Boston Guardian*, 28 February 1953, p. 12; *Agricultural Express*, 7 March 1930, p. 13.

75. Letters, *Manchester Guardian*, 13 November 1953, p. 6; Manchester also processed meat and offal into feeding products, and in 1958 the city's Public Cleansing Committee tendered bids on a dozen tons of MBM, plus tallow products, Classifieds, *Manchester Guardian*, 7 October 1958, p. 12.

76. 'Do Dogs Get Cleaner Meat than Family?', *Bury Free Press*, 30 January 1953, p. 1.

77. P. N. Hobson, S. Bousfield and R. Summers, *Methane Production from Agricultural and Domestic Wastes* (London, 1981), pp. 175–6.

78. HOC, *War on Waste*, p. 26.

79. J. W. Wilesmith and G. A. H. Wells, 'Bovine Spongiform Encephalopathy', in Bruce W. Chesebro (ed.), *Transmissible Spongiform Encephalopathies* (Berlin, 1991), p. 33.

80. Rosemary Collins, 'Abattoir Carve-Up Alleged by MMC', *Guardian*, 9 April 1985, p. 19.

81. Monopolies and Mergers Commission, Prosper De Mulder Ltd and Croda International plc, Department for Trade and Industry, August 1991, Cm 1611, pp. 1–2.

82. David Leigh, 'Animal Feed Firm Paid Out To Tories', *Observer*, 16 June 1996, p. 9; Opinion of the [European Union] Economic and Social Committee on 'The Bovine Spongiform Encephalopathy (BSE) Crisis and Its Wide-Ranging Consequences for the European Union', Official Journal C 295, 07/10/1996 P. 0055.

83. Colin Spencer, 'All the Food Fit To Eat', *Guardian*, 4 August 1990, p. 19. See also Martin Wainwright, 'Dead Cow Mountain Will Finally Be Burned and Returned to the Land', *Guardian*, 8 July 1998, p. 10.

84. Anthony Tucker, 'Grants Needed To Aid Waste Recycling', *Guardian*, 1 August 1973, p. 7; Philip Jordan, 'Recycling Gets a Minister', *Guardian*, 26 July 1974, p. 7.

85. Letters, *Times*, 10 July 1973, p. 10.

86. Burgess, 'Point Was Not Wasted'.

87. John Powell, 'Scrap Industry Has Little To Fear', *Birmingham Daily Post*, 30 January 1973, p. 95.

88. Murray, 'Waste Not, Want Not'.

3. Nellie Dark's Bundles (1930s–1950s)

1. Letters, *Times*, 30 June 1937, p. 12.
2. Letters, *Times*, 4 March 1940, p. 7.
3. 'Waste on Dumps', *Times*, 5 March 1940, p. 5.
4. The failure of some of the efforts can be partially explained by a failure of collectivism. See James C. Scott, *Seeing like a State* (London, 1998).
5. MOA, File Report 846, 'Twelfth Weekly Report', 25/08/1941, p. 28.
6. MOA, Topic Collection 43, 'Propaganda and Morale 1939–1944', Box 3–B (images 1442, 1443, 1463).
7. MOA, File Report 961, 'Beaverbrook Waste Paper Appeal', 12 November 1941, pp. 1–4; MOA, Topic Collection 43, 'Propaganda and Morale 1939–1944', Box 3–B (image number 1442 and 1459); Box 3–C (image 1502).
8. D. B. Halpern, 'The Salvage of Waste Material', *Oxford Bulletin of Economics and Statistics*, 4:12 (1942), pp. 235–42; MOA, File Report 1171–2, 'The Paper Salvage Campaign', 24 March 1942, p. 2.
9. 'Wealth from Waste', *Times*, 20 January 1945, p. 2.
10. 'Shirts Off to Berlin!', *Times*, 20 October 1943, p. 3; 'Save Rags, Rope and String', *Times*, 17 July 1942, p. 2; J. C. Dawes, 'Making Use of Waste Products', *Journal of the Royal Society of Arts*, 90:4613, 15 May 1942, p. 398.
11. 'More Salvage Essential', *Times*, 4 March 1941, p. 2.
12. 'Question Time', *Times*, 23 October 1942, p. 5; 'Waste Rubber Salvage', *Times*, 16 April 1942, p. 2.
13. Letters, *Gloucestershire Echo*, 29 March 1941, p. 3; MOA, File Report 845, 'Some Criticisms and Suggestions on Salvage Collection', 24 August 1941, p. 2.
14. MOA, Diarist 5427, 28 June 1940.
15. Many are reproduced in *Make Do and Mend* (London, 2007).
16. Gloucestershire Archives, K 1348/1, Gloucestershire County Council, Salvage Officer, letters of appreciation.
17. Boston R. D. C. Salvage Department leaflet, printed by the *Lincolnshire Chronicle & Leader*, 1941.
18. Advertisements, *Times*, 3 June 1942, p. 2; 27 August 1942, p. 3; 1 September 1942, p. 2. See also Dawes, 'Making Use of Waste Products', p. 392; and Sandra Trudgen Dawson, 'Rubber Shortages on Britain's Home Front', in Mark J. Crowley and Sandra Trudgen Dawson (eds), *Home Fronts* (Woodbridge, 2017), pp. 59–75.

19. Letters, *Times*, 11 September 1939, p. 4; 13 March 1940, p. 9.

20. Instructions disseminated in Berlin, 1813, collected in Ernst Müsebeck (ed.), *Gold gab ich für Eisen* (Berlin, 1913), p. 220; Volker Issmer (ed.), *Als 'Mitläufer' (Kategorie IV) entnazifiziert: die Memoiren meines Vaters* (Münster, 2001), p. 176. See also Letters, *Times*, 8 November 1914, p. 4.

21. News, *Times*, 1 March 1940, p. 7.

22. 'Pots and Pans Take Wings', *Newcastle Journal*, 13 July 1940, p. 6.

23. 'Love Letters Salvage', *Lincolnshire Echo*, 14 November 1940, p. 4; 'Salvage Drive Opens To-Day', *Cheltenham Chronicle*, 30 August 1941, p. 7; 'Romances in a Secret Sack', *Sunday Mirror*, 25 January 1942, p. 3; 'Love Letters in Salvage', *Cheshire Observer*, 19 February 1944, p. 8. 'Love Letters as Salvage', *Derby Daily Telegraph*, 6 November 1941, p. 12.

24. 'War on Waste', *Times*, 29 February 1940, p. 5.

25. For an account with a more 'top-down' approach the reader should procure a copy of Peter Thorsheim's excellent *Waste into Weapons* (Cambridge, 2015).

26. 'Waste on Dumps', *Times*, 5 March 1940, p. 5; see also, MOA, File Report 846, 'Twelfth Weekly Report', 25 August 1941, p. 27.

27. MOA, Topic Collection 43, 'Propaganda and Morale 1939–1944', Box 3–C (image 1486); MOA, File Report 1171/2, 'The Paper Salvage Campaign', 23 March 1942, pp. 3–4.

28. MOA, File Report 961, 'Beaverbrook Waste Paper Appeal', 12 November 1941, p. 1.

29. 'Leg for Scrap', *Hartlepool Northern Daily Mail*, 16 July 1940, p. 1; 'Pots and Pans Take Wings', *Newcastle Journal*, 13 July 1940, p. 6; Peter Thorsheim, *Waste into Weapons* (Cambridge, 2015), pp. 68–9.

30. MOA, draft article for *Manuscript*, appendix VIII.

31. MOA, Diarist 5427, 13 July 1940.

32. 'Aluminium from the Royal Family', *Times*, 19 July 1940, p. 7; MOA, Topic Collection 43, 'Propaganda and Morale 1939–1944', Box 3–G (images 1831 and 1868).

33. Cited in G. C. Peden, *The Treasury and British Public Policy, 1906–1959* (New York, 2000), p. 310.

34. MOA, Topic Collection 43, 'Propaganda and Morale 1939–1944', Box 3–G (images 1827–1854).

35. MOA, Topic Collection 43, 'Propaganda and Morale 1939–1944', Box 3–G (image 1841); MOA, draft article for *Manuscript*, p. 99; MOA, Report on 'Pans to Planes', 15 July 1940, p. 9.

36. MOA, Report on 'Pans to Planes', 15 July 1940, p. 8.

37. MOA, File Report 268, 'Pans to Planes', 15 July 1940, pp. 6–7, 10; MOA, Topic Collection 43, 'Propaganda and Morale 1939–1944', Box 3–G (image 1851).

38. MOA, Diarist 5024, September 1940.

39. MOA, Diarist 5427, 5 August 1940.

40. MOA, Draft article for *Manuscript*, pp. 74–5.

41. MOA, Topic Collection 43, 'Propaganda and Morale 1939–1944', Box 3–B (images 1448, 1459, 1468). See also MOA, File Report 961, 'Beaverbrook Waste Paper Appeal', 12 November 1941, pp. 1–4.

42. MOA, Diarist 5418, October 1941; January 1942; MOA, Topic Collection 43, 'Propaganda and Morale 1939–1944', Box 3–B (image 1446)]; MOA, File Report, 961, 'Beaverbrook Waste Paper Appeal', 12 November 1941, p. 4.

43. Dawes, 'Making Use of Waste Products', p. 395.

44. J. F. M. Clark, 'Dawes, Jesse Cooper (1878–1955)', *ODNB*.

45. Bristol Archives, M/BCC/SAN/14/1, Bristol City Council: committee records and minutes, Salvage Drive minute book, 1941–44, fols 1–5, 8, 9, 17, 38, 42, 51, 52 (i–iv); 'Give Old Books for Salvage', *Carluke & Lanark Gazette*, 13 June 1941, p. 4.

46. 'Bristol's New Book Drive', *Western Daily Press*, 6 September 1944, p. 2; 'Notes of the Day', *Western Daily Press*, 6 January 1949, p. 5; 'Book Salvage Drive in Bristol', *Western Daily Press*, 31 October 1942, p. 4; 'Salvage Books To Be Inspected', *Western Daily Press*, 11 February 1943, p. 3.

47. 'Twelfth Night Salvage', *Times*, 6 January 1942, p. 2; 'High-Speed Steel', *Times*, 20 April 1942, p. 2.

48. Advertisement, *Shields Daily News*, 6 February 1941, p. 3.

49. MOA, Topic Collection 43, 'Propaganda and Morale 1939–1944', Box 3–E (Image 1461).

50. 'New Salvage Orders', *Times*, 25 June 1940, p. 3; Bus as Paper Dump', *Times*, 22 January 1942, p. 2; 'Bone Salvage', *Western Morning News*, 2 April 1942, p. 4; MOA, File Report 846, 'Twelfth Weekly Report', 25/08/1941, p. 30.

51. MOA, Topic Collection 43, 'Propaganda and Morale 1939–1944', Box 3–C (images 1488, 1502, *passim*).

52. Dawes, 'Making Use of Waste Products', p. 406. This calculation, however, ignores the fact that notes were printed on watermarked paper made from linen.

53. MOA, Topic Collection 43, 'Propaganda and Morale 1939–1944', Box 3–C (images 1440, 1447, 1494, 1586); see also MOA, File Report 961, 'Beaverbrook Waste Paper Appeal', 12 November 1941, p. 5. See also MOA, Diarist 5427, 28 June 1940.

54. John Collings Squire, 'Review', *Illustrated London News*, 25 July 1942, p. 19.

55. Valerie Holman, *Print for Victory* (London, 2008), esp. pp. 66–105.

56. Letters, *Times*, 9 December 1942, p. 5; 11 December 1942, p. 5. Further correspondence was sparked: see Letters, *Times*, 17 December 1942, p. 5; 'Standard Styles', *Yorkshire Post & Leeds Intelligencer*, 9 February 1942, p. 2.

57. Emily Cockayne, 'Judging a Book by Its Covers', www.rummage. work/blog/dustjackets, 18 July 2019.

58. Author's own copy.

59. Peter Pirbright (pseud.), *Off the Beeton Track* (London, 1946).

60. 'Memoranda', *British Journal of Inebriety*, 15 (1917), p. 175.

61. 'News', *Newcastle Journal*, 25 November 1918, p. 7; 'Costly Joke of Millions of Economy Labels', *Leeds Mercury*, 23 March 1921, p. 9.

62. 'Re-Use Those Envelopes', *Hartlepool Northern Daily Mail*, 21 July 1942, p. 2.

63. 'An Ill Wind', *Nottingham Evening Post*, 18 September 1940, p. 4.

64. 'Forty Million Envelopes', *Times*, 6 January 1944, p. 5.

65. 'Salvage on the Railways', *Times*, 20 April 1942, p. 2.

66. 'Old Gas Bags Help To Cook Hitler's Goose', *Manchester Evening News*, 18 July 1940, p. 3.

67. 'Tudor Rose Mark Found on Old Guns', *Nottingham Journal*, 3 September 1940, p. 4; Howard Blackmore, *The Armouries of the Tower of London*, pt 1, *Ordnance* (London, 1976), p. 70; 'Georgian Group', *Sunderland Daily Echo & Shipping Gazette*, 15 October 1940, p. 2.

68. 'Treasures from Scrap Heaps', *Nottingham Evening Post*, 5 April 1940, p. 8.

69. 'Saved from Scrap', *Western Morning News*, 21 August 1940, p. 2.

70. Gloucestershire Archives, GBR/L6/23/B3262, Gloucester Borough Records, town clerk's files, 'Salvage': correspondence and other

papers, letter from the Ministry of War Transport to the Town
Clerk, Gloucester, 23 March 1942.

71. 'Railings', *Times*, 9 April 1938, p. 8.

72. 'Town Hall, Portmeirion', Young Roots Project, https://fdet.
wrexham-history.com/town-hall-portmeirion; Jan Morris,
Portmeirion (Woodbridge, 2006), p. 232; Jonah Jones, *Clough
Williams-Ellis: The Architect of Portmeirion* (Bridgend, 1996), p. 131.

73. 'Collecting Old Iron and Steel', *Times*, 15 August 1940, p. 9.

74. 'Iron and Steel for Weapons', *Times*, 21 June 1941, p. 2; '580,000
Tons of Railings Salved', *Times*, 17 June 1943, p. 2.

75. Northamptonshire Archives, ZA 2374, Defence Regulations to
requisition unnecessary iron or steel railings in county for scrap,
correspondence, papers 1941–43, letter from J. W. Hollyoak to Mrs
Clarke, 12 January 1943; letter from Mrs Clarke, 22 January 1943.

76. 'Inspectors Descend on Burton Latimer', *Northampton Mercury*, 13
March 1942, p. 1.

77. 'Scrap Railings Protest at Belper', *Derby Daily Telegraph*, 8
November 1941, p. 5.

78. 'New High Sheriff', *Northampton Mercury*, 8 March 1935, p. 7.
Thornby Hall documentation is from the author's own collection.

79. Northamptonshire Archives, ZA 2374, letter from Edwin Lutyens to
Sidney Harris, 18 November 1941; letter from Lady Wimborne, 22
November 1941; note by Sidney Harris, 19 January 1942.

80. Northamptonshire Archives, ZA 2374, letter from F. Brodie Lodge,
13 November 1942; letter from Sidney Harris, 27 November 1941.

81. Northamptonshire Archives, ZA 2374, letter from M. D. Allfrey, 19
March 1943; letter from C. M. Clarke, 2 April 1943; letter to C. M.
Clarke, 7 April 1943.

82. Gloucestershire Archives, GBR/L6/23/B3262, Gloucester Borough
Records, town clerk's files, 'Salvage': correspondence and other
papers, letter from Mrs E. Purcell, 1942; letter from Reginald J.
Bevan, 26 January 1942.

83. 'Woman's "Why My Railings?"', *Northampton Mercury*, 27
November 1942, p. 1.

84. Bristol Archives, LM/C/X16/5/8/2a–b, miscellaneous Lord Mayor's
correspondence with F. H. Miles and Miss M. H. Wetherman, 1942.
See also Bristol Archives, LM/C/X16/5/14a, miscellaneous Lord
Mayor's correspondence relating to salvaging materials, 1941.

85. MOA, Diarist 5259, 6 May 1942.

86. Letter, *Liverpool Daily Post*, 30 June 1942, p. 2.

87. MOA, Diarist 5335, February 1942.

88. 'Brute Force Used To Take Railings', *Northampton Mercury*, 9 October 1942, p. 7.

89. Gloucestershire Archives, GBR/L6/23/B3262, 'Report re Damage to Copings by Removal of Railings', 1942.

90. Gloucestershire Archives, GBR/L6/23/B3262, letter from M. V. Gransmore, 46 Cromwell Street, Gloucester, 28 January 1942.

91. Philip Ziegler, *London at War, 1939–1945* (London, 1995), p. 182; 'London Letter', *Belfast News-Letter*, 19 July 1941, p. 4.

92. 'Home News', *Times*, 15 December 1941, p. 2.

93. 'Scrap Railings Order Dismissed', *Western Daily Press*, 3 October 1941, p. 2.

94. 'City Depts Refuse To Give Railings', *Manchester Evening News*, 20 January 1942, p. 5.

95. 'Railings at County Buildings', *Dumfries & Galloway Standard*, 17 October 1942, p. 7.

96. Thorsheim, *Waste into Weapons*, pp. 228–9.

97. Letter to *London Evening Standard,* by Christopher Long, 24 May 1984, www.christopherlong.co.uk/pri/wareffort.html

98. Thorsheim, *Waste into Weapons*, pp. 8, 228–9, 231, 261; Hazel Conway concludes that dumping of park railings was 'most unlikely': 'Everyday Landscapes: Public Parks from 1930 to 2000', *Garden History*, 28:1 (2000), p. 123.

99. 'Iron and Steel for Weapons', *Times*, 21 June 1941, p. 2.

100. Gloucestershire Archives, K 1348/1, appendix B.

101. Bristol Archives, M/BCC/SAN/14/1, Salvage Drive Minute Book 1941–44, meeting, 28 July 1942.

102. Sidney Dark, *Not Such a Bad Life* (London, 1941), p. 300.

103. MOA, File Report 768, 'Demolition in London', July 1941, pp. 2, 56–8, 152; John Gale, *Clean Young Englishman* (London, 1988), p. 66.

104. Juliet Gardiner, *Wartime Britain, 1939–1945* (London, 2004), p. 399.

105. 'ZSL London Zoo during World War Two', Zoological Society of London, https://www.zsl.org/blogs/artefact-of-the-month/zsl-london-zoo-during-world-war-two, 1 September 2013.

106. 'Dustmen Pig-Keepers', *Daily Herald*, 4 December 1939, p. 3; 'War on Waste', *Times*, 29 February 1940, p. 5.

107. Advertisement, *Times*, 2 September 1943, p. 3.

108. 'Corporation Salvage Campaign', *Cheshire Observer*, 8 February 1941, p. 8.
109. '2,250,000 Tons of Salvage', *Times*, 16 April 1942, p. 2.
110. 'Salvage Collection', *Western Morning News*, 26 September 1942, p. 6. See, however, 'Canine Contraband', *Barnoldswick & Earby Times*, 13 February 1942, p. 4.
111. 'Valuable Salvage', *Chelmsford Chronicle*, 13 September 1940, p. 5.
112. See, for example, 'What Bones Salvage Means in Britain's War Effort', *Lincolnshire Echo,* 13 August 1940, p. 4.
113. '2,250,000 Tons of Salvage', *Times*, 16 April 1942, p. 2.
114. Advertisement, *Times*, 22 July 1943, p. 3.
115. Dawes, 'Making Use of Waste Products', p. 390.
116. Dawes, 'Making Use of Waste Products', p. 399.
117. 'For High Explosives', *Brechin Advertiser*, 19 May 1942, p. 2.
118. MOA, File Report 845, 'Some Criticisms and Suggestions on Salvage Collection', 24 August 1941, p. 2.
119. 'Rattle up the Bones', *Forfar Dispatch*, 15 July 1943, p. 2.
120. 'Bow Wow', *Yorkshire Evening Post*, 2 December 1942, p. 4; 'Dogs and Salvage Bones', *Portsmouth Evening News,* 6 March 1943, p. 3. Similar letters were also published in Liverpool and Birmingham.
121. MOA, File Report 1338, 'The Salvage of Household Bones', 3 July 1942.
122. Classifieds, *Bristol Evening Post*, 20 December 1939, p. 15; see also Suzanne Griffith, *Stitching for Victory* (Stroud, 2009), pp. 108–9.
123. 'Get Out Those Flour bags', *Sunday Post*, 4 February 1945, p. 2; 'Dolls from Flour Bags', *Liverpool Evening Express,* 17 December 1943, p. 2.
124. Glenna Hailey, *Sugar Sack Quilts* (Iola, WI, 2008), pp. 12–17.
125. 'Flour-Bag Pants Allegation', *Hull Daily Mail*, 15 April 1947, p. 1; 'Humour in Hansard', *Taunton Courier & Western Advertiser*, 3 May 1947, p. 2.
126. 'Iron Lung from Scrap', *Liverpool Evening Express,* 11 August 1943, p. 3.
127. 'Fire Engine from Scrap', *Evening Dispatch*, 6 July 1940, p. 5.
128. 'Kitchens Built from Scrap', *Gloucester Citizen*, 19 September 1941, p. 7.
129. 'Organ Built from Scrap', *Dundee Evening Telegraph,* 26 December 1949, p. 2.
130. 'Wastepaper Salvage Campaign', *Times*, 24 May 1947, p. 8.

131. Letters, *Times*, 26 April 1947, p. 5.

132. 'Unwanted Waste Paper', *Times*, 18 June 1949, p. 4; 'Combined Op by RDCs To Beat Drought', *Gloucestershire Echo*, 27 September 1949, p. 3.

133. Cf. Patricia Nicol, *Sucking Eggs* (London, 2010); Mark Riley, 'From Salvage to Recycling – New Agendas or the Same Old Rubbish?', *Area*, 40:1 (2008), pp. 79–89.

134. Bristol Archives, M/BCC/SAN/14/1, Salvage Drive Minute Book 1941–1944, minutes of meetings on 22 October 1942; 11 November 1942.

135. Henry Irving, 'Paper Salvage in Britain during the Second World War', *Historical Research*, 89:244 (2016), pp. 391–2.

136. MOA, Report on 'Pans to Planes', 15 July 1940, pp. 6–7.

137. Bristol Archives, LM/C/X16/5/3/5a, Bristol City Council: Lord Mayor's Office, general correspondence, 1941.

138. Bristol Archives, LM/C/X16/5/3a, Bristol City Council: Lord Mayor's Office, general correspondence relating to salvaging materials, 1942.

139. Bristol Archives, LM/C/X16/5/12, Bristol City Council: Lord Mayor's Office, correspondence with Messrs Gedye & Sons Ltd, 1942.

140. See, for example, Gloucestershire Archives, GBR/L6/23/B3262, Gloucester Borough Records, town clerk's files, 'Salvage': correspondence and other papers, 'Notes for Schools'.

141. See, for example, inside back cover of *The Dandy*, 13 November 1943.

142. 'Notes of the Week', *Northampton Mercury*, 12 July 1940, p. 3.

143. MOA, Diarist 5118, 26 February 1940.

144. MOA, Topic Collection 43, 'Propaganda and Morale, 1939–1944', Box 3–G (image 1827).

4. Darn (1900s–1930s)

1. Francis Fox, *Sixty-Three Years of Engineering* (London, 1934), pp. 263–5; 'New Linen from Old Plans', *Times*, 7 November 1917, p. 3; 'New Linen from Old Plans', *Times*, 6 September 1919, p. 6.

2. Visual Arts Data Service (VADS), Imperial War Museum: Posters of Conflict, 'Pick 'Em Up', IWM PST 13432, https://vads.ac.uk/large.php?uid=32949&sos=15; 'Salvage', 1912, IWM PST 13424, https://

vads.ac.uk/large.php?uid=32943&sos=20; 'Salvage!' IWM PST
13407, https://vads.ac.uk/large.php?uid=32934&sos=21

3. VADS, Imperial War Museum: Posters of Conflict, 'Salve now!',
1918, IWM PST 13413, https://vads.ac.uk/large.
php?uid=32938&sos=22.

4. Quartermaster-General's Branch, *Salvage*, rev. edn, France, October
1918.

5. 'Standard Ships from Old Tins', *Globe*, 4 October 1918, p. 4; Henry
J. Spooner, *Wealth from Waste* (London, 1918), pp. 11–12.

6. Theodor Koller, *Utilization of Waste Products* (London, 1918),
esp. pp. 60–64.

7. Frederick A. Talbot, *Millions from Waste* (London, 1920), pp. 5, 12,
19–29.

8. Spooner, *Wealth from Waste*, p. 238; Advertisement, *East London
Observer*, 11 October 1902, p. 8. See for examples: 'Sanitary
Business', *Lowestoft Journal*, 17 December 1904, p. 6; 'Improvement
Committee', *Hartlepool Northern Daily Mail*, 5 April 1906, p. 3;
Driffield Times, 15 September 1906, p. 3; 'A Use for Old Tins',
Torquay Times & South Devon Advertiser, 5 October 1906, p. 2;
'Waterworks and Lighting Committee', *West Sussex County Times*, 21
February 1914, p. 5. See also: 'Drainage and Health Committee',
Suffolk and Essex Free Press, 10 September 1913, p. 5; 'Sale of Old
Tins', *Staffordshire Advertiser*, 28 March 1914, p. 11; 'Salvage from
Refuse', *Birmingham Daily Post*, 16 April 1918, p. 6; 'Wealth in the
Dustbin', *Times*, 4 March 1919, p. 7.

9. 'Refuse of Value', *Sheffield Daily Telegraph*, 17 April 1918, p. 3.
Spooner, *Wealth from Waste*, p. 22; 'Drainage and Health
Committee', *Suffolk and Essex Free Press*, 12 May 1915, p. 5.

10. Talbot, *Millions from Waste*, p. 150; 'Gold in Old Tins', *Aberdeen
Press & Journal*, 9 September 1916, p. 2; Classifieds, *Birmingham
Daily Post*, 29 July 1918, p. 1; Spooner, *Wealth from Waste*, p. 239;
'Salvage of Waste', *Folkestone, Hythe, Sandgate & Cheriton Herald*, 6
April 1918, p. 7.

11. Letters, *Times*, 4 March 1916, p. 7.

12. Letters, *Surrey Advertiser*, 13 March 1916, p. 4.

13. Frances Balfour, 'Physicians, Heal Yourselves', *Times*, 3 March 1916,
p. 9.

14. Letters, *Times*, 7 March 1916, p. 7.

15. Talbot, *Millions from Waste*, pp. 38–43.

16. 'Demand for Waste Paper', *Times*, 30 November 1917, p. 3.

17. 'Value of Refuse', *Times*, 29 April 1918, p. 3.

18. 'Waste Products', *Milngavie & Bearsden Herald*, 26 April 1918, p. 1.

19. Spooner, *Wealth from Waste*, p. 20.

20. 'Earthquake in England', *Globe*, 22 April 1884, p. 5; David J. Blake, *Window Vision* (Bungay, 1989), pp. 15, 71.

21. F. H. Crittall, *Fifty Years of Work and Play* (London, 1934), pp. 10–11, 99–105.

22. 'National Salvage Council', *Liverpool Daily Post*, 21 February 1918, p. 5.

23. 'National Salvage Council', *Liverpool Daily Post*, 2 March 1918, p. 4.

24. 'Salvage from Refuse', *Birmingham Daily Post*, 16 April 1918, p. 6; 'Wealth in the Dustbin', *Times*, 4 March 1919, p. 7.

25. 'New Bones for Old', *Nottingham Evening Post*, 2 May 1918, p. 2; 'Kitchen Munitions', *Nottingham Journal*, 18 January 1918, p. 1.

26. 'Waste Utilisation in Birmingham', *Municipal Engineering and Sanitary Record*, 21 March 1918, p. 206; 'Birmingham's Success with Waste Paper', *Municipal Engineering and Sanitary Record*, 27 June 1918, p. 418.

27. 'Waste Tins Wanted in Birmingham', *Municipal Engineering and Sanitary Record*, 7 February 1918, p. 106; 'Death of Mr J. Jackson', *Birmingham Daily Gazette* 19 September 1933, p. 11.

28. 'Municipal Economy', *Sheffield Daily Telegraph*, 5 July 1915, p. 6; Wasteful Habits', *Sheffield Evening Telegraph*, 25 March 1918, p. 3; 'National Salvage in Earnest', *Municipal Engineering and Sanitary Record*, 14 March 1918, pp. 193–4; Census returns for 1911.

29. See: Timothy Cooper, 'Challenging the "Refuse Revolution": War, Waste and the Rediscovery of Recycling, 1900–50', *Historical Research*, 81:214 (2008); J. F. M. Clark, 'Dawes, Jesse Cooper (1878–1955)', *ODNB*.

30. Editorial, *Municipal Engineering and Sanitary Record*, 14 March 1918, pp. 177–8.

31. 'National Salvage Council Officers', *Municipal Engineering and Sanitary Record*, 11 April 1918, p. 254.

32. 'Aims of the National Salvage Council', *Liverpool Daily Post*, 14 March 1918, p. 3.

33. 'The Salvage Club', *Times*, 2 May 1918, p. 3.

34. Letters, *Municipal Engineering and Sanitary Record*, 30 May 1918, p. 348.

35. Letters, *Municipal Engineering and Sanitary Record*, 13 June 1918, p. 378.

36. 'Valuable Rubbish', *Times*, 20 April 1918, p. 3.

37. 'Gathering Up the Fragments', *Kirkintilloch Herald*, 3 April 1918, p. 7; 'Fortune in Paper', *Hartlepool Northern Daily Mail*, 22 April 1918, p. 4.

38. Talbot, *Millions from Waste*, pp. 121, 127, 154.

39. VADS, Imperial War Museum: Posters of Conflict, 'Sell Your Waste Paper – Lendrum's', IWM PST 13427, https://vads.ac.uk/large.php?uid=32946&sos=24; 'Sell your Waste Paper', IWM PST 13425, https://vads.ac.uk/large.php?uid=32944&sos=25

40. 'Home Paper Supplies', *Times*, 14 March 1918, p. 7.

41. 'The Waste of Paper', *Times*, 14 March 1918, p. 3.

42. 'Still Harping on Waste Paper', *Municipal Engineering and Sanitary Record*, 11 April 1918, p. 254.

43. 'More Government Waste', *Municipal Engineering and Sanitary Record*, 16 May 1918, p. 327.

44. Talbot, *Millions from Waste*, pp. 5, 11, 298–304.

45. Talbot, *Millions from Waste*, pp. 12–13.

46. Advertisement, *Bystander*, 25 February 1914, p. 48.

47. Talbot, *Millions from Waste*, pp. 12–13.

48. 'The Bristol Soapmakers', *Western Mail*, 3 April 1914, p. 18; 'New Companies', *Western Mail*, 14 May 1914, p. 7.

49. 'More New Works for Irlam', *Manchester Evening News*, 1 January 1914, p. 6.

50. 'Soap, Candle and Glycerine', *Western Mail*, 3 April 1914, p. 15; Andrew Wynter, *Curiosities of Toil*, 2 vols (London, 1870), vol. 1, pp. 24–5.

51. 'For the Toilet', *Southern Reporter,* 15 October 1914, p. 2.

52. Restrictions started in early 1916, limiting use to medicinal (and not for toiletries), and these were extended to medical use by 1917.

53. A.E. Musson, *Enterprise in Soap and Chemicals* (Manchester, 1965), p. 266; 'A Wonder Village', *Hampshire Telegraph*, 4 May 1923, p. 12.

54. 'Glycerine from Kitchen Waste', *Scotsman*, 13 August 1917, p. 3.

55. 'Glycerine Comes into Its Own', *Globe*, 16 May 1917, p. 5.

56. 'The Rise of Glycerine', *Diss Express,* 14 July 1911, p. 3.

57. 'Killing at a Distance', *Falkirk Herald*, 14 July 1915, p. 4.

58. 'Germany's Shortage of Glycerine', *Perthshire Advertiser*, 28 July 1915, p. 2.

59. See, for example, 'German Ghouls', *Western Daily Press*, 20 April 1917, p. 6; 'Germans Disgust Chinese', *Leeds Mercury*, 2 March 1917, p. 3.

60. 'Lever Bros Ltd', *Liverpool Daily Post*, 27 March 1918, p. 3. For Crosfield & Sons, see Musson, *Enterprise in Soap and Chemicals*, p. 267.

61. 'Glycerine Comes into Its Own', *Globe*, 16 May 1917, p. 5; 'Glycerine from Kitchen Waste', *Scotsman*, 13 August 1917, p. 3.

62. 'Local Notes', *Western Daily Press*, 2 July 1918, p. 3; VADS, Imperial War Museum: Posters of Conflict, 'Bone and Fat Bucket', IWM PST 13428, https://vads.ac.uk/large.php?uid=32947&sos=9

63. 'A Visit to the Ragman', *Yorkshire Evening Post*, 30 March 1918, p. 2; 'Panel Doctor's Glycerine Fine', *Huddersfield Daily Examiner*, 27 September 1918, p. 2; 'Restriction of Glycerine Supplies', *Scotsman*, 13 February 1917, p. 9; 'Substitute for Glycerine', *Dundee Reporter's Journal*, 10 March 1917, p. 4.

64. 'Lord Mayor Enjoys Whale Steak', *Western Times*, 14 May 1918, p. 3; 'Tale of a Stranded Whale', *Illustrated London News*, 1 June 1918, p. 11.

65. 'What Happened to the Whale', *Sheffield Evening Telegraph*, 23 May 1918, p. 4.

66. 'Whales and the War', *Southern Reporter*, 30 May 1918, p. 2.

67. 'Planters' Margarine Company', *Dundee Courier*, 7 November 1914, p. 5; 'Lever Bros Ltd', *Liverpool Daily Post*, 27 March 1918, p. 3.

68. William Clayton, *Margarine* (London, 1920), pp. 1–4, 16.

69. Talbot, *Millions from Waste*, pp. 184–5.

70. 'Planters' Margarine Company', *Dundee Courier*, 7 November 1914, p. 5; 'Lever Bros Ltd', *Liverpool Daily Post*.

71. Talbot, *Millions from Waste*, p. 185.

72. 'Extra', *Yarmouth Independent*, 22 June 1918, p. 2.

73. TNA, DSIR 37/238, Department of Scientific and Industrial Research: Ministry of Munitions, Chemical Waste Products Committee, 'Waste Products Considered of No Use: Questionnaires Returned by Manufacturers', 1918.

74. Talbot, *Millions from Waste*, pp. 214, 221; 'Gathering Up the Fragments', *Kirkintilloch Herald*.

75. Talbot, *Millions from Waste*, p. 57; Spooner, *Wealth from Waste*, pp. 221, 255–7.

76. 'News Briefs', *Burnley News*, 20 July 1918, p. 8; 'Nuts for Respirators', *Leeds Mercury*, 25 July 1918, p. 3; 'Shells and Fruit Stones', *Whitby Gazette*, 2 August 1918, p. 9; 'Collections of Fruit Stones and Nut Shells', *West Sussex County Times*, 21 September 1918, p. 3.

77. Talbot, *Millions from Waste*, p. 25.

78. 'Old Negatives Wanted', *Globe*, 1 August 1918, p. 4; Talbot, *Millions from Waste*, pp. 210–12.

79. Ian Hernon, *Riot!* (London, 2006), p. 133.

80. 'Collection of Wool', *Hampshire Advertiser,* 4 May 1918, p. 3.

81. 'Dogs' Wool Used for Garments Now', *Canadian Textile Journal*, 36 (1919), p. 137; 'Town, Port & Garrison', *Dover Express*, 10 May 1918, p. 2.

82. Mark Pomeroy, 'The Red Cross', https://www.royalacademy.org.uk/article/the-royal-academy-joins-the-war

83. 'Dogs' Wool', *Butchers' Advocate*, 65, 1918, p. 23; 'A New Association', *Dogdom*, 19:6 (August 1918), p. 283.

84. https://vads.ac.uk/large.php?uid=32936&sos=12; http://blog.maryevans.com/2013/06/the-dog-wool-spinners-of-the-royal-academy.html; 'Chelsea Gossip', *Chelsea News & General Advertiser*, 5 April 1918, p. 2; 'The New Dogs' Wool Industry', *Farringdon Advertiser & Vale of the White Horse Gazette*, 13 April 1918, p. 2.

85. 'The Development of the Textile Industries', *Journal of the Royal Society of Arts*, 66:3412, 12 April 1918, pp. 357–8.

86. 'Correspondence', *Dumfries & Galloway Standard*, 15 May 1918, p. 5; 'What Shall We Do with Our Dogs?', *Manchester Guardian*, 19 March 1918, p. 4; 'The British Dogs' Wool Association', *Sphere*, 30 March 1918, p. 23; 'Wool from Dog Combings', *Times*, 18 March 1918, p. 5.

87. 'Ladies' Kennel Association Notes', *Tatler*, 10 July 1918, p. 32.

88. 'Philokuon', 'War-Time Dog Devices', *Staffordshire Advertiser*, 9 November 1940, p. 2; 'The Chronicle Causerie', *Sevenoaks Chronicle & Kentish Advertiser*, 24 June 1921, p. 8; 'The Development of the Textile Industries', *Journal of the Royal Society of Arts*, pp. 357–8.

89. 'What Shall We Do with Our Dogs?', *Manchester Guardian*.

90. 'Notes of the Day', *Exeter & Plymouth Gazette*, 23 July 1918, p. 2.

91. Spooner, *Wealth from Waste*, pp. 3, 6–7, 15.

92. Talbot, *Millions from Waste*, pp. 5, 12, 19–29.

93. 'The Largest Salvage Dump in England', *Times*, weekly edition, 25 July 1919, illustrated section, p. 6.

94. Bella Sidney Woolf, 'The Quiver Army of Helpers', *Quiver*, April 1919, pp. 461–5.

95. 'Value of Refuse', *Cambridge Daily News*, 29 April 1918, p. 3; 'Wealth in the Dustbin', *Times*, 4 March 1919, p. 7.

96. 'National Salvage', *Daily Record*, 4 March 1918, p. 2.

97. C. Basil Barham, 'When Pig Keeping Pays', *Globe*, 26 February 1918, p. 6.

98. Talbot, *Millions from Waste*, pp. 163–4.

99. 'Our London Letter', *Cambridge Daily News*, 2 May 1918, p. 2.

100. 'Our London Letter', *Cambridge Daily News*, 30 October 1918, p. 2.

101. Neil MacGregor, *History of the World in 100 Objects* (London, 2010), pp. 524–8.

102. 'Suffragette Outrages', *Manchester Guardian*, 11 July 1913, p. 7; Roger C. Davies, 'Suffragette Bombs, 1912–1914', 'Standing Well Back', blog http://www.standingwellback.com/home/2018/2/8/suffragette-bombs-1912–1914.html; City of London Police Museum, https://www.cityoflondon.police.uk/about-us/history/museum/Pages/photo-gallery.aspx.

103. 'Thalia Campbell', http://www.birdchildsandgoldsmith.com/acatalog/Thalia_Campbell_Banners.htm

104. 'Meeting at Broadstairs', *Thanet Advertiser*, 22 December 1917, p. 2; '850 Women Workers', *Thanet Advertiser*, 27 July 1918, p. 3.

105. Classified, *Liverpool Echo*, 7 January 1914, p. 2.

106. Classified, *Edinburgh Evening News,* 10 March 1900, p. 5.

107. Classified, *Portsmouth Evening News,* 12 January 1926, p. 11.

108. See, for examples, Paul N. Hasluck, *Domestic Jobbing* (London, 1907), and Bernard E. Jones, *Household Repairs* (s.i, 1917). Both were reprinted often in the first half of the twentieth century. Harras Moore, *Odd Jobs around the House* (London, 1927); Adverts, *Yorkshire Evening Post*, 28 May 1923, p. 7; *Daily Herald*, 13 March 1923, p. 1.

109. E. Griffith, *Manual of Plain Needlework* (Oxford, 1930).

110. http://www.archives.thepenvro.com/Magazines-1930s.html; http://www.archives.thepenvro.com/Magazines-1940s.html

111. 'Blazing Fireworks', *Liverpool Echo*, 30 September 1929, p. 4.

112. 'Use for Old Flour Bags', *Hampshire Telegraph*, 10 October 1903, p. 12; 'Flour Bags in Lindean', *Southern Reporter,* 4 April 1929, p. 2;

'Diary of a Highland Housewife', *Aberdeen Press & Journal*, 25 April 1933, p. 2.

113. 'Not from a Flour Bag', *St Andrew's Citizen*, 25 May 1935, p. 11.

114. 'A New Use for Flour Bags', *Sheffield Weekly Telegraph*, 18 December 1915, p. 12. See also, 'Letters Out of Flour Bags', *Dundee People's Journal*, 4 September 1915, p. 4; 'Appeal for Flour Bags', *Aberdeen Press & Journal*, 29 May 1916, p. 8.

115. 'Waste Not, Want Not', *Illustrated War News*, 12 April 1916 [Part 88], p. 31.

116. 'Woman's Sphere', *Aberdeen Press & Journal*, 27 March 1918, p. 2.

117. Talbot, *Millions from Waste*, pp. 5, 11, 298–304.

118. Victor Schukov, 'Meet Frederick Talbot, One of Pointe-Claire's Long Forgotten Celebrities', *Montreal Gazette*, 17 November 2014, https://montrealgazette.com/news/local-news/west-island-gazette/victor-schukov-meet-frederick-talbot-one-of-pointe-claires-long-forgotten-celebrities

119. 'Deaths', *Sevenoaks Chronicle and Kentish Advertiser*, 10 October 1924, p. 18; *London Gazette*, 25 October 1907, p. 7151; UK, Outward Passenger Lists, 1890–1960 and Form 30A Ocean Arrivals (Individual Manifests), 1919–1924 [database on-line]. Provo, UT, USA: Ancestry.com Operations, Inc., 2012.

120. 'A Wonder Village', *Hampshire Telegraph*, 4 May 1923, p. 12; 'Northampton Grocers Visit Wonder Town', *Northampton Mercury*, 27 April 1923, p. 5.

5. Reconstructors and Destructors (1880s–1900s)

1. 'Jumble Sale', *The Graphic*, 19 November 1892, pp. 616, 619; 'Mr Winston Churchill in Oldham', *Manchester Courier and Lancashire General Advertiser*, 29 November 1900, p. 8.

2. 'Wollaston', *Northampton Mercury*, 5 January 1889, p. 7. See also Vivienne Richmond, 'Rubbish or Riches? Buying from Church Jumble Sales in Late-Victorian England', *Journal of Historical Research in Marketing*, 2:3 (2010), pp. 327–41.

3. 'Miscellanea', *St James's Gazette*, 25 February 1889, p. 12.

4. 'Jumble Sale', *The Graphic*, 19 November 1892, pp. 616, 619.

5. For items that have been inventively repaired, see Andrew Baseman's riveting blog 'Past Imperfect: The Art of Inventive Repair', http://blog.andrewbaseman.com

6. Advertisements, *Derby Mercury*, 16 May 1900, p. 1.

7. See, for examples, Classifieds, *Bury Free Press*, 17 August 1912, p. 4; Classifieds, *Lincolnshire Echo*, 28 May 1919, p. 1; 'Sureties of the Peace', *Essex Standard*, 3 November 1876, p. 8; advertisements, *Bury Free Press*, 7 June 1913, p. 4; advertisements, *Bath Chronicle & Weekly Gazette*, 12 May 1923, p. 14.

8. Advertisements in: *North Devon Journal*, 2 November 1922, p. 4; *Folkestone, Hythe, Sandgate & Cheriton Herald*, 30 April 1921, p. 7. See also advertisements, *Lincolnshire Echo,* 28 May 1919, p. 1, and Classifieds, *Daily Gazette for Middlesbrough*, 7 September 1888, p. 1.

9. 'A Swindle', *Gloucester Citizen*, 19 February 1887, p. 3.

10. 'Bampton', *Western Times*, 18 August 1885, p. 8. See also, 'Long Ashton Petty Session', *Western Daily Press*, 11 June 1864, p. 2.

11. 'Magisterial', *Leicester Chronicle*, 18 July 1885, p. 6.

12. 'A Woman Burnt to Death', *Leicester Chronicle*, 6 February 1875, p. 10; 'Godmanchester', *Cambridge Independent Press*, 27 February 1875, p. 4.

13. 'Some Leamington "Visitors"', *Leamington Spa Courier*, 20 September 1912, p. 7.

14. J. Howorth, *Repairing and Riveting Glass, China and Earthenware*, 3rd edn (London, 1908), p. 5.

15. 'Scene in the Shire Hall', *Leamington Spa Courier*, 30 November 1923, p. 5.

16. 'J. Nike', *Western Times*, 31 March 1884, p. 2; Census Returns 1861, 1871, 1881, 1891; William White, *History, Gazetteer & Directory of the County of Devon* (Sheffield, 1878), pp. 381, 646, 914; see also *Post Office Directory of Exeter* (Exeter, 1912), p. 52.

17. There were more Nike households in Devon in 1881 than in the rest of the country. A pedlar's certificate from January 1872 links Ellen Nike to John Nike's address in 1881 (Devon Heritage Centre, www.findmypast.co.uk).

18. 'John Nike', *Western Times*, 28 June 1884, p. 3.

19. 'Exmouth', *Western Times*, 10 July 1877, p. 5; 'Assaults on the Police', *Exeter & Plymouth Gazette*, 7 December 1877, p. 7.

20. 'Local News', *Western Times*, 4 April 1899, p. 5.

21. 'Refusing to Quit', *Western Times*, 11 May 1887, p. 2; 'To-Day's Exeter Police', *Exeter Flying Post*, 1 April 1899, p. 4; 'Assault on a Police Constable at Exeter', *Western Times*, 6 August 1914, p. 2; 'Deaths', *Western Times*, 29 October 1915, p. 4.

22. 'Black & Co.', Whistle Museum, http://www.whistlemuseum.
com/2017/12/18/black-co-whistle-maker-glasgow-history-whistles-a-
strauss/; Thomas Hiram Holding, *Uniforms of the British Army,
Navy and Court* (London, 1894), p. 58.

23. *Order of Business at James Platt and Company*, December 1869,
p. 40.

24. James Platt, *Economy* (London, 1882), p. 22.

25. 'Old Artificial Teeth Bought', *London Evening Standard,* 29
December 1888, p. 8.

26. 'What Can Be Done with Old Sardine Boxes', *Leisure Hour*,
February 1888, pp. 134–5.

27. Thomas Greenwood, 'The Utilisation of Waste IV', *Leisure Hour*, 35,
December 1886, p. 840.

28. https://carvingswithstories.blogspot.co.uk/2015/12/another-hidden-
carved-treasure-in.html

29. 'Old Papers and Parchments', *Western Daily Press*, 28 July 1877, p. 8;
'Old Papers & Parchments', *Bristol Times*, 29 August 1867, p. 1.
Thomas Cort ran a similar establishment in Manchester, appealing
in 1870 for 'Old Ledgers, Letters, Briefs, Bookbinders' Shavings,
Newspapers and all kinds of Waste Paper': 'Waste Paper',
Manchester Evening News, 22 July 1870, p. 4.

30. Bristol Archives, 37164/B/10/9, Messrs C. T. Jefferies & Sons, Stock
and Purchasing Records, catalogues of salvage stock, 1856.

31. Classifieds, *Bristol Times & Mirror*, 1 January 1866, p. 1; 'Topics of
the Day', *Western Daily Press*, 9 May 1866, p. 2; advertisement,
Publishers' Weekly, 183, 17 July 1875, p. 186.

32. 'Great Fire in Redcliff Street', *Bristol Mercury*, 10 October 1881, p. 3.

33. Alfred Edward Parkman, 'Canynge House, Radcliffe Street after the
Great Fire 1881', http://www.bonhams.com/auctions/20981/
lot/3020/

34. 'C. T. Jefferies & Sons', *Western Daily Press,* 15 October 1881;
Classifieds, *Bristol Mercury*, 3 December 1881, p. 5.

35. Charles Dickens, *Bleak House* (London, 1852); Somerset Heritage
Centre, Gibbs Mss, DD/GB, 37–156, muniments of the Gore
family of Barrow Court.

36. Francis T. Buckland, *Curiosities of Natural History* (London, 1860),
pp. 193–4.

37. Advertisements: *Tatler*, 29 May 1907, p. 19; 1 July 1908, p. 2; see also
'For Sale', *Kentish Mercury*, 14 September 1906, p. 7.

38. J. C. Martin, *Saint Conan's Kirk Loch Awe: Guide Book* (s.i., 1954), p. 7.

39. University College London (UCL), 'London's Lost Warships Rediscovered', https://www.ucl.ac.uk/media/library/LondonsLostWarships

40. Frank C. Bowen, 'The Shipbreaking Industry', *Shipping Wonders of the World*, 40 (1936), pp. 1272, 1274.

41. Ian Buxton and Ian Johnston, *Battleship Builders* (Barnsley, 2013), p. 306.

42. 'End of the HMS Dreadnought', *Illustrated London News*, 23 January 1875, p. 3.

43. Alison Adburgham, *Liberty's* (London, 1975), pp. 109–10; 'The New Tudor Liberty Building', *Sphere*, 6 May 1922, p. 6, 3 June 1922, p. 28.

44. Charles Tracy, *Continental Church Furniture in England* (Woodbridge, 2001), pp. 65–6.

45. Tracy, *Continental Church Furniture*, pp. 36–42, 65.

46. 'Piccadilly Hall', *Morning Post*, 21 May 1888, p. 1; Tracy, *Continental Church Furniture*, pp. 1, 81–2; 'Antique Carved Oak Stalls', *Times*, 17 December 1886, p. 2; 'Notes on Current Affairs', *British Architect*, 29:26, 29 June 1888, p. 462.

47. Kim Woods, *Imported Images* (Donington, 2007), pp. 178, 487, 490–91.

48. Frank Trentmann, *Empire of Things* (London, 2016), p. 227; Tracy, *Continental Church Furniture*, pp. 36–42; Mary B. Shephard, 'Our Fine Gothic Magnificence', *Journal of the Society of Architectural Historians*, 54:2 (June 1995), pp. 187, 203–4, fn19; David King, 'Medieval Glass-Painting', Carole Rawcliffe and Richard Wilson (eds), *Medieval Norwich* (London, 2004), ch. 5, p. 121.

49. Shephard, '"Our Fine Gothic Magnificence"', p. 186.

50. 'Sale of Contents', *Diss Express*, 12 December 1913, p. 5.

51. 'Ancient Chapel Again Dedicated', *Yarmouth Independent*, 18 August 1934, p. 10.

52. George Cubbit, *Costessey Hall next Norwich: Sale [Catalogue]* (Norwich, 1913), pp. 11–12.

53. Advertisement, *Times*, 20 September 1920, p. 20.

54. Shephard, '"Our Fine Gothic Magnificence"', p. 187.

55. 'The New Crystal Palace', *Times*, 14 September 1852, p. 5; *Cassell's Illustrated Exhibitor* (London, 1862), p. 7.

56. 'Sir Charles Fox', *Engineering*, XVIII, 17 July 1874, p. 53.
57. 'The Crystal Palace', *Times*, 1 December 1936, p. 16; The Crystal Palace Foundation, 'Disaster Strikes, 1936', http://www.crystalpalacefoundation.org.uk/history/disaster-strikes-1936–2; BBC A History of the World, 'Glass from the Crystal Palace', http://www.bbc.co.uk/ahistoryoftheworld/objects/VbK_iXNXQz2vLUYzndNKhw
58. Cf. Tom Licence, *What the Victorians Threw Away* (Oxford, 2015).
59. [Wynter], 'The Use of Refuse', *Quarterly Review*, 124 (1868), p. 340.
60. For example, 'Broken Glass', *Northwich Guardian*, 6 July 1872, p. 2; Classifieds, *West Somerset & Free Press*, 8 February 1873, p. 8. See also British Library (BL), Evanion Collection, Evan.6557, W. Wright & Co.'s rag, bone, bottle and miscellaneous warehouse, 1880.
61. Thomas Greenwood, 'The Utilisation of Waste', *Leisure Hour*, September 1886, p. 624; Classifieds, *Daily Gazette for Middlesbrough*, 7 September 1888, p. 1.
62. Emily Hobhouse, 'Dust-Women', *Economic Journal*, 10: 39, 1900, p. 412.
63. 'A Scientific Wonder', *London Journal*, 29:744, 19 March 1898, p. 247.
64. 'The Brickmaking Industry', *Liverpool Mercury*, 18 August 1900, p. 6.
65. Edward Ballard, *Report in Respect of the Inquiry as to Effluvium Nuisances* (London, 1882), pp. 6–7.
66. 'Utilisation of Waste Products', *Chambers' Journal*, 5th ser., 11:559, 15 September 1894, p. 591; Thomas Greenwood, 'The Utilisation of Waste IV', *Leisure Hour*, December 1886, p. 840.
67. 'Penny Wisdom', *Household Words*, 134, 16 October 1852, p. 99.
68. William Crookes, 'Chemical Products – the Application of Waste', *Popular Science Review*, 2:5 (January 1863), p. 59.
69. The arguments that profits gained from by-products outweighed the costs of production were disputed at the same time: see Desrochers, 'Does the Invisible Hand Have a Green Thumb?' *Geographic Journal*, 175:1, 2009, pp. 7–12.
70. [Wynter], 'The Use of Refuse', p. 346.
71. Pierre Desrochers has detailed the interplay between legislative demands to clean up and the urge to find creative ways to maximise profit: 'win-win outcomes': 'Does the Invisible Hand Have a Green Thumb?', at pp. 12–13. See also Pierre Desrochers, 'Promoting

Corporate Environmental Sustainability in the Victorian Era', *V&A Online Journal*, 3 (Spring 2011).

72. *Trades' Guide for Midland Counties* (Birmingham, 1879), p. 176.

73. Advertisement, *Sheffield Daily Telegraph*, 26 February 1857, p. 4; invoice for Walter Skull & Son, High Wycombe, 20 September 1900, in John P. Birchall, 'British Glues and Chemicals' (draft), http://www.themeister.co.uk/hindley_images/meggitts_invoice.jpg; see also www.themeister.co.uk/hindley/british_glues_chemicals.htm

74. Advertisement, *Sheffield Daily Telegraph*, 12 March 1859, p. 1; J. Thomas Way, *On Superphosphate of Lime* (London, 1851); 'To Agricultural Gentlemen', *Sheffield Independent*, 21 January 1871, p. 4.

75. 'Nottinghamshire', *Sheffield Daily Telegraph*, 30 July 1896, p. 9.

76. *London Gazette*, 15 April 1904, p. 2448; 'New Companies', *Sheffield Daily Telegraph*, 12 March 1903, p. 11; 'New Companies', *Sheffield Daily Telegraph*, 22 August 1903, p. 15.

77. 'The Drowning Mystery at Denaby', *Sheffield Daily Telegraph*, 23 May 1903, p. 12.

78. 'Fires in Sheffield', *Sheffield Daily Telegraph*, 18 July 1902, p. 8.

79. '£5,000 Blaze', *Sheffield Daily Telegraph*, 23 May 1908, p. 5.

80. 'Fires', *Bolton Chronicle*, 17 September 1853, p. 5; 'Fire in Independent St', *Bolton Evening News*, 2 February 1877, p. 3; 'Fire at Waste Warehouse', *Bolton Evening News*, 28 July 1882, p. 3; 'Fire at Bolton Cotton Waste Warehouse', *Bolton Evening News*, 5 June 1893, p. 2; 'Another Fire at a Waste Warehouse', *Bolton Evening News*, 10 June 1893, p. 3; 'Cotton Waste Fires', *Municipal Engineering and Sanitary Record*, 7 February 1918, p. 101.

81. 'Bolton, The Dangers of Cholera', *Bolton Evening News*, 6 April 1893, p. 3.

82. Bolton Archives and Local Studies Service, NMXB/1/14, Agreement: Chapel Trustees (named) and Leonard Wild of Bolton, cotton waste dealer, 16 March 1892.

83. 'Dispute between Cotton Waste Dealers', *Bolton Evening News*, 17 August 1881, p. 3; 'A Fracas between Bolton Waste Dealers', *Bolton Evening News*, 17 September 1881, p. 3; 'Action against a Cotton Waste Dealer', *Bolton Evening News*, 5 February 1870, p. 3.

84. For objects which have been through an incinerator see the database, 'What the Victorians Threw Away', http://www.whatthevictorians threwaway.com/?s=destructor

85. A. W. Neal, *Refuse Recycling and Incineration* (Stonehouse, 1979), p. 5.

86. Alfred Fryer, *Vic; The Autobiography of a Pomeranian Dog* (Manchester, 1880), p. 56.

87. Tom Crook, *Governing Systems* (Oakland, CA, 2016), pp. 190–91; 'Sugar Machinery', *Public Ledger & Daily Advertiser*, 2 July 1874, p. 6; Talbot, *Millions from Waste*, p. 142.

88. Ernest R. Matthews, *Refuse Disposal* (London, 1915), pp. v, 4, 52, 55.

89. 'Utilisation of Waste Products', *Chambers' Journal*, 11:559, 15 September 1894, p. 592.

90. 'Dead Lions and Tigers', *Lancashire Evening Post*, 9 September 1907, p. 4.

91. 'A Lead to the Nation', *Preston Herald*, 4 May 1918, p. 6; Matthews, *Refuse Disposal*, pp. 42, 52, 65, 95; Mary Mills, 'Electricity from Household Waste in Edwardian Woolwich', *Bygone Kent*, September 2000, pp. 539–44.

92. Talbot, *Millions from Waste*, p. 142.

93. Hamish Williams, 'Up in Smoke: Christchurch Destructor', Christchurch uncovered, http://blog.underoverarch.co.nz/2017/12/up-in-smoke-christchurch-destructor

94. Hobhouse, 'Dust-Women', pp. 413–14.

95. John Hinshelwood, 'Smithfield Square's Fascinating Past: Part Two', Hornsey Historical Society, https://hornseyhistorical.org.uk/smithfield-squares-fascinating-past-part-two

96. 'Scheme for Utilising Waste Materials', *Edinburgh Evening News*, 17 September 1895, p. 4; 'Novel Housing Scheme at Liverpool', *Manchester Courier & Lancashire General Advertiser*, 7 March 1903, p. 15; 'A House Made of Refuse', *Luton Times & Advertiser*, 25 August 1905, p. 3.

97. R. Goulburn Lovell, 'Coal Concreted from Dusts and Ashes', *Journal of the Society of Architects*, 10:4, October 1917, p. 76; Hobhouse, 'Dust-Women', pp. 413–14.

98. Talbot, *Millions from Waste*, pp. 22, 142–3, 147–9, 154, 283–4, 303; Crook, *Governing Systems*, pp. 190–91; 'Sugar Machinery', *Public Ledger & Daily Advertiser*, 2 July 1874, p. 6.

99. 'Material from Dust Bins', *Newcastle Journal*, 3 July 1918, p. 4.

100. J. F. M. Clark, '"The Incineration of Refuse Is Beautiful": Torquay and the Introduction of Municipal Refuse Destructors', *Urban History*, 34:2 (2007), p. 276.

101. 'Auctions', *London Evening Standard*, 29 March 1884, p. 8; 'Obituary', *American Art News*, 8:22, 12 March 1910, p. 4; William R. Johnston, *William and Henry Walters* (Baltimore, MD, 1999), p. 139. 'Certified Copy of Marriage Certificate for Julius Ichenhauser and Diamanté Adutt', 27 January 1904, application number 8385893–1: Paddington Registry Office, 1904.

102. TNA, HO 334/12/4128, Naturalisation of Maurice Ichenhauser, 1884; 'Re: Julius Ichenhauser', *London Evening Standard*, 11 July 1895, p. 3; 'Receiving Orders', *London Evening Standard*, 26 June 1895, p. 8; 'Police', *Times*, 2 February 1888, p. 13; 'Julius Ichenhauser', *Morning Post*, 3 March 1892, p. 1; see also TNA, HO 334/15/5407, 'Naturalization of Julius Ichenhauser, 1887'; 'Natalie Segalla', 'Certified Copy of Marriage Certificate for Natalie Segalla and Julius Ichenhauser', 26 August 1884, application number 8385923–1: Westminster Registry Office, 1884.

103. 'High Court of Justice', *Times*, 26 February 1892, p. 14; 'Lord Coleridge on Ugly China', *Aberdeen Weekly Journal*, 25 February 1892, p. 3.

104. The Library and Museum of Freemasonry, United Grand Lodge of England Freemason Membership Registers, 1751–1921 [database on-line]. Provo, UT: Ancestry.com Operations, Inc., 2015, Phoenix Lodge, Registers, fols 154, 159–160.

105. 'The Law Courts', *London Evening Standard*, 24 June 1895, p. 6; 'The Law Courts', *London Evening Standard*, 22 June 1895, p. 5; 'Re Julius Ichenhauser', *London Evening Standard*, 11 July 1895, p. 3.

106. Toovey Auctioneers https://www.tooveys.com/lots/251474/a-victorian-rosewood-and-inlaid-specimen-snuff-box

107. 'Commission on Ancient Scottish Monuments', *Scotsman*, 22 November 1912, p. 9; various snuff boxes have appeared for sale on eBay.

108. 'A Jumble Sale', *Express & Echo*, 17 January 1894, p. 1; 'Jumble Sale at Countess Weir', *Exeter & Plymouth Gazette*, 26 September 1895, p. 3; 'Kennerleigh', *Western Times*, 9 August 1895, p. 6; 'Ideford', *Express & Echo*, 19 September 1899, p. 4.

6. Impostor Instruments (1850s–1880s)

1. W. John Stonhill, 'Paper Pulp from Wood, Straw, and Other Fibres in the Past and the Present', in John Rattray and Hugh Robert Mill (eds), *Forestry and Forest Products* (Edinburgh, 1885), pp. 437–71;

'Novelties', *British & Colonial Printer & Stationer*, 6 November 1884, p. 12.

2. Roberta Montemorra Marvin, *The Politics of Verdi's Cantica* (Abingdon, 2014), pp. 34, 78.

3. William England, 'Recovery of Gold and Silver from Waste Photographic Materials', *Photographic News,* 7:245, 15 May 1863, p. 234.

4. Thomas Sutton, *Dictionary of Photography* (London, 1858), p. 347.

5. 'Waste Not, Want Not', *British Journal of Photography*, 839:23, 2 June 1876, p. 254; George Dawson, 'Reduction of Silver Residues and Other Matters', *Illustrated Photographer*, 17 July 1868, pp. 290–91; Charles Corey, 'Recovery of Waste Silver', *British Journal of Photography*, 9:176, 15 October 1862, p. 380; 'Mr Hadow's Method of Recovering Waste', *Photographic News,* 18:819, 15 May 1874, p. 233.

6. 'Waste Not, Want Not' *British Journal of Photography*, p. 254.

7. 'Society of Arts', *Athenaeum*, 2212, 19 March 1870, p. 393.

8. Wakefield Museums Collections Online, G. and J. Hall, 1978.80/1/19, studio portrait of a woman, http://collections.wakefield.gov.uk/photographs/; advertisement, *Ossett Observer*, 1 January 1879, p. 1.

9. Esther Leslie, *Synthetic Worlds* (London, 2005), p. 84.

10. William Reed, *History of Sugar and Sugar Yielding Plants* (London, 1866), p. 117.

11. Ernst Hubbard (trans. M. J. Salter), *The Utilisation of Wood Waste* (London, 1902), pp. 13, 86, 105; Thomas Greenwood, 'The Utilisation of Waste', *Leisure Hour*, 35, September 1886, p. 624.

12. Deborah Jean Warner, *Sweet Stuff* (Washington, DC, 2011), p. 181.

13. *A Vindication of Saccharin* (London, 1888), p. 4.

14. 'Metropolitan Notes', *Nottingham Evening Post*, 28 April 1888, p. 2.

15. 'Notes on News', *Preston Herald,* 2 May 1888, p. 4.

16. 'Metropolitan Notes', *Nottingham Evening Post*, 28 April 1888, p. 2.

17. Reed, *History of Sugar*, pp. 131, 138–40; Ballard, *Effluvium Nuisances*, pp. 128–9.

18. N. P. Burgh, 'A Visit to a Bone Boiling Factory', *Technologist*, 4 (1864), pp. 139–44; see also Simmonds, *Waste Products*, p. 96.

19. For a fascinating modern interpretation cautiously celebrating this ingenuity, see Leslie, *Synthetic Worlds*, pp. 7–8, 23, 84.

20. Simon Garfield, *Mauve* (London, 2002), p. 38; William Crookes, 'Chemical Products – the Application of Waste', p. 62.

21. Garfield, *Mauve*, p. 79.

22. 'Judson's Simple Dyes', *Suffolk Chronicle*, 18 January 1862, p. 2.

23. Garfield, *Mauve*, p. 84; '1858', *New Scientist*, 9 October 1958, p. 98.

24. Edwin Yates and Andrew Yates, 'Johann Peter Griess, FRS (1829–88)', *Notes & Records*, 70 (2016), pp. 65–81.

25. D. H. Soxhlet, *The Art of Dying and Staining* (London, 1902), p. 73.

26. 'Contemporary Science', *London Review*, 26 July 1862, p. 87.

27. Census returns for 1881 and 1891.

28. Anthony Camp, *On the City's Edge* (s.i., 2016), pp. 55–6; British Library (BL), Evanion Collection, Evan 7042, 'Crawshaw's Crystal Dyes', *c.*1890.

29. Simmonds, *Waste Products*, p. 8.

30. 'The Use of Refuse', *Quarterly Review*, 124, 1868, p. 181.

31. Lyon Playfair cited in [George Dodd], 'Penny Wisdom', *Household Words*, 134, 16 October 1852, p. 100; Andrew Wynter, 'The Use of Waste Substances. No. II', *Good Words*, December 1876, p. 307.

32. Crookes, 'Chemical Products', p. 70.

33. Patent 1027 (1772).

34. George Dodd, 'Paper: Its Applications', *The Curiosities of Industry* (London, 1853), p. 17.

35. 'The Paper Maché Works and Other Wonders of Birmingham', *Taunton Courier & Western Advertiser*, 9 October 1844, p. 8; 'Partnership Dissolved', *Globe*, 23 May 1827, p. 1.

36. C. F. Bielefeld, *Portable Buildings* (London, 1853); Richard H. Horne, 'The Pasha's New Boat', *Household Words*, 87 (22 November 1851), pp. 209–13.

37. 'Southampton Street and Tavistock Area: Wellington Street', *Survey of London*, http://www.british-history.ac.uk/survey-london/vo136/pp226–229; Mary L. Shannon, *Dickens, Reynolds, and Mayhew on Wellington Street* (London, 2016), p. 93.

38. 'Destruction of Bielefeld's Papier Mâché Works', *Morning Post*, 10 March 1854, p. 5.

39. *London Gazette*, 21 June 1861, p. 2606; 'Re Bielefeld', *Morning Post*, 29 August 1861, p. 7.

40. Patent Office, *Abridgements* [...] *Paper, Pasteboard, and Papier Mâché* (1858), p. 82.

41. Census returns for 1851.

42. *London Gazette*, 7 July 1857, p. 2389; 'Sales by auction', *Birmingham Journal*, 3 September 1859, p. 1.

43. 'Corn Markets &c', *Birmingham Journal*, 17 September 1859, p. 8; 'Fire at Mr Bettridge's Papier Mâché Works', *Aris's Birmingham Gazette*, 16 January 1860, p. 3.

44. 'Birmingham Manufactures for India', *Birmingham Daily Gazette*, 29 April 1863, p. 3.

45. 'Birmingham Art for Egypt', *Birmingham Daily Post*, 28 October 1865, p. 4; 'Papier Mâché Panels', *Hull & Eastern Counties Herald*, 9 February 1865, p. 7; 'A Novelty in Ship Decoration', *Aris's Birmingham Gazette*, 16 March 1861, p. 6; '3 Pianofortes', *Cambridge Chronicle & Journal*, 2 October 1869, p. 5.

46. Leslie, *Synthetic Worlds*, p. 11.

47. See Susan Mossman in 'Perspectives on the History and Technology of Plastics', *Early Plastics* (London, 1997), pp. 29–32.

48. Silvia Katz, *Early Plastics* (Princes Risborough, 1986), p. 19; 'Xylonite', *British & Colonial Printer & Stationer*, 24 January 1884, p. 11; 'How Celluloid Is Made', *British & Colonial Printer & Stationer*, 17 August 1884, p. 9.

49. Friedrich Böckmann, *Celluloid* (London, 1907), pp. 8–10, 39, 71–2.

50. Christine Garretson Persans, *The Smalbanac* (Albany, NY, 2010), p. 51.

51. *London Gazette*, 29 May 1868, p. 3069; Patent: 1614 (1868); 2531 (1871).

52. Mossman, 'Perspectives on the History and Technology of Plastics', p. 29.

53. Peter Ashlee, *Tusks and Tortoiseshell* (s.i., 2013), p. 10; Marianne Gilbert, 'Plastics Materials: Introduction and Historical Development', in Marianne Gilbert (ed.), *Brydson's Plastics Materials*, 8th edn (Oxford, 2017), p. 3.

54. *Post Office London Directory for 1882* (London, 1882), pt III, pp. 1435, 1780.

55. Mark Suggitt, 'Working with Plastics', in Mossman, *Early Plastics*, pp. 138, 145; 'British Xylonite Co', Grace's Guide, www.gracesguide.co.uk/British_Xylonite_Co.

56. J. L. Meikle, *American Plastic* (New Brunswick, NJ, 1995), pp. 12, 14, 18.

57. Abdul Sheriff, *Slaves, Spices and Ivory in Zanzibar* (London, 1987), pp. 2, 77; William Beinart and Lotte Hughes, *Environment and Empire* (Oxford, 2007), pp. 67–8.

58. John M. Mackenzie, *The Empire of Nature* (Manchester, 1988), p. 205.

59. Roger Newport, 'Plastics and Design', in Mossman, *Early Plastics*, p. 79.

60. 'Parkesine, Prize Medal', inside cover of British Library (BL), d6 8233, John Naish Goldsmith, 'Alexander Parkes, Parkesine, Xylonite and celluloid', typewritten, 1934.

61. 'Useful Shams', *Scientific America*, 62:19, 8 May 1880, p. 297.

62. J. Hamilton Fyfe, 'Rags for the Ragged', *Once a Week*, 3 May 1862, pp. 528–30.

63. Carolyn Steedman, *Dust* (Manchester, 2001), pp. 112–14, 128–36.

64. Benjamin Lambert. 'On the Paper Manufacture', *Technologist*, 4 (1864), p. 13– 21.

65. Dodd, 'Paper', p. 7.

66. Great Seal Patent Office, *Abridgements of the Specifications Relating to the Manufacture of Paper, Pasteboard, and Papier Mâché* (London, 1858), esp. pp. 44–6.

67. Commissioners of Patents, *Title of Patents of Invention, Chronologically Arranged from March 2, 1617 (14 James I) to October 1852 (16 Victoriae)*, 2 vols (London, 1854), vol. 2, p. 1083; Patent Office, *Abridgements   Paper, Pasteboard, and Papier Mâché* (1858), p. 117.

68. Sandwell Community History and Archives Service, BS6/5/1/9/4, Chance Brothers, Smethwick, copy of a specification for improvements in the manufacture of paper in the name of Henrik Zander of Middlesex, 1839; Census returns for 1841; see also Patent Office, *Abridgements   Paper, Pasteboard, and Papier Mâché* (1858), p. 88; Patent 360 (1852).

69. Dodd, 'Paper', p. 1.

70. M. C. Cooke, 'Paper Materials Patented since the Year AD 1800', *Technologist*, 1 (1861), pp. 51–5.

71. HOC, *Report from the Select Committee on Paper* (London, 1861), esp. pp. 12–15.

72. P. L. Simmonds, 'Animal Substances Used for Writing On', *Technologist*, 6 (1866), p. 13.

73. Alois Ritter Auer von Welsbach, 'On Maize Paper', *Technologist*, 3 (1863), p. 358.

74. Wynter, 'The Use of Waste Substances. No. II', p. 308.

75. George Dodd 'Done to a Jelly', *Household Words*, 222 (24 June 1854), p. 439.

76. *London Gazette*, 12 October 1877, p. 5614.

77. *Post Office London Directory for 1882*, vol. 2, p. 1236; vol. 3, p. 1740; *Business Directory of London 1884*, 2 parts (London, 1884), pt I, pp. 528–9, 574; pt II, p. 576; *London Gazette*, 10 July 1883, p. 3510; Stadtarchiv Oldenburg, Oldenburg, Niedersachsen, Heiratsregister (marriage records), 1876–1920, Schroeder, 20 November 1886, p. 121.

78. Census returns for 1871 and 1881; Classifieds, *Hendon & Finchley Times*, 1 November 1884, p. 8; see adverts in *Islington Gazette*, for example, 14 December 1881, p. 1; 27 September 1882, p. 1. A hundred years previously, Harman Viëtor had made small fortepianos in London. Harman does not appear to have been highly respected and left in disillusionment for Philadelphia in the 1770s following the unhappy outcome of a legal case concerning the rape of his daughter. See Margaret Debenham and Michael Cole, 'Pioneer Piano Makers in London, 1737–74', *Royal Music Association Research Chronicle*, 44:1 (2013), pp. 75–80; *OB*, t17710515–6, 15 May 1771; Geoffrey Lancaster, *The First Fleet Piano*, 2 vols (Canberra, 2015), vol. 1, pp. 47–8.

79. Marie E. Kent, 'Exposing the London Piano Industry Workforce (*c.*1765–1914)', unpublished PhD thesis, London Metropolitan University, 2013, 2 vols, vol. 1, p. 262.

80. Advertisement, *Musical Opinion & Music Trade Review*, 4:38, 1 November 1880, p. 70.

81. *London Gazette*, 1 March 1878, p. 1779; Patent, 616 (1878).

82. Patent Office, *Abridgements Paper, Pasteboard, and Papier Mâché* (1858), pp. 64–5; Patent 10064 (1844).

83. Patent Office, *Abridgements […] Paper, Pasteboard, and Papier Mâché* (1858), pp. 83–4; Patent 13861 (1851); *Repertory of Patent Inventions*, 20 (1852), pp. 166–70.

84. 'Furniture Exhibition, stand 166', *London Evening Standard*, 15 August 1881, p. 1; advertisement, *Musical Opinion & Music Trade Review*, 4:38, 1 November 1880, p. 70.

85. Edward and Eva Pinto, *Tunbridge and Scottish Souvenir Woodware* (London, 1970), pp. 121–2; 'The New Artificial Wood', *Leeds Mercury*, 23 April 1859, 6; Patent 2232 (1856).

86. Called various names, such as Sciffarin ware, wood-marble and Simili bois, see Hubbard, *Utilisation of Wood Waste*, pp. 116–19. In the 1860s Michael Dietrich Rosenthal and Solomon Gradenwitz made imitation wood handles for umbrellas by mixing ground sawdust with 'common glue', which was coloured and varnished to suit: Great Seal Patent Office, *Abridgements of the Specifications Relating to Umbrellas, Parasols, and Walking Sticks* (London, 1871), pp. 133–4; Patent 1682 and 3301 (1865).

87. Hubbard, *Utilisation of Wood Waste*, pp. 4, 123–129.

88. 'County Advertisements', in E. R. Kelly (ed.), *Kelly's Directory of Leicestershire and Rutland* (London, 1881), supplement, p. 32; advertisement, *Clerkenwell Press*, 15 July 1882, p. 2; 'Musical instruments at the Furniture Exhibition', *Musical Opinion & Music Trade Review*, 5:57, 1 June 1882, p. 368; Advertisement, *Printing Times & Lithographer*, 15 January 1882, supplement, p. 7.

89. 'The Furniture Exhibition at Islington', *Furniture Gazette*, 13 August 1881, p. 92.

90. A. W. N. Pugin, *True Principles of Pointed or Christian Architecture* (London, 1841), p. 30; see also Meikle, *American Plastic*, p. 13; I am grateful for coffin details from Rosemary Hill, email correspondence, 6 June 2017.

91. 'Fritz Viëtor & Co.', *Printing Times & Lithographer*, 15 January 1882, p. 7; R. J. Gettens and G. L. Stout, *Painting Materials* (London, 1942), p. 92; Mossman, 'Perspectives on the History and Technology of Plastics', p. 21; *Post Office London Directory for 1885*, p. 1341; Robert Hunt (ed.), *Ure's Dictionary of Arts, Manufacture, and Mines*, 6th edn, 3 vols (London, 1867), vol. 1, pp. 496–7.

92. 'Musical Instruments at the Furniture Exhibition', *Musical Opinion & Music Trade Review*, 7:81, 1 June 1884, p. 422.

93. Advertisement, *Musical Opinion & Music Trade Review*, 1 April 1883, p. 301.

94. *Post Office London Directory for 1882*, vol. 1, p. 198; vol. 2, p. 1236; vol. 3, pp. 1740, 1770; *Business Directory of London 1884*, vol. 1, pp. 528–9, 574; *London Gazette*, 10 July 1883; 'Sales of the Year 1880', *Times*, 28 February 1880, p. 16.

95. 'By Messrs C. & H. White', *Bucks Herald*, 22 May 1880, p. 4; 'Fires in London', *Reynold's Newspaper*, 2 May 1880, p. 1.

96. *Post Office London Directory for 1882*, vol. 1, pp. 198, 313; vol. 3, pp. 1770, 1780, 1883; *London Gazette*, 29 March 1887, p. 1879; *Business Directory of London 1884*, vol. 2, pp. 541, 608, 747.

97. 'Cardboard Manufacture from Stable Manure', *Pall Mall Gazette*, 3 April 1888, p. 12.

98. 'The Furniture Exhibition at Islington', *Furniture Gazette*, 13 August 1881, p. 92; 'Timber from Straw', *Furniture Gazette*, 13 August 1881, p. 93.

99. 'Legal Answers', *Sheffield Weekly Telegraph*, 23 April 1887, p. 5.

100. 'Bills of Sale', *Commercial Gazette*, 11 March 1880, p. 243; 'Bills of Sale', *Commercial Gazette*, 9 July 1885, p. 650; 'Re Cleasby', *London Evening Standard*, 8 December 1885, p. 8; census returns for 1881; 'Bills of Sale', *Commercial Gazette*, 14 August 1884, p. 746; census returns for 1881; *Business Directory of London 1884*, vol. 2, p. 747; *London Gazette*, 8 August 1913, p. 5699; 'Meeting of Creditors in Sheffield', *Sheffield Independent*, 15 November 1882, p. 3. Rayden was bankrupt in 1896: *Edinburgh Gazette*, 29 May 1896, p. 524; 'Sheffield County Court', *Sheffield Independent*, 12 October 1882, p. 5; 'The Claim by a Money Lender', *Sheffield Independent*, 21 October 1882, p. 15; http://www.merseyfire.gov.uk/Historical/liverpoolSalvageCorp.htm;http://discovery.nationalarchives.gov.uk/details/r/05dc5149–0deb-433e-97b4-e71aa2161f8d#0

101. 'Bills of Sale', *Commercial Gazette*, 3 December 1885, p. 1142; census return for 1881.

102. Jeanne Thompson, 'A Compilation of Reformed Church Indexes from the Town of Leer, Ostfriesland Covering the Years 1796–1873', 2001, http://www.ogsa.us/pdf-files/LeerReformedChurchIndex.pdf; TNA, BT 31/3704/23038, Company No: 23038; Fritz Viëtor and Company Ltd. Incorporated in 1886, 'Memorandum of Association', stamped 17 August 1886.

103. Census returns for 1881 and 1891; TNA, BT 31/3704/23038, 'Memorandum of Association'.

104. 'Clerkenwell', *Morning Advertiser*, 30 December 1871, p. 6; 'Universal Type-Writer Libel Case', *Derby Daily Telegraph*, 9 April 1889, p. 4; 'Dispute among Freemasons', *Canterbury Journal, Kentish Times and Farmers' Gazette*, 13 April 1889, p. 6; 'Mansion House', *London Evening Standard*, 14 February 1867, p. 7; see also 'Serving a

Summons Whilst Shaking Hands', *Dundee Evening Telegraph*, 13 March 1891, p. 2.

105. TNA BT 31/3704/23038, letter from Fritz Viëtor and others, dated 30 September 1886; a blank space was left where Crickmay's signature should have appeared; letter from Dunn & Palmer, Solicitors, to the Registrar of the Joint Stock Company, Somerset House, dated 6 October 1886.

106. Census returns for 1871 and 1891; *London Gazette*, 21 September 1888, p. 5302; 26 October 1888, p. 5857; *Post Office London Directory for 1882*, vol. 3, pp. 1402, 1881; *Business Directory of London 1884*, vol. 1, p. 157; 'Partnerships Dissolved', *Times*, 24 October 1888, p. 12; *Commercial Gazette*, 24 October 1888, p. 1016.

107. Census returns for 1881 and 1891; 'Re Edward Tauchert', *Morning Post*, 30 May 1892, p. 8.

108. Census returns for 1861 and 1871; 'Bankruptcy Proceedings', *Birmingham Daily Post*, 14 January 1891, p. 6; 'Birmingham County Court', *Birmingham Daily Post*, 12 June 1891, p. 6.

109. *Post Office London Directory for 1882*, vol. 2, p. 696; vol. 3, pp. 68, 1435, 1780; *Post Office London Directory for 1899*, vol. 3, p. 763; *London Gazette*, 5 April 1892, p.2033; census returns for 1881.

110. 'Extracts from the Register of County Courts' Judgements', *Commercial Gazette*, 11 September 1884, p. 836.

111. 'High Court of Justice', *Morning Post*, 1 June 1886, p. 6; 'High Court of Justice', *Morning Post*, 1 November 1886, p. 3; 'Re Julius Ichenhauser', *London Evening Standard*, 11 July 1895, p. 3; 'Antique Oak Panelling', *Times*, 2 October 1885, p. 2; Stadt Bibliotek Fürth, HB 26 71.64.8, *Addreß und Geschäfts – Handbuch der Fürth* (Fürth, 1884), p. 118.

112. 'Outer House', *Scotsman*, 19 February 1886, p. 3; 'R. N. Paterson Brothers & Co. *v.* Viëtor & Co.', *Glasgow Herald*, 6 November 1886, p. 6; 'Paterson's Manufactory Ltd', *Falkirk Herald*, 12 November 1887, p. 7; see also *OB*, t18820109–171, Deception, forgery, 9 January 1882.

113. Advertisement, *Paper Mills Directory*, 24 (1884), p. 85.

114. *London Gazette*, 6 August 1886, p. 3825.

115. TNA BT 31/3704/23038, 'Memorandum of Association of Fritz Viëtor & Company Limited'; letter dated 30 September 1886, letter dated 11 October 1886, from Nathaniel Whitcombe; list of book debts, town, Burdett old and new accounts; 'New Companies',

Manchester Courier & Lancashire General Advertiser, 21 August 1886, p. 4.

116. 'Bankruptcy Proceedings', *Birmingham Daily Post*, 14 January 1891, p. 6.

117. TNA, BT 31/3704/23038, list of book debts, town.

118. 'Bills of Sale', *Commercial Gazette,* 16 September 1886, p. 863 (see also p. 860); *Commercial Gazette*, 15 February 1888, supplement, p. 10; *Post Office London Directory for 1887*, p. 1858; *Post Office London Directory for 1888*, p. 1078.

119. Supplement to the *Commercial Gazette*, 15 February 1888, p. 10; another creditor was Wilhelm Stern & Co., a company dealing in coloured and fancy papers located on the same street as the business owned by Ichenhauser's family.

120. *London Gazette*, 20 December 1887, p. 7006; 10 January 1888, p. 332; 24 January 1888, p. 590.

121. 'Caution', *Scotsman*, 3 March 1886, p. 1; http://www.glasgowwestaddress.co.uk/1888_Book/Paterson_Sons_&_Co.htm

122. See Arjun Appadurai, 'Introduction', in Arjun Appadurai (ed.), *The Social Life of Things* (Cambridge, 1986), p. 7; Meikle, *American Plastic*, p. 13.

123. 'Furniture Trades' Exhibition', *Sewing Machine Gazette & Journal of Domestic Appliances*, 1 September 1881, p. 31.

124. 'Fifty Pounds Reward', *London Evening Standard,* 25 March 1882, p. 1.

125. 'Trade Jottings and Notes', *Musical Opinion & Music Trade Review*, 10: 112, 1 January 1887, p. 183.

126. Cyril Ehrlich, *The Piano* (London, 1976), p. 151; W. L. Sumner, *The Pianoforte* (New York, 1966), p. 91.

127. 'Piano forte', *Manchester Guardian*, 23 November 1880, p. 3.

128. 'H.T. HINCKS', *Leicester Chronicle,* 23 June 1883, p. 1; see also 8 November 1884, p. 1.

129. 'Sales by Auction', *Manchester Courier & Lancashire General Advertiser,* 10 November 1883, p. 8.

130. 'Central Auction Rooms', *Northampton Mercury,* 15 November 1884, p. 4; 'Sales by Mr Broadbent', *Sheffield Daily Telegraph*, 12 August 1893, p. 4; 'Sale This Day', *East Anglian Daily Times*, 29 June 1893, p. 2; 'Removed from Colne and Burnley Residences', *Burnley Express* 16 November 1940, p. 6; '69 Church-St, Burnley', *Burnley Express* 28 December 1901, p. 4; 'Going Abroad', *Burnley Express* 8

August 1903, p. 4; 'Sales by Auction', *Burnley Express* 19 November 1904, p. 4; 'Removed from Colne and Burnley Residences', *Burnley Express*, 16 November 1940, p. 6.

7. The Wonderland of Rework (1840s–1850s)

1. 'Great Fire in Blackstock Street', *Liverpool Daily Post*, 20 September 1858, p. 8; 'Great Fire in Liverpool', *Lloyd's Weekly Newspaper*, 26 September 1858, p. 9.

2. 'Fires', *Liverpool Daily Post*, 14 December 1857, p. 5; 'Latest Liverpool News', *Glasgow Herald*, 27 June 1859, p. 4; 'Destructive Fire in Liverpool', *Liverpool Daily Post*, 7 February 1862, p. 5; 'Great Fire in Liverpool Yesterday', *Liverpool Daily Post*, 15 July 1867, p. 5.

3. Letters, *Luton Times & Advertiser*, 12 February 1859, p. 4.

4. 'Miscellaneous', *Reynold's Newspaper*, 21 January 1866, p. 5; *Rules and Regulations of the London Salvage Corps Established 1866* (London, 1888), p. 5; 'Great Destruction of Property', *Morning Advertiser*, 23 July 1866, p. 2; 'Terrific Fire', *London Evening Standard*, 10 May 1866, p. 5; Cornelius Walford, *Insurance Cyclopaedia*, 6 vols (London, 1871–80), vol. 4, p. 159.

5. Walford, *Insurance Cyclopaedia*, vol. 4, p. 105; Clive Trebilcock, *Phoenix Assurance and the Development of British Insurance*, vol. 2 (Cambridge, 1998), p. 16.

6. 'The Crystal Palace', *Times*, 1 December 1936, p. 16; The Crystal Palace Foundation, 'Disaster Strikes, 1936', http://www. crystalpalacefoundation.org.uk/history/disaster-strikes-1936–2; BBC A History of the World, 'Glass from the Crystal Palace', http:// www.bbc.co.uk/ahistoryoftheworld/objects/ VbK_iXNXQz2vLUYzndNKhw

7. William Crookes, 'Chemical Products – the Application of Waste', *Popular Science Review*, 2:5, January 1863, pp. 58–70.

8. Henry Mayhew, *London Labour and the London Poor*, 4 vols (London, 1861–62), vol. 2, pp. 135–58, 452–61.

9. Mayhew, *London Labour and the London Poor*, vol. 2, p. 152; Emily Cockayne, 'Who Did Let the Dogs Out? Nuisance Dogs in Late-Medieval and Early Modern England', in Laura D. Gelfand (ed.), *Our Dogs, Our Selves* (Leiden, 2016), p. 47.

10. Mayhew, *London Labour and the London Poor*, vol. 2, pp. 152–76.

11. Mayhew, *London Labour and the London Poor*, vol. 1, p. 290.

12. 'Rubbish', *Chambers' Journal*, 4th ser., 12:621, 20 November 1875, p. 749.

13. Christopher Herbert. 'Rat Worship and Taboo in Mayhew's London', *Representations*, 23 (1988), pp. 1–24.

14. Charles Dickens, *Our Mutual Friend* (London, 1865).

15. Hounsell, *London's Rubbish*, pp. 16, 49, 54; Velis, Wilson and Cheeseman, '19th Century London Dust-Yards', pp. 1285–6.

16. See Desrochers, 'Promoting Corporate Environmental Sustainability in the Victorian Era', p. 477.

17. David Greysmith, 'Simmonds, Peter Lund (1814–1897), *ODNB*; 'Obituary', *Yorkshire Post & Leeds Intelligencer*, 7 October 1897, p. 8.

18. See, for example, Richard Herring, *Paper & Paper-Making*, 3rd edn (London, 1863), p. 71.

19. 'Rubbish', *Chambers' Journal*, 4th ser., 12:621, 20 November 1875, p. 749.

20. Emily Hobhouse, 'Dust-Women', *Economic Journal*, 10:39 (1900), p. 411–12.

21. 'Waste Materials', *Chambers' Journal*, 4th ser., 12:580, 6 February 1875, p. 96; Ballard, *Effluvium Nuisances*, pp. 83, 110–11.

22. 'The Manufacture of Arsenic', *Chambers' Journal*, 5th ser., 10:480, 11 March 1893, pp. 149–51.

23. *The Encyclopaedia Britannica*, 7th edn, 24 vols (Edinburgh, 1842), vol. 2, p. 662.

24. Timothy Cooper and Pierre Desrochers have provided accounts of the activities and influences of men who appeared to spearhead sustainability by lecturing and writing for scientific journals: Timothy Cooper, 'Peter Lund Simmonds and the Political Economy of Waste Utilisation in Victorian Britain', *Technology & Culture*, 52:1 (2011), pp. 21–44; Pierre Desrochers, 'Victorian Pioneers of Corporate Sustainability', *Business History Review*, 83:4 (2009), pp. 703–29.

25. 'Children's Books for Christmas', *Star*, 22 December 1881, p. 4; See also 'Literature', *Worcester Journal*, 5 November 1881, p. 6.

26. Asa Briggs, *Victorian Things* (London, 1988), pp. 47–50.

27. Clara Matéaux, *Wonderland of Work* (London, c.1880).

28. Census returns for 1851, 1861 and 1911; [Thomas] Milner Gibson, *Report from the Select Committee on the School of Design* (London, 1850), p. 73–4, 82–3; F. Graeme Chalmers, 'Fanny McIan and

London's Female School of Design, 1842–57', *Woman's Art Journal*, 16:2 (1996), pp. 3–9.

29. Charles Dickens, 'The Female School of Design', *Household Words*, 2:51 (1851), pp. 577–81; census returns for 1851; 'Prize Exhibition at the Government School of Design', *Illustrated London News*, 19 January 1850, p. 12.

30. Annie Carey, *The Wonders of Common Things* (London, 1873), pp. 90, 99, 101, 251–84.

31. Census returns for 1861 and 1871; Esther Carey, *Eustace Carey* (London, 1857), p. 383; Annie Carey (posth.), *School-Girls* (London, 1881), preface.

32. M. A. Paull, *May's Sixpence: or, Waste Not, Want Not* (Edinburgh, 1880), p. 65.

33. M. A. Paull, *Romance of a Rag* (London, 1876), dedication, pp. 53–4, 57, 60, 64, 68, 70–72, 74, 76.

34. Mayhew, *London Labour and the London Poor*, vol. 2, pp. 30–31; 'Shoddy', *Yorkshire Post & Leeds Intelligencer*, 30 July 1872, p. 3; Hanna Rose Shell, 'Images, Technology, and History. Shoddy Heap: a Material History between Waste and Manufacture', *History & Technology*, 30:4 (2014).

35. Spooner, *Wealth from Waste*, pp. 202–3; S. E. Fryer, rev. D. T. Jenkins, 'Lister, Samuel Cunliffe, first Baron Masham (1815–1906), *ODNB*.

36. 'Yorkshire', *Westminster Review*, 140 (April 1859), p. 191.

37. Mayhew, *London Labour and the London Poor*, vol. 2, pp. 30–31.

38. 'Varieties', *Leisure Hour*, 841, 1 February 1868, p. 96.

39. 'Waste in Wollen Factories', *Leeds Times*, 20 June 1857, p. 2; 'Shoddy', *Yorkshire Post & Leeds Intelligencer*, p. 3; Samuel Jubb, *The History of Shoddy* (London, 1860), pp. 43, 48, 52.

40. Jubb, *History of Shoddy*, pp. 21–2, 37.

41. 'Yorkshire', *Westminster Review*.

42. 'Shoddy', *Hull Packet*, 23 June 1843, p. 2; 'March of Impudence', *Northern Star & Leeds General Advertiser*, 24 June 1843, p. 4.

43. 'Frauds in Manufacture and the Truck System', *Spectator*, 23 April 1842, p. 392.

44. Cited in 'News of the Neighbouring Towns', *Manchester Courier & Lancashire General Advertiser*, 14 July 1847, p. 8.

45. 'Shoddy', *Yorkshire Post & Leeds Intelligencer*, 30 July 1872, p. 3.

46. 'March of Impudence', *Northern Star & Leeds General Advertiser*.

47. Jubb, *History of Shoddy*, pp. 2, 21–2, 24–5, 37; 'The Woollen Trade', *Leeds Intelligencer,* 3 July 1847, p. 5.

48. 'Contemporary Science' (*London Review*); *London Gazette*, 19 June 1857, p. 2127; Lyon Playfair, 'Paper Making – Ulmate of Ammonia', *Technologist* 3 (1863), pp. 513–18; Patent 1591 and 3023 (1857); British Library, Add. MS 37189, Correspondence of Charles Babbage, 20 vols, 8 fols 486–91, 1836–55; Seyed Behrooz Mostofi (ed.), *Who's Who in Orthopaedics* (London, 2005), p. 346; 'To Paper Makers', *London Evening Standard,* 19 March 1861, p. 1.

49. Jubb, *History of Shoddy*, p. 23.

50. Crookes, 'Chemical Products', p. 67.

51. 'Our Dust-Bins', *Leisure Hour,* 17, 2 November 1868, p. 720.

52. 'Waste in Wollen Factories', *Leeds Times.*

53. 'Shoddy', *Yorkshire Post & Leeds Intelligencer,* 30 July 1872, p. 3; Lyon Playfair, 'On the Utilisation of Waste', *Technologist,* 3 (1863), p. 403. Flock wallpaper was also available in the eighteenth century, made using 'the rough nap taken off by the Cloth-dresser': Joseph Collyer, *Parent's and Guardian's Directory* (London, 1761), p. 208.

54. Cf. Costas A. Velis, David C. Wilson and Christopher R. Cheeseman, '19th Century London Dust-Yards: A Case Study in Closed-Loop Resource Efficiency', *Waste Management*, 29:4 (2009), pp. 1282–90.

55. [Andrew Wynter], 'The Use of Refuse', *Quarterly Review*, 124 (1868), pp. 334–57.

56. Matéaux, *The Wonderland of Work*, p. 161; advertisement, *Musical Opinion & Music Trade Review*, 12:133 (October 1888), p. 42.

57. Advertisement, *Musical Opinion & Music Trade Review*, 11:130 (July 1888), p. 472; see also *London Gazette*, 29 May 1868, p. 3069.

58. Andrew Wynter, *Curiosities of Toil* (London, 1870), vol. 1, p. 4.

59. Ballard, *Effluvium Nuisances*, p. 89.

60. 'Waste Materials' [*Chambers' Journal*], p. 95; 'Utilization of Waste Products', *Chambers' Journal*, 5th ser., 5:217, 25 February 1888, p. 128.

61. For a more comprehensive study of types of potash, see, Paul Warde, 'Trees, Trade and Textiles: Potash Imports and Ecological Dependency in British Industry, c.1550–1770', *Past & Present*, 240 (2018), pp. 47–82.

62. H. Dussaunce, *A General Treatise on the Manufacture of Soap* (London, 1869), pp. 21, 98–100, 251, 467, 664, 701.

63. 'Utilization of Waste Products', *Chambers' Journal*, 5th ser., 5:217, 25 February 1888, p. 128; Simmonds, *Waste Products*, pp. 2, 208.

64. 'Pure Coal Tar Soap', *Nottingham Journal*, 12 February 1867, p. 1; 'Jubilaeus carbonis detergentis', *Chemist and Druggist*, 26 July 1913, pp. 135–6; Emily Cockayne, 'In Your Soap Box', www.rummage. work/blog/coaltarsoap, 23 March 2019.

65. N.P. Burgh, 'A Visit to a Bone Boiling Factory', *Technologist*, 4 (1864), pp. 139–44.

66. Ballard, *Effluvium Nuisances*, pp. 72, 116, 134.

67. Edward Dawson Rogers, *Rogers' Directory of Norwich and Neighbourhood* (Norwich, 1859), p. 20; 'Messrs Horsfield and Bagshaw', *Norfolk Chronicle*, 30 November 1867, p. 1.

68. Advertisement, *North Devon Journal*, 26 March 1857, p. 1.

69. Antony Gibbs & Sons and William Joseph Myers & Co., *Peruvian and Bolivian Guano* (London, 1844), pp. 85–8.

70. Dussaunce, *On the Manufacture of Soap*, p. 263.

71. Bodleian Library, Trade Cards 23 (47), Thomas Elliott, Nurseryman and Seedsman; Ballard, *Effluvium Nuisances*, p. 136.

72. 'The Value of Chemical Analysis', *Cornishman*, 19 December 1878, p. 6.

73. Justus Liebig (ed. Lyon Playfair), *Organic Chemistry* (London, 1840), p. 185.

74. Ballard, *Effluvium Nuisances*, p. 5.

75. *OB*, t18570706–817, 6 July 1857.

76. Ballard, *Effluvium Nuisances*, p. 117.

77. Lyon Playfair, 'On the Utilization of Waste', *Technologist*, 3 (1863), p. 406.

78. *OB*, t18570706–817, 6 July 1857.

79. John Booker, *Essex and the Industrial Revolution* (Chelmsford, 1974), pp. 182–3.

80. Frederick Falkner, *The Muck Manual* (London, 1843), pp. 163–4. British Museum, 1982.U.876, trade card of Benoney Hurd (1802–1840); Mark Jenner, 'Polite and Excremental Labour? London's Nightmen *c*.1600–*c*.1850', Institute of Historical Research/SOAS podcast, https://www.youtube.com/watch?v=fB3aOQIa4ik 2013

81. Leslie, *Synthetic Worlds*, p. 83.

82. Liebig (ed. Playfair), *Organic Chemistry*, pp. 197–9.

83. Peter Hounsell, *London's Rubbish* (Stroud, 2013), p. 15.

84. [Edwin Chadwick], *Inquiry into the Sanitary Condition of the Labouring Population of Great Britain* (London, 1842), pp. 379–81; Sidney Dark, *Not Such a Bad Life* (London, 1941), p. 15.

85. 'The London Concentrated Chemical Manure Company', *Exeter & Plymouth Gazette*, 8 April 1848, p. 2; Ballard, *Effluvium Nuisances*, p. 149 (Ballard identified Bilston in Staffordshire as one place where poudrette-making made use of the town's effluent after 1875); Mayhew, *London Labour and the London Poor*, vol. 2, p. 410; [Chadwick], *Inquiry*, p. 379; *London Gazette*, 9 March 1858, p. 1381; 'Poudrette de Bondy', *Yorkshire Gazette*, 23 August 1856, p. 2.

86. Select Committee on Sewage of Towns, *Second Report from the Select Committee on Sewage of Towns* (s.l., 1862), p. 469.

87. Letter to the Editor, *Times*, 18 August 1858, p. 7; see Stephen Halliday, *Year of the Great Stink* (Stroud, 1999).

88. Edmund Edward Fournier d'Albe, *Life of Sir William Crookes* (London, 1923), p. 257.

89. Tom Crook, 'Putting Matter in Its Right Place: Dirt, Time and Regeneration in Mid-Victorian Britain', *Journal of Victorian Culture*, 13:2 (2008), pp. 200–22; Nicholas Goddard, '"A Mine of Wealth?", The Victorians and the Agricultural Value of Sewage', *Journal of Historical Geography*, 22:3 (1996), pp. 277–90.

90. Lewis Angell, *Sanitary Science and the Sewage Question* (London, 1871), pp. 21–3, 25–6; Crookes, 'Chemical Products', p. 64; William Cameron Sillar et al., *The A.B.C. Sewage Process* (London, 1868), pp. 3–7; Fournier d'Albe, *Life of Sir William Crookes*, p. 257; 'Bucks', *Bucks Herald*, 28 January 1882, p. 1; see also 'Sewage of Towns', *Bolton Evening News* 22 April 1882, p. 4.

91. Angell, *Sanitary Science and the Sewage Question*, pp. 27–8, appendix, iii–iv; see also Goddard, '"A Mine of Wealth?"', p. 280.

92. Mayhew, *London Labour and the London Poor*, vol. 2, p. 109.

93. Gloucestershire Archives, D6059/1, account book and stock book of James Andrews of 50 Westgate Street, Gloucester, rag merchant and general dealer, 1835–1880.

94. Fyfe, 'Rags for the Ragged', pp. 529–30.

95. Benjamin Lambert, 'On the Paper Manufacture II', *Technologist*, 4 (1864), p. 20.

96. 'The Old Bookshop in Portsmouth-Street', *Illustrated London News*, 5 January 1884, p. 14; 'The Old Curiosity Shop', *Shields Daily Gazette*, 4 January 1884, p. 4.

97. Census returns for 1841, 1871, 1881 and 1891.

98. 'Robbing the Graphotyping Company', *Clerkenwell News*, 28 June 1871, p. 5.

99. 'The Marine Store Shop Nuisance Again!', *Devizes and Wiltshire Gazette*, 31 January 1861, p. 3.

100. 'The Marine Store Shop Nuisance', *Devizes and Wiltshire Gazette*, 17 January 1861, p. 3.

101. J. Hamilton Fyfe, 'Rags for the Ragged', *Once a Week*, 3 May 1862, p. 529.

102. PP 1860 (195) III.I, bill for regulating the business of dealers in marine stores, 1860.

103. 'The Marine Store Nuisances', *Cambridge Independent Press,* 2 April 1859, p. 8.

104. 'A New Industry', *Times*, 18 April 1889, p. 3; *London Gazette*, 27 November 1877, 6685; Patent 2765 (1877); census returns for 1891.

105. 'Sales by Auction', *Morning Advertiser*, 30 May 1859, p. 8.

106. 'Fire', *Norfolk News*, 2 June 1866, p. 6.

8. Rebaptisms of Fire (1830s–1840s)

1. 'News', *Globe,* 22 January 1836, p. 4; 'Singular Death', *Chelmsford Chronicle*, 29 January 1836, p. 1.

2. 'Middlesex Adjourned Sessions', *Morning Post,* 13 September 1834, p. 4. See also, TNA, HO 17/59/123, Home Office, petition on behalf of Henry Kirkman, 3 October 1832; Emily Cockayne, *Cheek by Jowl* (London, 2012), p. 52.

3. 'Marlborough Street', *Morning Advertiser,* 18 November 1825, p. 3.

4. 'Melting Pewter Pots', *Morning Advertiser,* 29 September 1825, p. 2.

5. 'Thames Police', *London Evening Standard,* 14 June 1832, p. 4; 'Nuisance', *Evening Mail,* 5 September 1832, p. 1; 'Gross Nuisance', *Morning Advertiser,* 26 August 1834, p. 3; London Metropolitan Archives, MJ/SP/1834/08/020, Henry Kirkman, indictment for nuisance, Middlesex Sessions of the Peace, 1834.

6. 'Mary-La Bonne', *Public Ledger & Daily Advertiser*, 30 October 1826, p. 4; 'Old Bailey', *Globe*, 9 December 1826, p. 4.

7. 'Police', *Evening Mail,* 12 October 1825, p. 4; 'Henry Kirkman', *Morning Chronicle*, 10 January 1826, p. 4.

8. 'Thames Police', *London Evening Standard,* 14 June 1832, p. 4.

9. 'Marylebone', *London Courier & Evening Gazette*, 27 October 1830, p. 4.

10. 'Thames Police', *Public Ledger & Daily Advertiser*, 6 January 1832, p. 3.

11. 'Marylebone', *Morning Post*, 23 April 1832, p. 4.

12. 'Thames Police', *London Evening Standard*, 14 June 1832, p. 4.

13. 'Thames-Office', *Bell's Weekly Messenger*, 17 June 1832, p. 6.

14. 'Middlesex Sessions', *London Courier & Evening Gazette*, 4 September 1832, p. 4; 'Middlesex Sessions', *Atlas*, 9 September 1832, p. 3; 'Shocking Death', *Bell's Weekly Messenger*, 24 January 1836, p. 5.

15. John Haggard, *Reports of Cases Argued and Determined in the Consistory Court of London*, 2 vols (London, 1822), vol. 1, pp. 409–14.

16. 'Much Excitement', *Evening Mail*, 22 January 1836, p. 6; London Metropolitan Archives, MJ/SP/C/W/1730, Coroner's report Henry Kirkman, Westminster district, St Pancras no. 33, 22 January 1836.

17. *OB*, t18250113–200, 13 January 1825.

18. Desrochers, 'Does the Invisible Hand?', pp. 5–6; Charles Slack, *Noble Obsession* (London, 2002), p. 61.

19. London Metropolitan Archives, CLC/W/JB/044/MS01018/001, St Dunstan in the West precinct: registers of presentments of the Wardmote Inquest, 2 vols, 1558–1870, vol. 1, fol. 200. Presentment of William Sturt, 1814.

20. Peter Thorsheim, 'The Paradox of Smokeless Fuels: Gas, Coke and the Environment in Britain, 1813–1949', *Environment and History*, 8:4 (2002), pp. 381–401.

21. Mary Mills, *The Early East London Gas Industry and its Waste Products* (London, 1999), pp. 19–26.

22. [Robert Stephenson], *Description of the Patent Locomotive Steam Engine* (London, 1838), p. 20; Samuel Charles Brees, *Appendix to Railway Practice* (London, 1839), p. 313; François Marie Guyonneau de Pambour, *Practical Treatise on Locomotive Engines upon Railways* (London, 1836), p. 268.

23. 'Devizes Gas and Coke Company', *Devizes and Wiltshire Gazette*, 8 February 1827, p. 3; 'Bristol Diocesan Association', *Bristol Mirror*, 7 February 1829, p. 3.

24. 'Price of Gas', *Kendal Mercury*, 15 June 1844, p. 2.

25. [Dodd], 'Penny Wisdom', p. 98.

26. M. R. Fox, *Dye-Makers of Great Britain, 1856–1976* (Manchester, 1987), p. 62. See also Samuel Clegg Jr, *Practical Treatise of the*

Manufacture and Distribution of Local Gas, 4th edn (London, 1866), p. 37.

27. Mills, *Early East London Gas Industry*, pp. 71–5; 'Imperial Gas Light & Coke Company', *Morning Advertiser*, 13 October 1824, p. 1; Matéaux, *The Wonderland of Work*, p. 28.

28. John Burridge, *The Naval Dry Rot* (London, 1824), p. 114.

29. Andrew Wynter, 'The Use of Waste Substances, no. II', *Good Words*, December 1876, p. 306.

30. Mills, *Early East London Gas Industry*, pp. 27–8, 33–4, 65.

31. *Account of the Qualities and Uses of Coal Tar and Coal Varnish* (Edinburgh, 1784), pp. 6–8.

32. P. F. Tingry, *The Painter's & Colourman's Complete Guide* (London, 1830), p. 180; see also T. Bentley, *Prospectus of the Various Paints for the Interior of Houses* (London, 1816), p. 4.

33. Mills, *Early East London Gas Industry*, pp. 25, 35, 56–7.

34. Benjamin Cook, 'Methods of Producing Heat, Light & Various Useful Products from Pit Coal', *Philosophical Magazine*, 37:157 (1811), pp. 332–9; see Ray Shill, *Workshop of the World* (Stroud, 2006), chapter 4.

35. V. E. Chancellor, 'Hume, Joseph (1777–1855)', *ODNB*; The Royal Society, EC/1817/17, certificates of election and candidature, Joseph Hume, 1817, https://collections.royalsociety.org

36. 'Discoveries and Improvements in Arts, Manufactures and Agriculture', *Belfast Monthly Magazine*, 11:64, 30 November 1813, pp. 388–90.

37. University of Cambridge Digital Library, RG0 14/31, 210r, Papers of the Board of Longitude, letter from papers on nautical travel, https://cudl.lib.cam.ac.uk/view/MS-RGO-00014-00031/475.

38. 'Letter to the Editor', *Technical Repository*, 1 (1822), p. 312; 'Art. X *The Naval Dry Rot &c, &c, &.*', *Quarterly Review*, 30 (1824), pp. 216–30.

39. 'Letter to the Editor', *Technical Repository* (1822), pp. 312–14.

40. 'Ships: An Account of the Names and Rate of Each Ship in the Royal Navy Launched since January 1815', in HOC, *Accounts and Papers*, 3 vols, *3 February–6 July 1825* (1825), Table 21, pp. 331–40.

41. 'Letter to the Editor', *Technical Repository* (1822), p. 312.

42. Mills, *Early East London Gas Industry*, pp. 39–41; 'Ships: An Account of the Names and Rate of Each Ship in the Royal Navy'.

43. 'Late Fire at the Devonport Dock-Yard', *London Evening Standard*, 16 October 1840, p. 4.

44. Henry Francis Whitfeld, *Plymouth and Devonport* (Plymouth and Devonport, 1900), pp. 421–2.

45. Mills, *Early East London Gas Industry*, pp. 45, 49, 67–8.

46. Thomas Allen Britton, *A Treatise on the Origin, Progress, Prevention and Cure of Dry Rot in Timber* (London, 1875), pp. 130–32; see also Henry Potter Burt, 'On the Nature and Properties of Timber', *Minutes of the Proceedings of the Institute of Civil Engineers*, 12 (1853), p. 228.

47. George J. Stevenson, *The Methodist Hymn-Book and Its Association* (London, 1870), p. 307.

48. Mills, *Early East London Gas Industry*, p. 53; Hancock, *Personal Narrative*, p. 21.

49. *London Gazette*, 5 November 1841, p. 2737; 'Destructive Fire' *Morning Post*, 5 July 1841, p. 4; Mills, *Early East London Gas Industry*, p. 54.

50. HOC, *Report from the Select Committee on Turnpike Trusts and Tolls* (London, 1836), p. 54.

51. John Henry Cassell, *Treatise on Roads and Streets* (London, 1835), p. 20.

52. Mills, *Early East London Gas Industry*, p. 53.

53. Cassell, *Treatise on Roads*, pp. 22–4, 26–9; see London County Council, *Survey of London*, vols 43 and 44, *Poplar, Blackwell and Isle of Dogs* (London, 1994), pp. 427–9.

54. Ruth Richardson, *Death, Disease and the Destitute*, 2nd edn (London, 1987), pp. 132–41; Cassell, *Treatise on Roads*, pp. 28–9.

55. TNA, PROB/11/1883/480, will of John Henry Cassell, 11 September 1837; TNA, RG/4/4247, nonconformist burials, St Ann's Limehouse, 11 March 1837; cf. Mills, *Early East London Gas Industry*, p. 54.

56. Cassell, *Treatise on Roads*, p. 20.

57. Cassell, *Treatise on Roads*, pp. 9–10; HOC, *Report from the Select Committee on Turnpike Trusts*, p. 54.

58. HOC, *Report to her Majesty's Principal Secretary of State for the Home Department, from the Poor Law Commissioners, on an inquiry into the sanitary condition of the labouring population of Great Britain* (London, 1842), p. 380.

59. HOC, *Report from the Select Committee on Turnpike Trusts*, pp. 54–5; Cassell, *Treatise on Roads*, pp. 11–13, 15.

60. 'Manchester Asphaltum Company', *Manchester Times*, 5 May 1838, p. 2.

61. Cassell, *Treatise on Roads*; Stephen Inwood, *City of Cities* (London 2005), pp. 238–9.

62. Gordon Airey, 'Recycling of Bituminous Materials', in Peter Domone and John Illston (eds), *Construction Materials*, 4th edn (London, 2010), pp. 241–4.

63. Mills, *The Early East London Gas Industry*, p. 57.

64. *Encyclopaedia Britannica*, 8th edn, 22 vols (Edinburgh, 1853–60), vol. 8, p. 285.

65. Thomas Hancock, *Personal Narrative of the Origin and Progress of the Caoutchouc or India rubber Manufacture in England* (London, 1857), p. v; Slack, *Noble Obsession*, p. 62.

66. Slack, *Noble Obsession*, p. 63. Patent 4804 (1823).

67. Hancock, *Personal Narrative*, pp. v–vi; Slack, *Noble Obsession*, p. 64.

68. John Loadman and Francis James, *The Hancocks of Marlborough* (Oxford, 2010), p. 21.

69. Slack, *Noble Obsession*, pp. 57–8.

70. H. I. Dutton, *The Patent System* (Manchester, 1984), pp. 159–60.

71. Hancock, *Personal Narrative*, pp. 14–16.

72. Frederick Walton, 'On the Introduction and Use of Elastic Gums and Analogous Substances', *Journal of the Royal Society of Arts*, 10:489, 4 April 1862, p. 326; Hancock, *Personal Narrative*, pp. 23, 72; Slack, *Noble Obsession*, pp. 64–5.

73. Hancock, *Personal Narrative*, pp. 73–4, 81–2; Loadman and James, *Hancocks of Marlborough*, pp. 57, 60.

74. Slack, *Noble Obsession*, pp. 5–7, 128–34; Hancock, *Personal Narrative*, pp. 91–92, 107.

75. Hancock, *Personal Narrative*, pp. 88–90, 92.

76. Goldsmith, 'Alexander Parkes, Parkesine, Xylonite and Celluloid', p. 12; Patent 2359 and 11147 (both 1855); Hancock, *Personal Narrative*, p. 123.

77. Hancock, *Personal Narrative*, p. 95.

78. Slack, *Noble Obsession*, p. 236; The Science Museum Library holds Alexander Parkes's copy of Hancock, *Personal Narrative*.

79. W. T. Brannt, *India Rubber, Gutta Percha and Balata* (London, 1900), pp. 291–4; Hancock, *Personal Narrative*, p. 127.

80. J. M. Ball, 'Development of Reclaiming Industry, part 1', in A. Nourry (ed.), *Reclaimed Rubber* (London, 1962), p. 4.

81. Henry Murray, 'Recent Improvements in Minor British Art Industries', *Art Journal*, new ser. 8 (1869), p. 9; John R. Jackson, 'Cork and Its Uses', *Technologist*, 5 (1865), p. 196.

82. 'Letter to the Editor', *Monthly Magazine or British Register*, 39, 1 June 1815, pp. 402–3.

83. Jackson, 'Cork and Its Uses', pp. 193, 196.

84. Census returns for 1841; 'Kamptulicon', *Building News*, 24 October 1862, p. 319.

85. Patent 10054 (1844).

86. 'Infringement of Patent – Vice Chancellor's Court', *Patent Journal and Inventors' Magazine*, 87, 22 January 1848, pp. 205–6.

87. Advertisement, *Builder*, 23, 15 July 1843, p. 271.

88. 'Kamptulicon', *Building News*.

89. For example, advertisement, *Builder*, 21, 1 July 1843, p. 248; 'Patent Elastic Pavement & Kamptulicon Company', *Kentish Mercury*, 29 November 1845, p. 4.

90. 'Effects of Shot on Iron Vessels, Kamptulicon', *American Railroad Journal*, 20:559 (6 March 1847), p. 151.

91. Fred T. Jane, *The British Battle Fleet*, 2 vols (London, 1915), vol. 1, pp. 219–20.

92. 'New Uses of India Rubber', *Chambers' Edinburgh Journal*, 6 July 1844, p. 105.

93. 'Weekly Compendium', *Newcastle Journal*, 23 September 1843, p. 4.

94. 'Condensed News', *Oxford Chronicle & Reading Gazette*, 12 August 1843, p. 2; 'Fatal Accident', *Illustrated London News*, 12 August 1843, p. 3; 'Inquest on the Body of the Late Mr Ancona', *Morning Advertiser*, 11 August 1843, p. 3.

95. 'The Kamptulicon', *Evening Mail*, 8 May 1844, p. 5; 'Kamptulicon Life-Boat', *Artizan*, vol. 2, 1845, p. 250; 'The Patent Kamptulicon Life-Boat', *Year-Book of Facts* (London, 1845), pp. 52–3; 'New Life-Boat', *Morning Post*, 26 September 1844, p. 8.

96. [Tom Taylor] *Diogenes and His Lantern* (London, 1850), p. 28.

97. 'The History of Philosophy and Puffing', *Hogg's Instructor*, 5 (1850), p. 27.

98. *British Museum New Reading Room and Libraries* (London, 1867), p. 14; 'Kamptulicon', *Globe*, 26 May 1851, p. 1. See also, 'Burnley Union', *Burnley Advertiser*, 22 February 1868, p. 2.

99. 'Destructive Fires', *Exeter & Plymouth Gazette*, 17 November 1849,
     p. 7; 'Diary of Events for the Year 1849', *Era*, 6 January 1850, p. 10.
100. Census returns for 1851.
101. *London Gazette*, 31 October 1851, p. 2851; advertisement, *Investor's
     Manual*, 28 October 1865, p. 305.
102. *London Gazette*, 16 October 1866, p. 5490; Patent 1564 (1866).
103. Murray, 'Recent Improvements in Minor British Art Industries',
     p. 9.
104. 'Kamptulicon', *Mechanics' Magazine*, 31 October 1862, p. 274.
105. Walton, 'On the Introduction and Use of Elastic Gums', pp. 332–4.
106. 'Destructive Fires', *Essex Herald*, 15 November 1864, p. 8; 'Immense
     Fire at Deptford', *Morning Post*, 19 September 1870, p. 7.
107. 'Fires', *Gloucester Citizen*, 12 June 1877, p. 2.
108. *London Gazette*, 7 February 1865, p. 604.
109. Patent 911 (1870). This product is similar to one designed by Henry
     Trappes, another Mancunian, in 1855: Great Seal Patent Office,
     *Abridgements of the Specifications relating to the Manufacture of Paper,
     Pasteboard, and Papier Mâché* (London, 1858), p. 127.
110. 'Boulinikon or Buffalo Hide Floor Cloth', *Newcastle Journal*, 20 July
     1871, p. 1; 'The Town Hall Floor Coverings', *Bolton Evening News*, 4
     November 1874, p. 4.
111. 'Great Fire in Salford', *Sheffield Independent*, 8 March 1880, p. 3.
112. Murray, 'Recent Improvements in Minor British Art Industries',
     p. 9.
113. 'Southwark', *London Evening Standard*, 25 March 1845, p. 3;
     'Infringement of Patent – Vice Chancellor's Court', pp. 205–6;
     Patents 748 (1855); 2657 (1870); 'The Late Mr Elijah Galloway,
     C.E.', *Mechanics' Magazine*, 24 May 1856, p. 491; census returns for
     1861, 1881 and 1891; Murray, 'Recent Improvements in Minor British
     Art Industries', p. 10; Classifieds, *Clerkenwell News*, 11 August 1871,
     p. 8; Bodleian Library, Broadside Ballads Online, Harding B. 12
     (214), 'The Ladies' Argument' (n.d.). http://ballads.bodleian.ox.ac.
     uk/search/roud/V27423
114. [George Dodd], 'Penny Wisdom', *Household Words*, 134, 16 October
     1852, p. 98; Commissioners for the Exhibition of 1851, *Official
     Descriptive and Illustrated Catalogue*, 3 vols (London, 1851), vol. 3,
     p. 128.
115. Lee Jackson, *Dirty Old London* (London, 2014), chapter 5.

116. Cited in 'Desecration of the Tomb', *Leicester Chronicle*, 10 October 1846, p. 1.

117. Patent Office, *Abridgements of the Specifications   Papier Mâché*, pp. 25, 44.

118. *London Gazette*, 29 May 1863, p. 2808.

119. Simmonds, 'Animal Substances Used for Writing On', p. 15.

9. Dustwomen and a Straw Man (1780s–1830s)

1. John Millington, *Elements of Civil Engineering* (Philadelphia, PA, 1839), p. 263.

2. Peter Hounsell, *London's Rubbish* (Stroud, 2013), pp. 15–16; Matéaux, *Wonderland of Work*, p. 308; Millington, *Elements of Civil Engineering*, p. 263.

3. News, *Western Times*, 8 December 1827, p. 2; 'Serious Accident', *Bell's Life in London & the Sporting Chronicle*, 9 December 1827, p. 2.

4. News, *Exeter & Plymouth Gazette*, 28 March 1829, p. 3.

5. 'The Alchymy [*sic*] of the Dust-Hole', *Globe*, 2 June 1836, p. 1; Hounsell, *London's Rubbish*, pp. 13–15.

6. Millington, *Elements of Civil Engineering*, p. 263; Dickens, *Our Mutual Friend*.

7. Costas A. Velis, David C. Wilson and Christopher R. Cheeseman, '19th Century London Dust-Yards: A Case Study in Closed-Loop Resource Efficiency', *Waste Management*, 24:4 (2009), p. 1284.

8. Census returns for 1851.

9. Hounsell, *London's Rubbish*, p. 16; Millington, *Elements of Civil Engineering*, p. 263.

10. Joseph Lockwood, 'Bricks and Brickmaking', *Builder*, 115, 19 April 1845, p. 183.

11. G. C. Greenwell, 'The Manufacture of Various Types of Bricks', *Lectures Delivered at the Bristol Mining School 1857* (Bristol, 1859), pp. 282–3; Ken Gurke, *Bricks and Brickmaking* (Moscow, ID, 1987), pp. 12, 127–8; Emily Cockayne, *Hubbub* (London, 2007), p. 132. For an embedded bone fragment in a Hampton Court brick, see Twitter, HRP Surveyors @HRP_Surveyors, 23 February 2017, https://pbs.twimg.com/media/C5YsgwSWAAMqqEz.jpg

12. Adolphe Smith, 'Flying Dustmen', in John Thompson, *Victorian London Street Life* ([1877] repr. New York, 1994), p. 126.

13. John Lawrence Hammond and Barbara Bradby Hammond, *The Village Labourer, 1760–1832* (London, 1913), p. 116.

14. James Mill (ed.), review of *Malcolm's Agriculture of Surrey*, in *Literary Review* 1:5 (May 1806), p. 485.

15. Board of Agriculture, *On the Subject of Manures* (London, 1795); Frederick Falkner, *The Muck Manual* (London, 1843), p. 106.

16. Charles Babbage, *On the Economy of Machinery and Manufacture*, 4th edn (London, 1846), p. 218.

17. 'Agricultural Chemistry', *Chester Chronicle*, 28 January 1814, p. 4; Falkner, *Muck Manual*, p. 95.

18. 'At the Durham Assizes', *Oxford University and City Herald*, 20 April 1811, p. 4.

19. Falkner, *Muck Manual*, p. 95; 'Soot Manure', *Chester Courant*, 30 October 1821, p. 4; 'Universal Chimney Doctor', *Stamford Mercury*, 26 January 1827, p. 3.

20. See Leslie, *Synthetic Worlds*, p. 83.

21. Board of Agriculture, *On the Subject of Manures* (London, 1795); Falkner, *Muck Manual*, pp. 90–91, 106–7, 146–7; See also James Mill (ed.), review of *Malcolm's Agriculture of Surrey*, in *Literary Review* 1:5 (May 1806), p. 485.

22. Abraham Rees, *Cyclopædia; or, Universal Dictionary of Arts, Sciences, and Literature*, 39 vols (London, 1802–19), vol. 4, s.v. 'bone'.

23. [George Dodd], 'Done to a Jelly', *Household Words*, 222 (24 June 1854), p. 439.

24. Matéaux, *Wonderland of Work*, p. 308.

25. London Metropolitan Archives, MJ/SP/1778/09/055, Middlesex Sessions of the Peace, Sessions Papers 1778, 'John Oliver and James Allan, both of Shoreditch, prosecuted by John Thompson for setting up pan furnaces and boilers'. See also MJ/SP/1834/06/095, Middlesex Sessions of the Peace, Sessions Papers 1834, 'John Hyde, of 102 White Cross Street, Saint Luke, appeals against his conviction for causing a nuisance'.

26. London Metropolitan Archives, MJ/SP/1832/09/079, Middlesex Sessions of the Peace, Sessions Papers 1832, 'Petition and other papers, of John Barber, of Baldwin Place, Saint Andrew, Holborn'.

27. 'Charge of Nuisance', *Bury and Norwich Post*, 11 February 1835, p. 3; 'Misdemeanour', *Norfolk Chronicle*, 14 February 1835, p. 4; William White, *History, Gazetteer, and Directory of Norfolk* (Sheffield, 1836), p. 179.

28. Dodd, 'Done to a Jelly', p. 439.

29. Wakefield Archives (WA), QS1/150/9, quarter sessions rolls, Wakefield, indictment, public nuisance, October 1811.

30. WA, QS1/160/4, quarter sessions rolls, Pontefract, indictment, public nuisance, April 1821.

31. Rees, *Cyclopædia*, vol. 4, s.v. 'bone'; 'French Patent Granted to J. F. Boby for Making Glue', *Retrospect of Philosophical, Mechanical, Chemical and Agricultural Discoveries*, 8 (London, 1815), p. 249.

32. Dodd, 'Done to a Jelly', p. 439; Matéaux, *Wonderland of Work*, p. 286.

33. Rees, *Cyclopædia*, vol. 4, s.v. 'bone'.

34. 'Bone Manure', *Hull Advertiser & Exchange Gazette*, 14 December 1821, p. 2; 'W. Beckett', *Hull Advertiser & Exchange Gazette*, 19 April 1822, p. 2. See also, Advertisement, *Stamford Mercury*, 13 January 1826, p. 3.

35. 'Capital Mills at Castleford, Yorkshire', *Leeds Intelligencer*, 8 December 1794, p. 2; 'Nothing is Useless', *Chambers' Edinburgh Journal*, 36: 131 (July 1846), pp. 19–21.

36. 'Human Bones', *Westmorland Gazette*, 16 November 1822, p. 2.

37. 'Traffic in Human Bones', *Western Times*, 7 November 1829, p. 3.

38. 'Importation of Human Bones', *Cheltenham Chronicle*, 12 April 1832, p. 4.

39. 'Bone-Dust for Cultivation of Grain', *Repertory of Inventions*, 13:68 (London, 1832), pp. 214–25.

40. The discovery of a medieval skull remnant on the site does, however, raise the possibility that exhumed human bones from cemeteries were milled at the Narborough bone mill, but I am sceptical that this was a regular source of bone supplies. Email correspondence with Peter Goulding (9 May 2017); Narborough Bone Mill, www.bonemill.org.uk.

41. Rees, *Cyclopædia*, vol. 26, s.v. 'paper'; George Dodd, 'Paper: Its Applications', *The Curiosities of Industry* (London, 1853), pp. 7–8.

42. Benjamin Lambert, 'On the Paper Manufacture', *Technologist*, 3 (1863), p. 387.

43. Peter Bower, *Turner's Papers* (London, 1990), pp. 18. 123.

44. Alan Crocker and Martin Kane (eds), *The Diaries of James Simmons Paper Maker of Haslemere, 1831–1868* (Guildford, 2015), pp. 24, 27, 40, 45, 49, 58.

45. Richard Herring, *Paper and Paper Making*, 3rd edn (London, 1863), p. 88.

46. 'London', *Stamford Mercury*, 7 January 1791, p. 2.

47. A. D. Mackenzie, *The Bank of England Note* (Cambridge, 1953), pp. 8–9.

48. Samuel Knafo, *The Making of Modern Finance* (Abingdon, 2013), p. 115.

49. Matthias Koops, *Historical Account of the Substances Which Have Been Used To Describe Events, and To Convey Ideas, from the Earliest Date to the Invention of Paper*, 2nd edn (London, 1801), pp. 12–13, 229–30.

50. 'News', *Caledonian Mercury*, 6 December 1790, p. 3.

51. Koops, *Historical Account*, 2nd edn, p. 250; 'Winchester', *Salisbury & Winchester Journal*, 15 March 1802, p. 4.

52. 'London', *Hampshire Chronicle*, 23 May 1803, p. 2.

53. 'Friday Night's Post', *Chester Courant*, 3 November 1801, p. 2; Koops, *Historical Account*, 2nd edn, p. 13.

54. House of Commons, *Report on Mr Koops' Petition*, 10 June 1801, p. 6 [602].

55. 'Friday Night's Post', *Chester Courant*, 3 November 1801, p. 2.

56. 'Monthly Commercial Report', *Hereford Journal*, 10 August 1808, p. 4; 'Literary Intelligence, English and Foreign', *Scots Magazine*, 1 March 1808, p. 45.

57. 'Monthly Commercial Report', *Hereford Journal*, 5 April 1809, p. 2.

58. Rees, *Cyclopædia*, vol. 26, s.v. 'paper'.

59. John Murray, *Observations and Experiments on the Bad Composition of Modern Paper* (London, 1824), pp. 7–9, 11.

60. Patent Office, *Abridgements Paper, Pasteboard, and Papier Mâché*, p. 11; 'Specification of the Patent Granted to Mr Koops', *Repertory of Arts and Manufactures*, 14 (1801), p. 225; Patent 2392 (1800).

61. Koops, *Historical Account*, 2nd edn, pp. 16, 261.

62. Patent Office, *Abridgements Paper, Pasteboard, and Papier Mâché*, pp. 11–12.

63. Koops, *Historical Account*, 2nd edn, pp. 233, 250–53; Keri Davies, 'William Blake and the Straw Paper Manufactory at Millbank', in Karen Mulhallen (ed.), *Blake in our Time* (Toronto, 2010), p. 237.

64. 'To the Public', *Sun*, 4 July 1800, p. 1; 'To the Public', *True Briton*, 17 July 1800, p. 1; 'Important Discovery', *Caledonian Mercury*, 27 September 1800, p. 1.

65. Hunter, *Papermaking*, p. 338; 'To Brewers, Distillers &c', *Morning Advertiser*, 25 July 1807, p. 1.

66. Davies, 'William Blake and the Straw Paper Manufactory', p. 237; 'Bankrupts', *Literary Magazine & British Review*, 5 (July 1790), p. 79.

67. *London Gazette*, 2 June 1801, p. 626; 'Hesse *v* Stevenson', *Morning Post*, 29 November 1803, p. 3.

68. 'Parliamentary Intelligence', *Times*, 17 June 1801, p. 2.

69. George III, ch. 125, Act for Enabling Matthias Koops, Gentleman, To Assign the Benefit of an Invention of Making Paper from Straw', 27 June 1801.

70. Davies, 'William Blake and the Straw Paper Manufactory', p. 238.

71. Matthias Koops, *Thoughts on a Sure Method of Annually Reducing National Debt* (London, 1796); *Critical Review*, 20, May 1797, p. 93; Davies, 'William Blake and the Straw Paper Manufactory', p. 239. See the British Library catalogue entry for *To the Public, by the Proprietors of the Minerva Office. Pall Mall, for Fire, Lives, Annuities, &c.* (London, 1797), www.bl.uk.

72. Koops, *Historical Account*, 2nd edn, p. 8; Dard Hunter, *Papermaking* (London, 1943), p. 332.

73. *Encyclopaedia Britannica*, 8th edn, vol. 8, p. 293.

74. 'Law Intelligence, Westminster Hall', *Times*, 15 December 1790, p. 4.

75. 'Law Report', *Times*, 30 November 1797, p. 3.

76. H. I. Dutton, *The Patent System* (Manchester, 1984), p. 163.

77. 'Hesse *v.* Stevenson', in John Bernard Bosanquet and Christopher Puller (eds), *Reports of Cases Argued and Determined in the Courts of Common Pleas*, 2nd edn, 3 vols (London, 1814), vol. 3, pp. 565–78.

78. Hunter, *Papermaking*, p. 338; Davies, 'William Blake and the Straw Paper Manufactory', p. 250; 'Patent Straw Paper Manufactory shares', *Morning Chronicle*, 28 October 1802, p. 4.

79. Davies, 'William Blake and the Straw Paper Manufactory', p. 250.

80. James Greig (ed.), *The Farington Diary*, 8 vols (London, 1922–28), vol. 2, p. 251.

81. 'Messrs Robins', *Morning Advertiser*, 10 June 1806, p. 3; Patent Office, *Abridgements of the Specifications [...] Papier Mâché*, pp. 14–15; Patent 2840 (1805); Hunter, *Papermaking*, p. 201, 513, 526.

82. 'Kitchen Garden Earth', *Morning Post*, 15 October 1805, p. 1; 'To Brewers, Tavern and Coffee-House Keepers', *Morning Advertiser*, 1

July 1806, p. 1; 'To Brewers, Distillers &c', *Morning Advertiser*, 25 July 1807, p. 1.

83. 'To Builders, Papermakers, and Others', *Morning Advertiser*, 28 March 1808, p. 4; Hunter, *Papermaking*, p. 340; TNA, PROB 11/1569/246, will of Elizabeth Jane Koops of Westminster, Middlesex, 8 June 1815.

84. 'On the Superiority of the Straw-Paper Made by Mr Louis Lambert's New Patent Process', *Technical Repository*, 8 (1826), p. 248.

85. Lambert, 'On the Paper Manufacture', p. 392; *London Gazette*, 23 October 1860, p. 3805; 22 May 1863, 2695; 27 July 1866, 4252; census returns of 1861.

86. Deborah Lutz, *The Brontë Cabinet* (London, 2015), p. 26.

87. Pocketbook made from William Burke's skin, 1829, Surgeons' Hall Museum, Edinburgh, HC.AB.10.4.

88. Lutz, *The Brontë Cabinet*, pp. 1–35; see Asa Briggs, *Victorian Things* (London, 1988).

89. 'English Patents', *Halifax Courier*, 12 November 1853, p. 4; Patent Office, *Abridgements of the Specifications* [...] *Papier Mâché*, p. 161; see also 'Mansion House', *North London News*, 30 July 1864, p. 3.

90. Henry Constantine Jennings, *Abuses in the Existing Mode of Executing the Public Printing* (London, 1822).

91. 'House of Commons, July 12', *Canterbury Journal*, 14 July 1820, p. 4; 'Parliamentary Paper', *Morning Post*, 30 September 1822, p. 1; 'London, Friday Oct 4', *Morning Post*, 4 October 1822, p. 2.

92. *London Gazette*, 10 May 1859, p. 1911.

93. *OB*, t18020918–158, 18 September 1802; t18241202–82, 2 December 1824; t18381217–304, 17 December 1838; t17861719–45, 19 July 1786.

94. John Crawfurd, *Taxes on Knowledge* (London, 1836), p. 19.

95. *OB*, t18021027–47, 27 October 1802.

96. *OB*, t18270913–88, 13 September 1827.

97. *OB*, t18291203–124, 3 December 1829.

98. *OB*, t18310908–136, 8 September 1831; t18120701–27, 1 July 1812.

99. *OB*, t18020217–88, 17 February 1802.

100. TNA, H09/3, fols 5v, 19; *OB*, t18320517–111, 17 May 1832; 'Exeter, October 27, 1827', *Western Times*, 27 October 1827, p. 1.

101. *OB*, t183200216–13, 16 February 1832.

102. *OB*, t18201028–24, 28 October 1820.

103. *OB*, t18120115–9, 15 January 1812; *OB*, t18120701–79, 1 July 1812; *OB*, t18100221–15, 21 February 1810.

104. Leah Price has written about the connectedness of paper and the body, focusing on the reuse of books as food wrappings and for wiping body parts: showing how books could be absorbent in more ways than one: *How To Do Things with Books in Victorian Britain* (Princeton, NJ, 2012), esp. pp. 219–39.

105. John Nichols, *Illustrations of the Literary History of the Eighteenth Century*, 8 vols (London 1817–58), vol. 6, p. 270; vol. 8, pp. 289–90.

106. Tate Galleries, 'Joseph Mallord William Turner, *George IV's Departure from the 'Royal George'*, *c.*1822, http://www.tate.org.uk/art/artworks/turner-george-ivs-departure-from-the-royal-george-1822-n02880

107. R. J. B. Knight, 'The Introduction of Copper Sheathing into the Royal Navy, 1779–1780', *Mariner's Mirror*, 59:3 (1973), p. 303.

108. 'Improvements in Naval Architecture', *Naval Chronicle for 1808*, 20 (1808), pp. 305–6; Mills, *The Early East London Gas Industry*, p. 38.

109. Mills, *Early East London Gas Industry*, pp. 28–9; Cyril Arthur Townsend, 'Chemicals from Coal', Greater London Industrial Archaeology Society, www.glias.org.uk/Chemicals_from_Coal/PART_1.html; *Account of the Qualities and Uses of Coal Tar and Coal Varnish* (Edinburgh, 1784), pp. 3–4, 6–7, 23; 'Review: *Account of the Qualities and Uses of Coal Tar and Coal Varnish*', *Critical Review*, May 1785, pp. 375–6.

110. F. B. Laidlaw, 'The History of the Prevention of Fouling', US Naval Institute, *Marine Fouling and its Prevention* (Annapolis, MD, 1952), pp. 212–14; Townsend, 'Chemicals from Coal'.

111. Frank Johnson, *The Royal George* (London, 1971), pp. xxi, 3–4, 26. Not all the timber went into the ship; waste chips of wood and larger parts remained.

112. Knight, 'The Introduction of Copper Sheathing', pp. 301, 306.

113. 'List of Patents for New Inventions', *Hereford Journal*, 13 April 1825, p. 4; Thomas Hancock, *Personal Narrative of the Origin and Progress of the Caoutchouc or India Rubber Manufacture in England* (London, 1857), p. 21; Loadman and James, *The Hancocks of Marlborough*, p. 26; Hancock, *Personal Narrative*, pp. 154, 180.

114. Knight, 'The Introduction of Copper Sheathing', p. 303.

115. [Henry Slight and Julian Slight], *A Narrative of the Loss of the Royal George at Spithead*, 5th edition (Portsmouth, 1842), p. 50.

116. [Slight and Slight], *Narrative*, pp. 16, 83–5, 90, 97; Johnson, *Royal George*, pp. 154–5.

117. [Slight and Slight], *Narrative*, pp. 108–10.

118. Nicholas Tracy, *Who's Who in Nelson's Navy* (London, 2006), pp. 131–2; David W. London, 'The Spirit of Kempenfeldt', in Ann Veronica Coates and Philip MacDougall (eds), *Naval Mutinies of 1797* (Woodbridge, 2011), pp. 95–6.

119. Burghley House, https://www.burghley.co.uk/360ipad; 'George Wombwell and his Menagerie', *The Wesley Banner and Revival Record,* January 1851, pp. 66–7; 'New Iron Lighthouse', *Civil Engineer and Architect's Journal*, May 1844, p. 208; *London Gazette*, 11 December 1835, p. 2483; 16 October 1840, p. 2286; 16 July 1841, p. 1860; Basil Hall and Augustin Crueze, *Description of the Capstan Lately Recovered from the Royal George* (London, 1839).

120. [Slight and Slight], *Narrative*, pp. vi, III, 131.

121. *London Gazette*, 7 July 1840, p. 1602.

122. *London Gazette*, 5 July 1803, p. 812.

123. Geoffrey Warren, *Vanishing Street Furniture* (Newton Abbot, 1978), p. 124. See also, Martin H. Evans, 'Old Cannon Re-Used as Bollards', University of Cambridge, Personal Web Page Service, http://people.ds.cam.ac.uk/mhe1000/bollards/cannonbollards.htm

124. John Wallis (ed.), *London*, 3rd edn (London, 1810), p. 681; William J. Pinks, additional material by Edward J. Wood, *History of Clerkenwell* (London, 1881), p. 281.

125. Edward Mogg, *Mogg's New Picture of London* (London, 1848), p. 208.

126. *London Gazette,* 27 July 1838, p. 1693.

127. 'The Téméraire', *Times*, 12 October 1838, p. 7.

128. Stuart Rankin, 'Shipbuilding in Rotherhithe – Bull Head Dock to The Pageants', *Rotherhithe Local History Paper*, 4a (March 2000), p. 9.

129. 'Obituary Notices', *Annals of Military and Naval Surgery* (London, 1863), p. 360.

130. TNA, H09/3, Home Office: Convict Prison Hulks: Registers and Letter Books, Convict Hulks Moored at Woolwich and Devonport: Captivity, Ganymede, Discovery: Register of Prisoners, 1821–33; *London Gazette*, 1 January 1836, p. 10; Nicholas Pevsner, *The Buildings of England* (London, 1989), p. 681; David Cordingly, *Billy Ruffian* (London, 2003), pp. 300–01.

131. Thorsheim, *Waste into Weapons*, p. 161.

132. 'Nothing is Useless', *Chambers' Edinburgh Journal*, pp. 19–21.

133. 'Traffic in Human Bones', *Western Times*, 7 November 1829, p. 3.

10. Once More unto the Breeches (1720s–1780s)

1. *OB*, t17390117–45, 17 January 1739; Thomas Leach, 'Westbeer's Case', *Cases in Crown Law* (London, 1789), pp. 13–17; Dominus Rex *v.* Westbeer, *English Reports*, 176 vols (London, 1900–1932), XCIII (King's Bench), 1083–1085; CLXVIII (Crown Cases), 108–10.

2. TNA, PROB 11/720/262, will of Gabriel Westbeer 1 September 1742.

3. John Noel Balston, *The Elder James Whatman : England's Greatest Paper Maker (1702–1759)* (London, 1992); Marmite Museum, www.marmitemuseum.co.uk/marmite-history

4. 'Country News', *Derby Mercury*, 5 February 1762, p. 4.

5. 'For Makeing Crown Soap', in Carl Van Doren (ed.), *Letters of Benjamin Franklin and Jane Mecom* (Princeton, NJ, 1950), pp. 129–32.

6. John Hawkins, *A General History of the Science and Practice of Music*, 5 vols (London, 1776), vol. 2, p. 249.

7. Babbage, *Economy of Machinery and Manufacture* (1846), p. 11; BM, Prints & Drawings, Banks 85, 169, draft trade card of Philip Tuten, *c.*1750; BM, Prints & Drawings, Heal 52, 54, trade card of Samuel Grover, cutler, *c.*1690.

8. BM, Prints & Drawings, Heal 91, 16, trade card of Thomas Collett, *c.*1800.

9. B. Courtenay, *Courtenay, Perfumer* (London, 1780?); News, *Penny London Post*, 20–22 December 1749, p. 2; News, *Whitehall Evening Post*, 9–12 August 1760, p. 1.

10. *OB*, t17691018–1, 18 October 1769; History of Science Museum, Oxford, Collection Database, inventory number 48095, 'Terrestrial and Celestial Globe, *c.*1775', https://www.mhs.ox.ac.uk/collections/imu-search-page/record-details/?thumbnails=on&irn=2764&TitInventoryNo=48095; H.D.S, *Portable Instructions* (London, 1779), p. 84.

11. Richard L. Hills, *Papermaking in Britain, 1488–1988* (London, 1988), p. 52; Amélia Jumqua, 'Unstable Shades of Grey', in Ariane Fennetaux, Amélie Junqua and Sophie Vasset (eds), *The Afterlife of Used Things* (London, 2015).

12. David Vaisey (ed.), *The Diary of Thomas Turner* (Oxford, 1985): for example, pp. 10, 166; G. H. Dannatt, 'Bicester in the Seventeenth and Eighteenth Centuries', *Oxoniensia*, 26–7 (1961–2), pp. 252–4.

13. Collyer, *Parent's and Guardian's Directory*, p. 235.

14. Aaron Hill, *Essays for the Month of January, 1717* (London, 1717), pp. 21, 24–5, 27.

15. 'Bristol, March 28, 1772', *Bath Chronicle & Weekly Gazette*, 2 April 1772, p. 1.

16. Hills, *Papermaking in Britain*, p. 54.

17. Great Seal Patent Office, *Abridgements of the Specifications Relating to the Manufacture of Paper, Pasteboard, and Papier Mâché* (London, 1858), p. 4.

18. Patent Office, *Abridgements [...] Paper, Pasteboard, and Papier Mâché*, p. 5.

19. Patent Office, *Abridgements [...] Paper, Pasteboard, and Papier Mâché*, pp. 8–9.

20. Mark Kurlansky, *Paper* (New York, 2016), p. 171; Rees, *Cyclopædia*, vol. 26, s.v. 'paper'.

21. Patent Office, *Abridgements [...] Paper, Pasteboard, and Papier Mâché*, pp. 6–7.

22. Patent 1872 (1792).

23. Robert Kerr, *Memorial Relative to the Invention of Bleaching* (Edinburgh, 1792).

24. R. Campbell, *The London Tradesman* (London, 1747), pp. 176, 194; 'Buckram', *Encyclopaedia Perthensis*, 2nd edn, 23 vols (Edinburgh, 1816), vol. 4, p. 458.

25. [Anon.], *A General Description of all Trades* (London, 1747), p. 44.

26. H. Lowndes, *London Directory* (London, 1797), p. 48; *London Gazette*, 6 April 1822, p. 581; Mark Jenner, 'Polite and Excremental Labour? The Nightmen of London c.1600–c.1800', UEA Symposium 'Early Modern Orientations', 17 May 2013; BM, Prints & Drawings, Heal 36, 40, trade card of Robert Stone, 1761; Heal 36, 2, trade card of John Bates, 1763.

27. David Barnett, 'The Structure of Industry in London 1775–1825', unpublished PhD thesis, Nottingham University, 1996, p. 187; BM, Prints & Drawings, Banks, 70, 49, trade card of Robert Ledger; James Raven, *Publishing Business in Eighteenth Century England* (Woodbridge, 2014), p. 61; 'Buckram Manufactory, London', *Norfolk Chronicle*, 15 November 1788, p. 3.

28. Simmonds, *Waste Products*, p. 19; Charles Babbage, *On the Economy of Machinery and Manufacture*, 4th edn (London, 1846), p. 11.

29. Larvik Museum, Norway, trunk by Henry Nickles, https://digitaltmuseum.no/021026961918/kiste

30. Bodleian Library, Oxford, Trade Cards 28(54), Samuel Pratt, trunk-maker.

31. Mary Thale (ed.), *Autobiography of Francis Place* (London, 1972), pp. 47–9.

32. Ronald E. Wilson, *Sheffield Smelting Company Limited 1760–1960* (London, 1960), n.p.

33. *The Sportsman's Dictionary; of the Gentleman's Companion* (London, 1800), s.v. 'SHO'; T. B. Johnson, *The Shooter's Companion* (London, 1819), p. 92; review, 'William Greener, *The Gun*', *New Sporting Magazine*, 8:46 (February 1835), p. 212.

34. Henry W. Robinson and Walter Adams (eds), *The Diary of Robert Hooke, 1672–1680* (London, 1968), p. 84.

35. W. Watt, *Remarks on Shooting: In Verse* (London, 1839), p. 16.

36. V&A Museum, M.230&A-1930, silver bowl and saucer, *c.*1710, http://collections.vam.ac.uk/item/O103850/bowl-and-saucer-unknown

37. See also, Sara Pennell, 'Invisible Mending? Ceramic Repair in Eighteenth-Century England', in Fennetaux, Junqua and Vasset (eds), *Afterlife of Used Things*, pp. 107–21.

38. Isabelle Garachon, 'From Mender to Restorer: Some Aspects of the History of Ceramic Repair', in Hannelore Roemich (ed.), *Glass & Ceramics Conservation*, Interim Meeting of the ICOM-CC Working Group, 3–6 October 2010, Corning, NY (s.i., 2010), p. 25; Thomas Lupton, *A Thousand Notable Things* (London, 1579), p. 296.

39. John White, *Art's Treasury of Rarities*, 5th edn (London, *c.*1710), p. 24.

40. *Schole's Manchester & Salford Directory*, 2nd edn (Manchester, 1797), p. 50.

41. BM, Prints & Drawings, Heal 37, 37, trade card of Edmund Morris.

42. 'Richard Wright', *Penny London Post or The Morning Advertiser*, 13–15 March 1745, p. 4. See also extract of a letter from Josiah Wedgwood, in Robert Somerville, *On the Subject of Manures* (London, 1795), p. 34.

43. TNA, PROB 11/981/117, will of Peter Johnan, china-riveter, 17 September 1772.

44. East Sussex Record Office, RYE/9/35, 16 October 1765, Rye Corporation, assemblies, informations, examinations, depositions and associated papers.

45. Bernard and Therle Hughes, *English Porcelain and Bone China 1743–1850* (London, 1955), pp. 20–25, 73–7.

46. Beverly Lemire, 'Consumerism in Preindustrial and Early Industrial England: The Trade in Secondhand Clothes', *Journal of British Studies*, 27:1 (1988), pp. 1–24; Laurence Fontaine (ed.), *Alternative Exchanges* (New York, 2008); Fennetaux, Junqua and Vasset (eds), *Afterlife of Used Things*.

47. Avril Hart and Susan North, *Seventeenth- and Eighteenth-Century Fashion in Detail* (London, 2009), p. 34.

48. 'This Is To Give Notice', *Stamford Mercury*, 5 April 1744, p. 4.

49. Collyer, *Parent's and Guardian's Directory*, p. 215.

50. Ninya Mikhaila and Jane Malcolm-Davies, *The Tudor Tailor* (London, 2006), p. 42; [Tom Taylor], *Diogenes and His Lantern* (London, 1850), p. 25; [E. B.], *A New Dictionary of the Canting Crew* (London, 1699), C3r.

51. *The Three Merry Coblers* (s.i., 1634); see also Thale (ed.), *Autobiography of Francis Place*, p. 230.

52. Nancy Cox and Karin Dannehl, *Dictionary of Traded Goods and Commodities, 1550–1820* (Wolverhampton, 2007), *BHO*, http://www.british-history.ac.uk/no-series/traded-goods-dictionary/1550–1820

53. Thale (ed.), *Autobiography of Francis Place*, pp. 74, 78, 80, 110, 118.

54. Bodleian Library, Oxford, John Johnson Collection, trade card of Thomas Parker, 1791, https://www.bodleian.ox.ac.uk/johnson/online-exhibitions/a-nation-of-shopkeepers/cries-itinerants-and-services#gallery-item=161232.

55. Advertisement, *Manchester Mercury*, 20 May 1777, p. 4.

56. Advertisements, *Hampshire Chronicle*, p. 8 September 1777, 3; *Kentish Gazette*, 21 January 1778, p. 1.

57. Advertisement, *Manchester Mercury*, 29 February 1780, p. 2.

58. Advertisement, *Shrewsbury Chronicle*, 20 June 1778, p. 3.

59. Advertisement, *Morning Chronicle & London Advertiser*, 2 January 1777, p. 2.

60. Advertisement, *World*, 19 December 1787, p. 3.

61. 'Mr Spence', *Morning Post and Daily Advertiser*, 25 January 1783, p. 3; advertisement, *World*, 11 May 1789, p. 4.

62. Advertisements, *Oxford Journal*, 13 December 1777, p. 1; 12 December 1778, p. 1; Stephen D. Behrendt, A. J. H. Latham and David Northrup (eds), *The Diary of Antera Duke* (Oxford, 2010), p. 265; 'Mr Moor', *Manchester Mercury*, 14 August 1787, p. 4.

63. Advertisement, *World*, 18 January 1788, p. 13.

64. Advertisement, *Morning Post*, 21 May 1804, p. 1.

65. 'A Novelty – Teeth Transplanted', *Cheltenham Chronicle*, 2 April 1829, p. 4.

66. Collyer, *Parent's and Guardian's Directory*, p. 161.

67. Thomas Fuller, *Pharmacopoia Extemporanea* (London, 1710), p. 264; Joseph Warner, *Cases in Surgery, with Remarks* (London, 1754), pp. 128–9.

68. Marie Roberts, *Gothic Immortals* (London, 1990), pp. 99–100; Pierre Joseph Macquer, *A Dictionary of Chemistry*, 2 vols (London, 1771), vol. 2, p. 461; Richard Pearson, *A Practical Synopsis of the Materia Aliementaria, and Materia Medica*, 2nd edn (London, 1808), p. 462; Rees, *Cyclopædia*, vol. 4, s.v. 'bone'. Sarah Lowengard *The Creation of Color in Eighteenth-Century Europe* (New York, 2008), chapter 23.

69. 'Blue Manufactory', *St James's Chronicle or the British Evening Post*, 14–16 August 1787, p. 1.

70. Babbage, *Economy of Machinery and Manufacture* (1846), p. 11; Ballard, *Effluvium Nuisances*, p. 80. Ballard notes that the 'carbonaceous matter' left behind in the making of prussiate of potash was used as a deodoriser in sewers and manure works.

71. Matéaux, *Wonderland of Work*, p. 28.

72. *Chrysal; or, The Adventures of a Guinea* (Dublin, 1760), pp. viii–x; Collyer, *Parent's and Guardian's Directory*, pp. 96–7.

73. Thomas Dibdin, 'Bibliographiana', *Director*, II (London, 1807), pp. 318–19; John Nichols, *Literary Anecdotes of the Eighteenth Century*, 6 vols (London, 1812), vol. 3, p. 621.

74. Kristian Jensen, *Revolution and the Antiquarian Book* (Cambridge, 2011), p. 127.

75. 'To Be Sold at Auction by Mr Christie', *Public Advertiser*, 2 April 1776, p. 4.

76. 'Anecdotes of John Ratcliffe', *Gentleman's Magazine*, LXXXII, February 1812, p. 114.

77. Christie's sale 6012, Wentworth, 8 July 1998, http://www.christies.com/lotfinder/Lot/

le-doctrinal-de-sapience-translated-from-french-998509-details.aspx; Jensen, *Revolution and the Antiquarian Book*, p. 81.

78. TNA, PROB 11/1015/299, will of John Ratcliffe of Bermondsey, 29 January 1776.

79. 'News', *Morning Chronicle & London Advertiser*, 13 February 1779, p. 3.

80. Benjamin Adams supplies an affidavit to Samuel Ayres's will, confirming his handwriting; TNA, PROB 11/1081/478, will of Samuel Ayres, mariner, 27 September 1781.

81. TNA, PROB 11/1224/197, will of Benjamin Adams, merchant of Red Lion Square, Middlesex, 2 November 1792.

82. 'News', *Public Advertiser*, 12 March 1779, p. 3.

83. Mariners' Pre-Voyage Wills, 1799, http://expedition-archive.com/naval-history/mariners-pre-voyage-wills-1799.

84. 'News', *General Evening Post*, 10–12 August 1779, p. 3; 'London', *Public Advertiser*, 8 September 1779, p. 2; Classifieds, *Morning Chronicle & London Advertiser*, 15 October 1779, p. 2.

85. 'Intelligence from Lloyds', *Caledonian Mercury*, 22 January 1780, p. 2.

86. R. Macpherson, *Dissertation on the Preservative from Drowning* (London, 1783), p. 94; for examples see: TNA, PROB 11/1073/89, will of John Niles, now belonging to the Terrible Privateer, 8 January 1781; TNA, PROB 11/1073/151, will of Thomas Beare, of the Terrible Private Ship of War, 13 January 1781.

87. TNA, PROB 11/1081/478.

88. *London Gazette*, 16 December 1780, p. 5.

89. 'For Makeing Crown Soap', Carl Van Doren (ed.), *Letters of Benjamin Franklin and Jane Mecom* (Princeton, NJ, 1950), pp. 129–32.

90. BM, 'The Botching Taylor Cutting His Cloth to Cover a Button', 1868,0808.4620, http://www.britishmuseum.org/research/collection_online/collection_object_details.aspx?assetId=74546001&objectId=1449253&partId=1; M. Dorothy George, *English Political Caricature to 1792* (Oxford, 1959), pp. 156–9.

91. BM, James Gillray, 'The State Tinkers', 1780, 1868,0808.4650, http://www.britishmuseum.org/research/collection_online/collection_object_details.aspx?objectId=1577553&partId=1

92. Edwin Roffe, 'Salmon in the Thames', *Notes & Queries*, 3rd ser., 6:144, 1 October 1864, p. 275.

93. Cockayne, *Hubbub*, pp. 125–6.

94. *A Pleasant New Song about a Joviall Tinker* (s.i., 1616).

95. John Dunton, *The Informer's Doom* (London, 1683), p. 144; *Epipapresbyter, Grand-Child to Smectymnuus* (London, 1685).

96. *Room for a Jovial Tinker, Old Brass to Mend* (s.i., 1680–82).

97. *The Jovial Tinker, or, the Willing Couple* (s.i., 1680).

98. 'London', *Kentish Gazette*, 1 June 1768, p. 2.

99. For charter boxes, see A. J. B. Wace, 'Lining Papers from Corpus Christi College', *Oxoniensia*, 11 (1937), pp. 166–170; [Thomas Brown], *The Reasons of Mr Joseph Haines* (London, 1690), p. 32.

100. Alexander Pope, *The Dunciad Variorum* (Dublin, 1729).

101. Henry Fielding, in W. B. Coley (ed.), *Contributions to The Champion and Related Writings* (Oxford, 2003), p. 449 (6 September 1740).

11. Uncivil Speculations (1630s–1720s)

1. Robert Aiken (ed.), *Letters and Journals Written by the Deceased Mr Robert Baillie*, 2 vols (Edinburgh, 1775), vol. 1, p. 14; George W. Sprott (ed.), *The Booke of Common Prayer* (Edinburgh, 1871), pp. xliii–lxix; Joong-Lak Kim, 'The Scottish-English-Romish Book: The Character of the Scottish Prayer Book of 1637', in Michael Braddick and David L. Smith (eds), *Experience of Revolution in Stuart Britain and Ireland* (Cambridge, 2011), pp. 14–32; Leonie James, *The Great Firebrand* (Woodbridge, 2017), p. 113; William Prynne, a prominent Puritan opponent of Laud, made a reference to printed articles being used as waste paper or to stop mustard pots in 'Brief Instructions for Churchwardens', *A Breviate of the Prelates Intollerable Usurpations* (Amsterdam, 1637), after p. 325.

2. John Dowden, 'Fragments of a Destroyed Edition of (So Called) "Laud's Scottish Prayer Book"', *The Athenaeum*, 3336, 3 October 1891, pp. 450–51.

3. Joong-Lak Kim, 'The Scottish-English-Romish Book', pp. 14–32; Gordon Donaldson, *The Making of the Scottish Prayer Book of 1637* (Edinburgh, 1954); Peter and Fiona Somerset Fry, *The History of Scotland* (London, 1982), pp. 170–71; Matthew Ward, *Pure Worship* (Eugene, OR, 2014), pp. 40–42.

4. N. R. Ker, *Fragments of Medieval Manuscripts Used as Pastedowns in Oxford Bindings* (Oxford, 1954). For more recent work see: Anna Reynolds, 'Such Dispersive Scattredness: Early Modern Encounters

with Binding Waste', *Journal of the Northern Renaissance*, 8 (2017), pp. 1–43; Adam Smyth, *Material Texts in Early Modern England* (Cambridge, 2018), esp. chapter 4; David Pearson, *Oxford Bookbinding, 1500–1640* (Oxford, 2000), p. 40.

5. Winchester Cathedral Library, 'Chases's Calendar of Muniments, 1623–50', fol. 84, cited in Paul Yeats-Edwards, 'The Winchester Malory Manuscript: An Attempted History', in Bonnie Wheeler, Robert L. Kindrick and Michael N. Salda (eds), *The Malory Debate* (Woodbridge, 2000), pp. 67–70.

6. Bulstrode Whitelocke, *Memorials of the English Affairs*, 4 vols (Oxford, 1853), vol. 1, pp. 188–9.

7. William Dugdale, *The History of St Paul's Cathedral in London* (London, 1716), p. 148.

8. Brian Sandberg, 'The Magazine of All Their Pillaging', in Laurence Fontaine (ed.), *Alternative Exchanges* (New York, 2008), p. 82.

9. David Cressy, 'Saltpetre, State Security and Vexation in Early Modern England', *Past & Present*, 212:1 (August 2011).

10. Peter Edwards, 'Turning Ploughshares into Swords', *Midland History*, 27:1 (2002), pp. 53, 58–9, 61, 64; Donald Woodward, 'Straw, Bracken and the Wicklow Whale', *Past & Present* 159:1 (1998), p. 71.

11. Roger Burt, 'The Transformation of the Non-Ferrous Metals Industries in the Seventeenth and Eighteenth Centuries', *Economic History Review* 48:1, February 1995, pp. 30–32; H. E. Salter (ed.), *Oxford Council Acts 1583–1626* (Oxford, 1928), p. 252n; *A History of the County of Oxford*, vol. 4, *The City of Oxford*, Victoria County History (London, 1979), www.british-history.ac.uk/report. asp?compid=22812 &strquery=lead

12. Edwards, 'Turning Ploughshares into Swords', p. 63.

13. Charles Carlton, *Going to the Wars* (London, 1992), pp. 277, 284.

14. A. P. Baggs et al., 'Parishes: Deddington', in Alan Crossley (ed.), *A History of the County of Oxford*, vol. 11, *Wootton Hundred (Northern Part)* (London, 1983), BHO, http://www.british-history.ac.uk/vch/oxon/vo111/pp81–120

15. John Rushworth, *5 Julii, 11 at Night: A Letter from a Leaguer before Colchester* (London, 1648), pp. 4–5; T. S., *A True and Exact Relation of the Taking of Colchester* (London, 1648), p. 2.

16. Adrienne Corless, 'A Hoard of 16th and 17th Century Children's Toys', http://irisharchaeology. ie/2013/02/a-hoard-of-16th-and-17th-century-childrens-toys

17. Jenny Mann (ed.), *Finds from the Well at St Paul-in-the-Bail, Lincoln* (Oxford, 2008), pp. ix, 26, 57–8, 63, 66, 76, 81.

18. Portable Antiquities Scheme Database, www.finds.org.uk, unique ID numbers IOW-41FB50; NLM-E0EA77; IOW-A3C45E; Hazel Forsyth and Geoff Egan, *Toys, Trifles & Trinkets* (London, 2005), pp. 387–91.

19. Forsyth and Egan, *Toys, Trifles & Trinkets*, pp. 67, 121, 388–9.

20. Edwards, 'Turning Ploughshares into Swords', p. 54.

21. William Dunn Macray, *A Register of the Members of St Mary Magdalen, Oxford*, new ser., 8 vols (London, 1894–1915), vol. 3, p. 57.

22. W. J. Alldridge, *The Goldsmith's Repository* (London, 1789), p. ix.

23. Matéaux, *The Wonderland of Work*, p. 308.

24. Rice Bush, *The Poor Mans Friend* (London, 1649), A2v, D1v; Boyd's inhabitants, in www.findmypast.co.uk.

25. Ming-Hsun Li, *The Great Recoinage of 1696 to 1699* (London, 1963), esp. chapters 4, 6, 7.

26. John Craig, *Newton at the Mint* (Cambridge, 1946), pp. 2, 188, 198; Stephen Dowell, *History of Taxation and Taxes in England*, 3rd edn, 4 vols, (New York, 1965), vol. 2, p. 211.

27. *The English Manufacture Discouraged* (London, 1697).

28. H. J. Powell, *Glass-Making in England* (Cambridge, 1923), pp. 153–4.

29. 'Adam Fitz-Adam', *The World*, 3 vols (London, 1770), vol. 1, p. 179; *The Plate-Glass Book* (London, 1757), Table IV.

30. 'M.P.', *The Figure of Five* (London, 1645), B1r.

31. Collyer, *Parent's and Guardian's Directory*, p. 258.

32. Randall Monier-Williams, *Tallow Chandlers of London*, 4 vols (London, 1970–77), vol. 4, pp. 9–12, 33, 58.

33. Harold Evan Matthews (ed.), *Proceedings, Minutes and Enrolments of the Company of Soapmakers, 1562–1642* (Bristol, 1940), p. 5.

34. 'M.P.', *The Figure of Five* (London, 1645), B1r.

35. See, for example, London Metropolitan Archives, CLA/047/ LJ/13/1779/002, City of London Sessions, Sessions Papers: Gaol Delivery and Peace, information of Thomas Kirby, 1779; Sean Shesgreen (ed.), *Criers and Hawkers of London* (Aldershot, 1990), pp. 154–5; Randall Monier-Williams, *The Tallow Chandlers of*

London, 4 vols (London, 1970–77), vol. 4, p. 22; A. H., *An Exact Legendary Compendiously Containing the Whole Life of Alderman Abel* (London, 1641).

36. John Dunton, *The Informer's Doom* (London, 1683), L3v.

37. *The Complaisant Companion* (London, 1674), p. 15.

38. [Anon.], *A Pack of Patentees. Opened. Shuffled. Cut. Dealt. And Played* (London, 1641), pp. 3–6.

39. Anon, *Hogs Caracter of a Projector* (London, 1642).

40. F. Bastian, *Defoe's Early Life* (London, 1981), pp. 136–61; Paula Backscheider, *Daniel Defoe* (Baltimore, MD, 1989), p. 40–61; Daniel Defoe, *An Essay upon Projects* (London, 1697).

41. Aaron Hill, *Essays, for the Month of December, 1716* (London, 1716), pp. 19–24. These balls sound like a late sixteenth-century concoction: Hugh Plat, *A Discoverie of Certain English Wants* (London, 1595), A3r–v.

42. Hill, *Essays,* pp. 9–14.

43. John Lord, *Modern Europe* (London, 1860), p. 224.

44. *Hogs Caracter of a Projector*, p. 3.

45. 'A.S.', *The Husbandman's Instructor, or, Countryman's Guide* (London, 1707), p. 78; *Some Remarks on the Royal Navy* (London, 1760), p. 29; Thomas Tryon, *The Merchant, Citizen and Country-man's Instructor* (London, 1701), p. 90; J. T. of Bristol, *An Impartial Inquiry into the Benefits and Damages Arising to the Nation from the Present Very Great Use of Low-Priced Spiritous Liquors* (London, 1751), pp. 10–11.

46. George Harris, *Life of Lord Chancellor Hardwicke*, 3 vols (London, 1847), vol. 1, pp. 266–70.

47. Edward Ward, 'South Sea Ballad', appended to *The Delights of the Bottle* (London, 1720), p. 56.

48. Susan Briscoe, *The 1718 Coverlet* (Newton Abbot, 2014).

49. V&A Museum, Number 1475–1902, bed-cover, unknown, http://collections.vam.ac.uk/item/O148231/bed-cover-unknown

50. Tina F. Smith and Dorothy Osler, 'The 1718 Silk Patchwork Coverlet: Introduction', *Quilt Studies*, 4/5 (2003), pp. 24–30.

51. L. Mare and W.H. Quarrell (eds), *Lichtenberg's Visits to England* (Oxford, 1938), p. 109.

52. Karen N. Thompson and Michael Halliwell, 'Who Put the Text in Textiles?', in Maria Hayward and Elizabeth Kramer (eds), *Textiles and Text* (London, 2007), pp. 237–43.

53. Randle Holme, *The Academy of Armory* (Chester, 1688), Book 3, p. 97.

54. John Styles, 'Patchwork on the Page', in Sue Pritchard (ed.), *Quilts, 1700–2010* (London, 2010), p. 51.

55. John Beresford (ed.), *James Woodforde. The Diary of a Country Parson*, 5 vols (Oxford, 1968), vol. 3, pp. 150, 308; vol. 5, p. 103.

56. One substance thought to give good protection against the ravages of the moth was Russian leather. This tough, birch-oiled, water-resistant, strong-smelling tanned young bovine leather was thought to protect against bookworms; scraps left by the bookbinders were valued as moth repellents, Middleton, *History of English Craft Bookbinding Technique*, p. 120.

57. Alexander Pope, *Memoirs of the Extraordinary Life, Works, and Discoveries of Martinus Scriblerus* (Dublin, 1741), pp. 96–7.

58. Edward Ward, *The Secret History of Clubs* (London, 1709), p. 90.

59. See Helen Smith, 'A Unique Instance of Art: The Proliferating Surfaces of Early Modern Paper', *Journal of the Northern Renaissance*, 8 (2017), pp. 1–39.

60. Bryan Christie, 'Back from the Brink', *Discover*, 33 (2016), pp. 15–18.

61. Cheshire Archives, ZCR/119/1079/11, 107 and ZCR/119/1079/12, 107, Frank Simpson Manuscripts, 'Chester Past & Present', photograph of a Jacobean mantelpiece and panelling, plus fragments of manuscript found behind.

62. [John Green], *The Construction of Maps and Globes* (London, 1717), pp. 120–21; G. R. Crone, 'A Note on Bradock Mead, alias John Green', *Library*, 5th ser., 6:1 (June 1951), pp. 42–3; 'John Green, *née* Bradock Mead, a.k.a. Rogers (d.1757)', Osher Map Library and Smith Center for Cartographic Education, http://www.oshermaps.org/special-map-exhibits/percy-map/john-green

63. 'You'd swear the pasteboard was the better man./"The dev'l", says he, "the head is not so full!"/ Indeed it is – behold the paper skull', Thomas Sheridan, 'Another [Picture of Dan]' 1718, in Jonathan Swift, *Miscellanies*, 10 vols (London, 1745), pp. 10, 205–6; BM, complete pack of cards, 1720, 1896,0501.923. See also *The Bubblers Mirrour* (London, c.1720).

64. Bernard C. Middleton, *A History of English Craft Bookbinding Technique*, 3rd edn (London, 1988), pp. 63–5, 68.

65. Robert Fleming, *The Petition of Robert Fleming bookseller in Edinburgh* (Edinburgh, 1771), p. 1.

66. Alfred Henry Shorter, *Paper Making in the British Isles* (Newton Abbot, 1971), p. 84.

67. Alexander Pope, *The Dunciad* (Dublin, 1728) p. 15; see also William Noblett, 'Cheese, Stolen Paper and the London Book Trade 1750–99', *Eighteenth-Century Life*, 38:3 (2014), pp. 100–10; Margaret Spufford, *Small Books and Pleasant Histories* (London, 1981), p. 49.

68. Price, *How To Do Things with Books in Victorian Britain*, p. 9.

69. Margaret Spufford, *Small Books and Pleasant Histories* (London, 1981), p. 48. William Cornwallis, *Essayes* (London, 1600), pp. I6v-I7v.

70. See also Spufford, *Small Books and Pleasant Histories*, p. 49; Alexander Brome, *Bumm-Foder, or Waste-Paper Proper to Wipe the Nation's Rump with, or Your Own* (London, 1660).

71. 'A New Protestant Litany', *The Muses Farwel to Popery & Slavery* (London, 1690), pp. 80–83.

72. Cambridge University Library, Newton's Waste Book, MS Add. 4004, http://cudl.lib.cam.ac.uk/view/MS-ADD-04004/1, esp. fols 73r–75v, 128r, 306r–v.

73. R. A. Foakes and R. T. Rickert (eds), *Henslowe's Diary* (Cambridge, 1961), pp. xi–xii; British Library, Harley MS 1957, sketchbook and ledger of Randle Holme and John Holme, c.1688–92.

74. [Anon.], *A Pack of Patentees*, esp. pp. 12–13.

75. *Statutes of the Realm, vol. 5, 1628–80*, ed. John Raithby (s.i, 1819), pp. 885–6, *BHO*, http://www.british-history.ac.uk/statutes-realm/vol5

12. Leading the Reforms (1530s–1630s)

1. Edward Peacock, *English Church Furniture* (London, 1866), pp. 53–4; Andrew White, 'Lincolnshire Museums Information Sheet, Archaeology Series 23, The Norman Manor House at Boothby Pagnell', 1981, https://www.thecollectionmuseum.com/assets/downloads/IS_arch_23_the_norman_manor_house_at_boothby_pagnell.pdf

2. Peacock, *English Church Furniture*, pp. 80–81, 94, 127.

3. Emily Cockayne, 'Re-Formations of the Reformation', www.rummage.work/blog/altars, 15 July 2019.

4. Peacock, *English Church Furniture*, pp. 37, 39, 93, 110, 112, 141.

5. Peter Marshall, *Beliefs of the Dead in Reformation England* (Oxford, 2002), p. 172; see also J. F. Fowler (ed.), *The Rites of Durham*,

Surtees Society, 107 (1902), pp. 59–60. See also Eamon Duffy, *Stripping of the Altars*, 2nd edn (London, 2005), pp. 480–87.

6. Peacock, *English Church Furniture*, pp. 65, 107, 167.

7. Peacock, *English Church Furniture*, pp. 29, 50, 53, 55–7, 59, 84, 114, 117, 120, 169–70.

8. Emily Cockayne, 'Experiences of the Deaf in Early Modern England', *Historical Journal*, 46:3 (2003), p. 496.

9. Margaret Aston, 'English Ruins and English History', *Journal of the Warburg and Courtauld Institutes*, 36 (1973), 231; Margaret Aston, *Broken Idols of the English Reformation* (Cambridge, 2016), p. 180.

10. Robinson and Adams (eds), *The Diary of Robert Hooke, 1672–1680*, p. 269.

11. Cited in Howard Colvin, *Essays in English Architectural History* (London, 1999), pp. 56–7.

12. V&A, Communion cup and paten cover, 1571–74, John Jones, Museum number 4636–1858; Robert Whiting, *Reformation of the English Parish Church* (Cambridge, 2010), p. 64.

13. Peacock, *English Church Furniture*, p. 41.

14. John Page Phillips, *Palimpsests* (London, 1980), pp. 1, 18.

15. Marshall, *Beliefs of the Dead*, pp. 88, 105; Mill Stephenson, *A List of Palimpsest Brasses in Great Britain* (London, 1903).

16. Stephenson, *A List of Palimpsest Brasses*, p. 215.

17. Whiting, *The Reformation of the English Parish Church*, p. 213.

18. Stephenson, *A List of Palimpsest Brasses*, pp. 6, 78–9.

19. BM, 1990,0105.1, Palimpsest, http://www.britishmuseum.org/research/collection_online/collection_object_details.aspx?objectId=46155&partId=1

20. Phillip McFadyen, *Norwich Cathedral Despenser Retable* (Cromer, 2015).

21. John Weever, *Ancient Funerall Monuments within the United Monarchie of Great Britaine* (London, 1631), pp. 51, 322.

22. Thomas Fuller, *Church-History of Britain* (London, 1655), p. 417.

23. Edmund Gibson (ed.), *Camden's Britannia Newly Translated into English* (London, 1695), B2.

24. G. W. O. Woodward, *The Dissolution of the Monasteries* (London, 1966), pp. 125–7; A. G. Dickens (ed.), *Tudor Treatises* (Leeds, 1959), pp. 123–4.

25. Elisabeth Leedham-Green, 'University Libraries and Book-Sellers', in Lotte Hellinga and J. B. Trapp (eds), *Cambridge History of the Book in Britain,* vol. 3, *1400–1557* (Cambridge, 1999), pp. 345–6.

26. Tessa Watt, *Cheap Print and Popular Piety, 1550–1640* (Cambridge, 1991), p. 141.

27. Thomas Nashe, *The Unfortunate Traveller* (London, 1594), A3v.

28. Thomas Nashe, *The Apologie of Pierce Pennilesse* (London, 1592), D4v.

29. Strype (ed), *Memorials*, vol. 1, pp. 189–90; Henry Barker, *English Bible Versions* (New York, 1911), p. 112.

30. Woodward, 'Swords into Ploughshares', p. 181.

31. Henry Ellis, *Original Letters Illustrative of English History*, 3rd ser., 4 vols (London, 1846), vol. 3, pp. 268–9; see also F. M. Stephenson, 'The Decline and Dissolution of the Gilbertine Order', unpublished PhD thesis, Coventry University, 2011, p. 173.

32. Colvin, *Essays in English Architectural History*, p. 59.

33. George Henry Cook, *English Monasteries in the Middle Ages* (London, 1961), p. 257.

34. Roger Burt, 'The Transformation of the Non-Ferrous Metals Industries in the Seventeenth and Eighteenth Centuries', *Economic History Review* 48:1 (February 1995), pp. 30–32; Brian Sandberg, 'The Magazine of All Their Pillaging', in Laurence Fontaine (ed.), *Alternative Exchanges* (New York, 2008), p. 82; Stephen Bull, *The Furie of the Ordnance* (Woodbridge, 2008), Appendix II.

35. Donald Woodward, '"Swords into Ploughshares": Recycling in Pre-Industrial England', *Economic History Review*, 2nd ser., 38: 2 (1985), p. 191.

36. John Strype (ed), *Memorials of the Most Reverent Father in God Thomas Cranmer*, 3 vols (Oxford, 1848), vol. 1, p. 327. See also 'Allowances of the Expenses of Lord Lansdowne, Envoy-Extraordinary to the Court of Spain, from 1st November 1684 to 3rd March 1688–9', TNA, SP 44/340 f.63, 447, 5 February 1690.

37. Madeleine Ginsburg, 'Rags to Riches', *Costume*, 14 (1980), p. 125; see Margaret Spufford and Susan Mee, *The Clothing of the Common Sort, 1570–1700* (Oxford, 2017), esp. p. 76.

38. John Fitzherbert, *Boke of Husbandry* (London, 1540), fols 60v-63v.

39. Gervase Markham, *The Countrey Farme* (London, 1616); Wendy Wall, 'Renaissance National Husbandry: Gervase Markham and the

Publication of England', *Sixteenth Century Journal*, 27:3 (1996), pp. 767–85, esp. pp. 771–80.

40. Thomas Nashe, *Christ's Tears over Jerusalem* (London, 1593), p. 52.

41. Thomas Tusser, *A Hundreth Good Pointes of Husbandrie* (London, 1570), fols 23, 32.

42. See also Simon Werrett, 'Recycling in Early Modern Science', *British Journal for the History of Science*, 46:4 (2013), esp. p. 642.

43. Lorna Hutson, *Thomas Nashe in Context* (Oxford, 1989), pp. 18–19, 21.

44. [Thomas Smith], *A Discourse of the Commonweal of this Realm of England*, ed. Elizabeth Lamond (Cambridge, 1929), p. 63.

45. Cited in Joan Thirsk, *Economic Policy and Projects* (Oxford, 1978), p. 91; Wallace Notestein, Frances Helen Relf and Hartley Simpson (eds), *Commons Debates 1621*, 7 vols (New Haven, CT, 1935), vol. 7, pp. 513–15.

46. Thirsk, *Economic Policy and Projects*, pp. 84–9.

47. R. H. Tawney and Eileen Power (eds), *Tudor Economic Documents*, 3 vols (London, 1924), vol. 2, pp. 269–95.

48. John Farquar Fraser (ed.), *Reports of Sir Edward Coke*, 6 vols (London, 1826), vol. 5, p. 508.

49. Bodleian Library, Oxford, MS.Eng.hist/c.479, 'An agreement between John Spilman and William Heyricke', 21 June 1604, fols 215–16, 222–23; State Papers Online, Gale, Centage Learning, 2018; R. Lemon (ed.), *Calendar of State Papers, Domestic Series, of the Reign of Elizabeth, 1581–1590* (London, 1865), 556, TNA, SP 12/217, fol. 114, 30 October 1588, http://go.galegroup.com.ueaezproxy.uea.ac.uk:2048/mss/i.do?id=GALE|MC4304205306&v=2.1&u=univea&it=r&p=SPOL&sw=w&viewtype=Calendar

50. William Hyde Price, *The English Patents of Monopoly* (Cambridge, 1913), p. 8; Thirsk, *Economic Policy and Projects*, p. 75.

51. Hills, *Papermaking in Britain*, p. 50; For a detailed account of sixteenth-century paper mills see Helen Smith, '"A Unique Instance of Art": The Proliferating Surfaces of Early Modern Paper', *Journal of the Northern Renaissance*, 8 (2017), pp. 1–39.

52. Thomas Churchyard, *A Sparke of Frendship and Warme Goodwill* (London, 1588).

53. Tawney and Power (eds), *Tudor Economic Documents*, vol. 2, pp. 253–4; William Turner Berry and Herbert Edmund Poole, *Annals of Printing* (London, 1966), pp. 109–10.

54. See also Reid Barbour, *Deciphering Elizabethan Fiction* (Newark, DE, 1993), p. 69.

55. Georgia Brown, *Redefining Elizabethan Literature* (Cambridge, 2004), pp. 65, 71.

56. Henry Bedel, *A Sermon Exhorting to Pitie the Poore* (London, 1573), fol. C1v.

57. All extracts from Thomas Nash[e], *Pierce Penilesse His Supplication to the Diuell* (London, 1592), fols B1r–v.

58. Frederick W. Thornsby (ed.), *Dictionary of Organs and Organists*, 2nd edn (London, 1921), pp. 37–8; Andrew Ashbee (ed.), *Records of English Court Music*, vol. 6, *1558–1603* (Aldershot, 1992), pp. 31, 42; Hyde Price, *English Patents*, p. 11; Ann Morton (ed.), *Calendar of the Patent Rolls Preserved in the Public Record Office. Elizabeth I*, vol. 9, *1580–1582* (London, 1986), p. 106.

59. Cited in Woodward, 'Swords into Ploughshares', p. 179.

60. Edmund Lodge, *Illustrations of British History*, 2nd edn, 3 vols (London, 1791), vol. 3, pp. 159–60; *Acts of the Privy Council of England Volume 15, 1587–1588*, ed. John Roche Dasent (London, 1897), p. 164, *BHO*, http://www.british-history.ac.uk/acts-privy-council/vo115

61. Hyde Price, *English Patents*, p. 143.

62. John Mottley, *A Survey of the Cities of London and Westminster*, 2 vols (London, 1733–35), vol. 2, p. 386; Tawney and Power, *Tudor Economic Documents*, vol. 2, p. 283 – Simon 'Farmor'.

63. *Acts of the Privy Council of England, vol. 22, 1591–1592*, ed. John Roche Dasent (London, 1901), pp. 483–4, *BHO*, http://www.british-history.ac.uk/acts-privy-council/vo122

64. Hutson, *Thomas Nashe in Context*, p. 187.

65. Richard Porder, *A Sermon of Gods Fearful Threatenings for Idolatrye Mixing of Religioun* (London, 1570), fol. 90v.

66. Hutson, *Thomas Nashe in Context*, p. 185.

67. Hutson, *Thomas Nashe in Context*, p. 185; Thirsk, *Economic Policy and Projects*, pp. 44–5, 60–61.

68. Tawney and Power, *Tudor Economic Documents*, vol. 2, p. 275.

69. David Landreth, 'Wit without Money in Nashe', in Stephen Guy-Bray, Joan Pong Linton and Steven Mentz (eds), *The Age of Thomas Nashe* (Farnham, 2013), pp. 135–52.

70. Hutson, *Thomas Nashe in Context*, pp. 18, 21.

71. William Page (ed.), 'Parishes: Guisborough', *A History of the County of York North Riding*, vol. 2 (London, 1923), pp. 352–65.

72. Jennifer Stead, 'Uses of Old Urine', *Old West Riding*, 1:2 (Spring 1982), pp. 1–9; Joseph Wright, *The English Dialect Dictionary*, 6 vols (London, 1898–1905), vol. 3, p. 520; Daniel Colwall, *Alum Works*, quoted in John Lowthrop (ed.), *The Philosophical Transactions and Collections to the End of the Year 1700*, 10 vols (London, 1721–56), vol. 2, p. 1052; Woodward, 'Straw, Bracken and the Wicklow Whale', pp. 70–71; W. R. Baumann, *The Merchant Adventurers and the Continental Cloth Trade* (Berlin, 1990), p. 28; James Haigh, *The Dyer's Assistant* (York, 1787), p. 18.

73. Hyde Price, *English Patents*, pp. 83–4.

74. Adam Hart-Davis and Emily Troscianko, *Taking the Piss* (Stroud, 2006), p. 74.

Conclusion: To Singe a Goose

1. eBay, https://www.ebay.co.uk/itm/Reclaimed-semi-circular-mirror-upcycled-from-a-crittall-window-burnished-steel/202021981780?hash=item2f0972ce54:g:JEwAAOSwjg1ZkvXS

2. Woodward, 'Swords into Ploughshares, p. 191.

3. See Gay Hawkins, *Ethics of Waste* (Oxford, 2006), p. 99.

4. Nothing Is Useless', *Chambers' Edinburgh Journal*, 36: 131 (July 1846), p. 19.

5. [George Dodd], 'Penny Wisdom', *Household Words*, 134, 16 October 1852, p. 101.

6. [Andrew Wynter], 'The Use of Refuse', *Quarterly Review*, 124 (1868), pp. 334–57.

7. [Robert Hogarth Patterson], 'The Economy of Capital', cited in James Platt, *Economy* (London, 1882), p. 29.

8. 'The Manufacture of Arsenic', *Chambers' Journal*, 10:480, 11 March 1893, p. 149.

9. [Dodd], 'Penny Wisdom', p. 97; Thomas Greenwood, 'The Utilisation of Waste II', *Leisure Hour*, 37, October 1886, p. 707.

10. Leslie, *Synthetic Worlds*, p. 11.

11. Spooner, *Wealth from Waste*, p. 197.

12. Talbot, *Millions from Waste*, pp. 5, 10–11.

13. 'Utilising Waste Paper', *South Shields Gazette*, 21 February 1891, p. 5.

14. MOA, Diarist 5427, 5 August 1940.

15. Letters, *Times*, 15 September 1886, p. 11.

16. See R. G. Stokes, R. Köster and S. C. Sambrook, *The Business of Waste* (Cambridge, 2013).

17. Talbot, *Millions from Waste*, pp. 19, 32, 35; see also Thomas Greenwood, 'The Utilisation of Waste II', *Leisure Hour*, 35, October 1886, pp. 707–9.

18. Ian Buxton and Ian Johnston, *Battleship Builders* (Barnsley, 2013), p. 306.

19. Michael Thompson, *Rubbish Theory* (originally 1979, reprint edn, London, 2017), p. 115.

20. Babbage, *On the Economy of Machinery and Manufacture*, pp. 148–9.

21. 'London', *Hampshire Chronicle*, 23 May 1803, p. 2. See also 'Gathering Up the Fragments', *Kirkintilloch Herald*, 3 April 1918, p. 7. See, for example, Thomas Tusser, *A Hundreth Good Pointes of Husbandry* (London, 1570), fol. 33v.

22. Strasser, *Waste and Want*.

23. Talbot, *Millions from Waste*, pp. 221–2.

24. 'Charitable Agencies and Poor Relief', *Times*, 2 December 1869, p. 5.

25. Bernard London, 'Ending the Depression through Planned Obsolescence', self-published pamphlet, 1932.

26. See various examples on the AvE YouTube Channel, for example, 'BOLTR: KitchenAid Mixer. SURPRISE!', https://www.youtube.com/watch?v=oqKp-oh9P18; https://www.youtube.com/watch?v=oqKp-oh9P18&feature=youtu.be&t=2282

27. James Strachan, *The Recovery and Re-Manufacture of Waste Paper* (Aberdeen, 1918), preface, pp. v–vi, see also pp. 3–8, 17, 40–41.

28. John Carvel, 'At £25 a Ton There's Money in Waste Paper', *Guardian*, 30 March 1977, p. 16.

29. Timothy Cooper, 'Rags, Bones and Recycling Bins', *History Today*, 56:2 (2006), pp. 17–18.

30. Clapp, *Environmental History of Britain*, p. 192.

31. Robert Somerville, *On the Subject of Manures* (London, 1795), p. 16.

32. Ballard, *Effluvium Nuisances*, p. 13.

33. 'Book Salvage Drive Bristol', *Western Daily Press*, 31 October 1942, p. 4.

34. Price, *How To Do Things with Books in Victorian Britain*, p. 195.

35. Robinson and Adams (eds), *Diary of Robert Hooke, 1672–1680*, p. 26.

36. George Dodd, 'Paper and Its Novelties', in *Curiosities of Industry* (London, 1853), pp. 17–18.

37. 'Papers', *Bradford Observer*, 17 September 1873, p. 3.

38. 'Where Cork Comes from', *Hampshire Advertiser*, 21 November 1903, p. 4.

39. Thomas Doubleday, 'To the Right Honourable Lord Althorp: Letter II', *Cobbett's Weekly Political Register*, 25 May 1833, columns 480–83.

40. Bristol Archives, 7955/2, inventory of goods, utensils, belonging to Farrell, Vaughan & Co., fol. 2.

41. *The Printers Case* (London, *c.*1711), p. 1.

42. Lewis Angell, *Sanitary Science and the Sewage Question* (London, 1871), p. 17.

43. For example, BL, Burney MS 408, Palimpsest, the upper text being homilies of St John Chrysostom over lower fragments of a tenth-century Gospel lectionary, https://www.bl.uk/manuscripts/FullDisplay.aspx?ref=Burney_MS_408

44. BL, Egerton 3245, Richard Rolle of Hampole, *Prick of Conscience*.

45. Frank C. Bowen, 'The Shipbreaking Industry', *Shipping Wonders of the World*, 40 (1936), pp. 1272, 1274.

46. 'Waste Paper', *Leisure Hour*, 30 (1881), p. 410.

47. Bristol Archives, M/BCC/SAN/14/1, Salvage Drive Minute Book 1941–1944, minutes for the meeting 28 July 1941.

48. [Edwin Chadwick], *An Inquiry into the Sanitary Condition of the Labouring Population of Great Britain* (London, 1842), p. 381.

49. Desrochers, 'Does the Invisible Hand?', pp. 8–11.

50. Simmonds, *Waste Products*, p. 460.

51. Bashley Britten, 'Utilisation of Blast-Furnace Slag for the Manufacture of Glass', *British Architect & Northern Engineer*, 6:13, 29 September 1876, p. 201.

52. Simmonds, *Waste Products*, pp. 2, 464–5; Greenwood, 'The Utilisation of Waste IV', p. 840; James Winter, *Secure from Rash Assault* (Berkeley, CA, 1999), pp. 146–7.

53. Britten, 'Utilisation of Blast-Furnace Slag', p. 201; Desrochers, 'Does the Invisible Hand?', pp. 9–11; Thomas Greenwood, 'The Utilisation of Waste IV', *Leisure Hour*, 35, December 1886, p. 840.

54. C. T. Flower, 'Manuscripts and the War', *Transactions of the Royal Historical Society*, 4th ser., 25 (1943), p. 28.

55. MOA, 'Salvaging History', 11/09/1941, pp. 1–2, 9.

56. William Downing Bruce, *A Letter […] on the Condition and Unsafe State of Ancient Parochial Letters* (London, 1950), pp. 2, 5.

57. 'On Parish Registers', *Notes & Queries*, 2nd ser., 2:34, 23 August 1856, p. 152; 'Parish Registers', *Notes & Queries*, 2nd ser., 3:69, 25 April 1857, pp. 321–2; Ralph Bigland, *Observations on Marriages, Baptisms, and Burials* (London, 1764), p. 61.

58. 'China Receives Letter from Queen Elizabeth I – 383 Years Too Late', New Zealand China Friendship Society, http://nzchinasociety. org.nz/china-receives-letter-from-queen-elizabeth-i-383-years-too-late, 6 August 2016; Lancashire Archives, DDSH/15/3/1–5, letter: Elizabeth to the Emperor of Cathay, 4 May 1602.

59. 'Redoubled Efforts for Salvage', *Times*, 17 March 1943, p. 2.

60. Flower, 'Manuscripts and the War', pp. 29–30.

61. TNA, Discovery Catalogue, Rutland Court of Quarter Sessions and Associated Records, http://discovery.nationalarchives.gov.uk/details/r/e5836a42-a906-43ae-8afd-732f5e59afa3 (accessed 20 September 2018). See also Peter Thorsheim, 'Bombs and Recycling Drives: The Double Threat to Books and Documents in Wartime Britain', in Mark J. Crowley and Sandra Trudgen Dawson (eds), *Home Fronts* (Woodbridge, 2017), p. 239.

62. East Sussex Record Office, SHE/2/8/10, Memorandum that the coronership records for 1868–1926 were handed to the postmaster at Battle [for salvage] on 6 March 1942.

63. Gloucestershire Archives, GBR/L6/23/B3262, Gloucester Borough Records, Town Clerk's Files, Salvage: correspondence and other papers, Letter from the Home Office, 24 February 1942.

64. Wiltshire and Swindon History Centre, 212A/19, selection of poor law records of the Chapelry made by Canon E. H. Goddard from papers put out for salvage in 1945.

65. 'Historical Treasures Saved from Scrap', *Northampton Mercury,* 29 May 1942, p. 5; Northamptonshire Record Office, Guilsborough Benefit Club Subscription book 1791–1894, ML/050.

66. 'Finds at Bristol', *Times*, 9 January 1943, p. 2.

67. Letters, *Times*, 14 May 1947, p. 5.

68. Nick Britten, 'Romance of the Open Road', *Telegraph*, 18 December 2003, p. 11.

69. Paul N. Hasluck (ed.), *Domestic Jobbing* (London, 1907), p. 19.

70. *OB*, t19071119–53, 19th November 1907.

71. John Roberts, *The Game of Billiards* (London, 1905), advert, opposite contents page.

72. Department for Environment, Food & Rural Affairs, 'Deposit Return Scheme in Fight against Plastic', https://www.gov.uk/government/news/deposit-return-scheme-in-fight-against-plastic, 28 March 2018.

73. Darrel Moore, 'Euro Closed Loop Recycling Placed Into Administration', *CIWM Journal Online*, 16 May 2016; Waste Partner, Bulletin Board, 16 May 2016, http://www.wastepartner.co.uk/bulletin-board/view/euro-closed-loop-recycling-goes-into-administration

74. See, for example, 'Hide and Seek! Dressed-Down Duchess Sports an Entirely Recycled Outfit', *Daily Mail*, 28 March 2019, Femail, p. 10.

75. Talbot, *Millions from Waste*, p. 11; See Martin O'Brien, *A Crisis of Waste* (London, 2007), p. 71, for a critique.

76. Francis Fox, *Sixty-Three Years of Engineering* (London, 1924), p. 264.

77. Letters, *Times*, 21 January 1941, p. 5.

78. 'Railways' Salvage Campaign', *Times*, 3 August 1944, p. 7.

79. Scrap Railings Order Dismissed', *Western Daily Press*, 3 October 1941, p. 2.

80. Brian Neville and Johanne Villeneuve, 'In Lieu of Waste', in Neville and Villeneuve (eds), *Waste-Site Stories* (Albany, NY, 2002), pp. 7, 19. Walter Benjamin started to explore the values of cast-offs in his unfinished *Arcades Project*, which considered what happens to objects spat out by capitalism and no longer regarded with the desire or approval they once enjoyed. See Davis Gross, 'Objects from the Past', in Neville and Villeneuve (eds), *Waste-Site Stories*, p. 36; Benjamin, *Arcades Project*. See also Sophie Gee, *Making Waste* (Princeton, NJ, 2010), p. 4.

81. N. W. Alcock and Nancy Cox (eds), *Living and Working in Seventeenth-Century England*, British Library CD-ROM (2000).

82. Goldsmith, 'Alexander Parkes, Parkesine, Xylonite and Celluloid', p. 22.

83. News, *Norwich Mercury*, 19 November 1881, p. 3.

84. Böckmann, *Celluloid*, p. 103.

85. Hills, *Papermaking in Britain*, p. 128.

Acknowledgements

In her acknowledgements to *Contrasting Communities* (1974), Margaret Spufford wrote: 'Financial help for a married woman with a young family, who remains bent on doing part-time research, is never easy to obtain.' I recalled those words in 2014, when Margaret died, and felt sad that the situation had not much improved. The culture that prevails in British universities is suppressing numberless important books and bright ideas. That loss can never be calculated and will never be compensated. To be able to write, on top of life's complications (including divorce, full-time teaching, burglary and six months of living in a shed), I have had to give up on anything approaching a social life for the past few years. I missed lots of people in this time, especially Mary Laven, Jason Scott-Warren, Jordan Claridge, Suzy Copping and the pub quiz crew.

Difficult circumstances reveal the genuinely supportive people, and Helena Carr, Jayne Gifford, Joel Halcomb and Gesine Oppitz-Trotman all went over and above being colleagues by reading chapters and providing good-humoured support. Alex Schofield, John Trotman and Amanda Dillon all passed their expert eyes over a full earlier draft, and their input has been valuable. My mother, Yvonne, assisted with genealogical research and, decades ago, taught me to sew and economise; my father, David, read drafts and photographed a Womble. One Womble was harmed in the making of this book.

I would also like to thank Emma Griffin, Camilla Schofield, Tony Howe, Rosemary Hill, Catherine Hastings and Frank Vickers, Ben Jones, Matthew Taunton, Tom Williamson, Bartek Witek, Izzy Carter, Michael Newstead, Cóilín Nunan, Keith Benton, Robert and Linda Tait and Lara Maiklem. Many thanks to Taryn Everdeen and Toby Sleigh-Johnson for their expert photography skills, and to archivists across the country (and also in Fürth, Germany), as well as countless eBay sellers for listing miscellaneous cast-off items which I draw together here. I have been fortunate to have not one but two conscientious and enthusiastic editors: many thanks to Rebecca Gray and Louisa Dunnigan. Matthew Taylor caught some last-minute slips. My agent, Clare Alexander, first recognised the value of this research, and I will be for ever in her debt.

Maud and Ned, I never did give you the childhoods I imagined I would, filled with the passing on of mending skills that I learned from my mother, and her mother too. Regardless, you have both weathered disruptions in your humorous ways and developed into amazing resourceful people. Kevin-the-recycled-cat demanded that I stop sometimes to fuss him. This helped more than I knew it could, so I thank the Dereham branch of Cats Protection. Most important has been the support of George, who got this book from the moment I talked to him about it and encouraged me to continue despite the adverse circumstances. Without him *Rummage* would be more superficial and have fewer boats in it, and I would be miserable.

Index

Note. An italic page reference indicates pages with relevant illustrations but without a matching indexed text.